Cold Chain Management for the Fresh Produce Industry in the Developing World

World Food Preservation Center Book Series

Series Editor:
Charles L. Wilson

For more information about this series please visit: http://worldfoodpreservationcenter.com/crc-press.html

Cold Chain Management for the Fresh Produce Industry in the Developing World

Edited by
Vijay Yadav Tokala
Majeed Mohammed

CRC Press is an imprint of the
Taylor & Francis Group, an **informa** business

First edition published 2022
by CRC Press
6000 Broken Sound Parkway NW, Suite 300, Boca Raton, FL 33487-2742

and by CRC Press
2 Park Square, Milton Park, Abingdon, Oxon, OX14 4RN

ISBN: 978-0-367-49819-1 (hbk)
ISBN: 978-1-032-12689-0 (pbk)
ISBN: 978-1-003-05660-7 (ebk)

DOI: 10.1201/9781003056607

Typeset in Times
by codeMantra

Contents

SECTION I   Cooling and Cold Chain

SECTION II   Cooling Systems

SECTION III   Cold Chain Development, Capacity Building and Case Studies

Foreword

With pleasure, I have written, read, or reviewed the chapters in this book on cold chain development for fresh produce. I believe that the editors have succeeded in making the case for their closing remarks in Chapter 18, where they stated, "Cold chain management is a requirement to achieve a global reduction of postharvest losses and food waste."

During my 30-year career as a consultant, trainer, and researcher in postharvest management, I have enjoyed working with many of the chapter authors. Some of them were partners on research projects, development projects, or food loss assessment work funded by USAID, FAO, the World Bank, or USDA. Others are board members or supporters of the nonprofit work of the Postharvest Education Foundation. All of our work included aspects of the cold chain and its development – whether we were assessing a food supply chain to determine its status (or absence), managing field trials on new cooling or cold storage technologies, introducing new postharvest technologies in extension or training programs, or reviewing the literature to search for evidence of findings that could be applied to the postharvest issues faced by smallholder farmers, storage managers, traders, and marketers.

Many of the chapter authors, no matter from which region of the world, point to the high levels of fresh produce losses and highlight the need for more research, more training, and more investments in the industry in order to achieve cold chain development for developing regions. It can be discouraging for some to read about these many needs and gaps, but I believe that whenever and wherever we find gaps in the cold chain, we know that there is more work to be done.

Recently, there is the good news of projects reporting success in their cold chain development efforts, including the "Zero Loss Aggregation and Training Centers" for mango farmers in Kenya, funded by the Rockefeller Foundation's Yieldwise Project. ColdHubs in Nigeria is just one example of a private sector investment and innovative business model. India is highlighted in several chapters as a national model of cold chain development efforts, led by the National Centre for Cold-chain Development (NCCD). Everywhere it seems, sustainable cold chain technologies and affordable options for smallholders are on the rise. With a new US$ 3.5 million grant from the UK Government and the University of Rwanda's Rubirizi Campus allocation of land and facilities valued at US$ 3 million, the Africa Centre for Sustainable Cooling and Cold Chain was launched in 2021. Currently, they are hiring world-class cold chain experts, procuring cutting-edge equipment, designing new educational and outreach opportunities, and developing innovative business models focused on cold chain development.

With the publication of this new book, the authors have done their part in gathering the most current information and sharing their recommendations for future cold chain development efforts. I thank them and commend their efforts. Let us continue to identify the gaps and do the work of reducing postharvest losses in the fresh produce industry.

Dr. Lisa Kitinoja
The Postharvest Education Foundation

Preface

The authors and their contributions to the 18 chapters in this book did recognize that a more integrated evaluation of cold chain performance is key for developing a resource-efficient, energy-smart food supply chain. They also concede that the overall purpose of the cold chain, for all the stakeholders in the fresh produce industry, is to maintain the quality and safety of the food products during handling, transporting, and storing in their journey from producer to consumer. Several chapters focus on the development of cold chain infrastructure as crucial in reducing postharvest losses and wastages, increasing farmer income, generating employment opportunities, and improving the livelihood of partakers along the supply chain. A thorough review of basic and contemporary literature is now available on the present status and challenges of cold chain collaborators.

The information is to alert stakeholders toward an urgent need to develop intelligent cold chain infrastructure which is often considered as a major impediment and a weak link in the horticulture supply chain in developing countries. The underlying principles across the themes in each of the three sections emphasize that an efficient cold chain system and its supporting infrastructure would promote developing countries to meet food safety and quality standards. The increasing capacity is driven by greater reliance on the cold chain to meet the growing trade and consumption of better quality perishables, the retail boom, and a growing middle class in developing countries. In identifying cold chain gaps, the authors also provide historical evidence to highlight the importance of the cold chain in developing countries and descriptions of several successful case studies.

Several cold chain options and technologies to produce cold conditions for handling, transport, and storage of perishables are addressed at various levels, with some being relatively simple and inexpensive meant for short-term storage, while other technologies are more sophisticated and complex to store the produce for as long as a few months. Recommendations are articulated for precooling operators to choose and adapt according to simple farm-based methods such as using ice to more complex systems like forced air, hydrocooling, or vacuum cooling (Kitinoja, 2014). For storage, options for food handlers that range from small walk-in cold rooms to large-scale commercial refrigerated warehouses are examined. Detailed reports are given on small-scale cold room designs with traditional mechanical refrigeration systems, low-cost CoolBot™ with air-conditioner base systems, or evaporative cool chambers. The improvement in efficiency through the measurement, analysis, and management of time-temperature conditions is reviewed, along with the accompanying technical and practical challenges delaying the implementation of such methods.

The total duration of the cold chain is described as being highly dependent on the specific product and the target market, with some cold chains being as short as a few hours and others lasting several months. The distribution center is identified as a critical control point in many cold chain management systems, as it provides the opportunity to sort and combine shipments that are received from many suppliers and to schedule the shipments' departure according to retailer demand, product arrival time, and quality.

We hope that using the outreach methods typically for linking growers, traders, storage operators, transporters, and marketers would incorporate the recommendations and guidelines relating to cold chain development and management. Introducing designs for synergistic collaborations for successful cold chain development with different sectors, and developing appropriate curriculum and training programs would cater for different food supply chain actors and service providers, and effective implementation of various outreach methods, including training-of-trainers programs and local postharvest innovation platforms.

We wish to convey our sincere gratitude to all authors for their time and valuable efforts in their contribution to the chapters in this book. We also want to express our appreciation to the Board of Directors of the Postharvest Education Foundation for their written and editorial commitments. We would like to specially acknowledge Dr. Kitinoja and Dr. Holcroft for their constant support and technical inputs. Many thanks to Dr. Wilson, Founder and CEO of World Food Preservation Center® LLC, for initiating this book series and also to all the staff members at the CRC Press for their relentless support throughout the book project.

REFERENCES

Kitinoja, L. (2014). Exploring the potential for cold chain development in emerging and rapidly industrializing economies through liquid air refrigeration technologies. Liquid Air Energy Network, UK.

Kitinoja, L., Tokala, V.Y. and Mohammed, M. (2019). Clean cold chain development and the critical role of extension education. *Agriculture for Development 36*(3): 19–25.

Editors

Dr Tokala earned his PhD in Environment and Agriculture with a specialization in postharvest physiology at Curtin University, Perth, Western Australia. He was awarded the 2016-International Postgraduate Research Scholarship (IPRS) and the Australian Postgraduate Award (APA) through a merit-based international competition to pursue his PhD. Dr Tokala completed his Bachelor's degree with a specialization in Horticulture at Dr. Y.S.R. Horticultural University, Andhra Pradesh (India) and was the top scorer in the college (Dux of the college). His postgraduate research was on fruit and vegetable processing and was awarded the University Gold Medal for securing the university's top ranking at S.K. Rajasthan Agricultural University, Rajasthan (India).

Dr Tokala is currently employed at Amity University as an Assistant Professor, since January 2020. Previously, he held the position of Teaching Assistant and Laboratory Demonstrator at Curtin Horticulture Research Laboratory, Australia. Before that, he worked as Horticulture Officer in the Government of Andhra Pradesh (India) with duties as a field consultant and extension officer in both rural and urban areas, with the main objective of enhancing quality horticulture production. He is also voluntarily serving as President /Board of Directors for The Postharvest Education Foundation, a US-based nonprofit organization committed to reducing global postharvest losses by providing innovative training programs to participants from more than 30 different developing nations.

He has edited two books on postharvest capacity building, published more than 30 research papers in peer-reviewed journals, and presented papers at several national and international platforms. His international consulting experience includes desktop studies and coordination with local teams for postharvest management of horticulture crops in Ethiopia, Nepal, Rwanda, and Senegal.

Majeed Mohammed is a professor of postharvest physiology at the University of The West Indies (UWI), with 31 years of experience in teaching, research, and outreach activities. He earned his PhD in postharvest physiology at UWI, MSc in postharvest physiology at the University of Guelph, and BSc in agriculture at UWI. His research is centered on the physiology and biochemistry of ripening of tropical fruits, effects of ethylene antagonists in delaying senescence of cut-flowers, alleviation of physiological disorders of minimally processed fruits and vegetables, and cold chain management and development of value-added food products. He has developed four schools of research: the first in Postharvest Physiology, the second in Commodity Utilization, the third in Food Quality Management, and the fourth in Postharvest Extension. His latest work focuses on assisting Caribbean countries with developing sustainable, efficient, and inclusive mechanisms to produce, transform, and deliver healthy and safe food to consumers. He is currently Board Director with the Postharvest Education Foundation (PEF), USA, and a member of the UN/FAO Panel of Experts from Latin America and the Caribbean on the Prevention and Reduction of Food Losses and Waste.

Contributors

Puran Bridgemohan
Biosciences, Agriculture & Food Technologies
   Department
The University of Trinidad and Tobago
Arima, Trinidad and Tobago

Amanda Brondy
Global Cold Chain Alliance
Arlington, Virginia

Angelos Deltsidis
Department of Horticulture
University of Georgia
Athens, Georgia

Neeru Dubey
Ernst and Young
Delhi, India

Rashmi Ekka
Agribusiness Associates Inc.
Davis, California

Gurbinder Gill
Agribusiness Associates Inc.
Davis, California

Wynand Groenewald
Future Green Now
Johannesburg, South Africa

Deirdre Holcroft
Holcroft Postharvest Consulting
Lectoure, France
and
The Postharvest Education Foundation
La Pine, Oregon

Nnaemeka C. Ikegwuonu
ColdHubs Nigeria Limited
Owerri, Nigeria

Madison Jaco
Global Cold Chain Alliance
Arlington, Virginia

Eduardo Kerbel
Carrier Transicold
San Jose, Costa Rica

Lisa Kitinoja
The Postharvest Education Foundation
La Pine, Oregon

Pawanexh Kohli
National Centre for Cold-chain Development
   (NCCD)
New Delhi, India

Bal Vipin Chander Mahajan
Punjab Horticultural Postharvest Technology
   Centre
Punjab Agricultural University
Ludhiana, India

Ramadhani O. Majubwa
Department of Crop Science and Horticulture
Sokoine University of Agriculture
Morogoro, Tanzania

Majeed Mohammed
The Postharvest Education Foundation
La Pine, Oregon

Theodosy J. Msogoya
Department of Crop Science and Horticulture
Sokoine University of Agriculture
Morogoro, Tanzania

Hosea D. Mtui
Department of Crop Science and
   Horticulture
Sokoine University of Agriculture
Morogoro, Tanzania

Swarajya Laxmi Nayak
Division of Food Science and Postharvest
   Technology
ICAR-Indian Agricultural
   Research Institute
New Delhi, India

Divine Njie
United Nations Food and Agriculture
    Organization
Rome, Italy

Olubukola M. Odeyemi
Department of Horticulture
Federal University of Agriculture
Abeokuta, Nigeria

Toby Peters
Cold Economy
University of Birmingham
Birmingham, United Kingdom

Leyla Sayin
Toby Peters Consultancy
United Kingdom

Eleni Pliakoni
Department of Horticulture and Natural
    Resources
Kansas State University
Manhattan, Kansas

Lowel Randell
Global Cold Chain Alliance
Arlington, Virginia

Steven A. Sargent
Horticultural Sciences Department
University of Florida
Gainesville, Florida

Shruti Sethi
Division of Food Science and Postharvest
    Technology
ICAR-Indian Agricultural Research Institute
New Delhi, India

Mandeep Sharma
Agribusiness Associates Inc.
Davis, California

R.R. Sharma
Division of Food Science and Postharvest
    Technology
ICAR-Indian Agricultural Research Institute
New Delhi, India

Vijay Yadav Tokala
The Postharvest Education Foundation
La Pine, Oregon

Eric Verploegen
MIT D-Lab
Massachusetts Institute of Technology
Boston, Massachusetts

Charles Wilson
World Food Preservation Center LLC
Shepherdstown, West Virginia

Section I

Cooling and Cold Chain

1 Cold Chain and Its Importance—Current Global Status

Lisa Kitinoja, Vijay Yadav Tokala, and Majeed Mohammed
The Postharvest Education Foundation

Bal Vipin Chander Mahajan
Punjab Agricultural University

CONTENTS

1.1 INTRODUCTION

While global commodity chains are fairly modern expansions in the fresh produce transportation and distribution industry, the refrigerated movement of temperature-sensitive commodities is a practice that dates back to 1797 with the use of naturally formed ice blocks (Rodrigue and Notteboom, 2016). Cold storage was also a key component of the food trade between colonial powers and their colonies. The first reefer ship for the banana trade was introduced in 1902 by the United Food Company. This enabled the banana to move from the status of an exotic fruit that had a small market to one of the world's most consumed fruits. Before this, the banana fruit used to arrive in retail markets too ripe and experienced huge losses. The impacts of mechanical refrigeration systems on the reefer transportation industry during the 1950s were monumental.

The term cold chain or cool chain denotes a temperature-controlled supply chain with the series of actions and equipment applied to maintain a product within a specified low-temperature range

DOI: 10.1201/9781003056607-2

3

from its harvest until it is consumed. An unbroken cold chain is an uninterrupted series of refrigerated handling, storage and distribution activities, along with associated equipment and logistics, which maintain a desired low-temperature range (Kitinoja et al., 2019). It is used to preserve and to extend and ensure the shelf life of products, such as fresh agricultural produce, seafood, frozen and processed foods. Such products, during transport and when in transient storage, are sometimes called cool cargo. Unlike other goods or merchandise, cold chain goods are perishable and always *en route* toward end-use or destination, even when held temporarily in cold stores, and hence are commonly referred to as cargo during its entire logistics cycle.

1.2 COLD CHAIN MARKET SIZE, STATUS, GLOBAL OUTLOOK 2019 TO 2025

The increasing trade of perishable products and rising government support for the development of infrastructure for cold chain facilities are anticipated to drive markets growth from 2019 to 2025 (Market Insights Report, 2019). Currently, 40% of food produced globally is refrigerated at some point in local, regional and international trade, and 15% of global energy consumption is devoted to refrigeration. Providers and stakeholders in the cold chain service are currently using cutting-edge technologies to cater to the increasing demand for food safety in fresh and processed food products. This is expected to create opportunities, which would boost the proliferation of numerous multinational vendors into the market, to provide better cold chain solutions. Stringent food safety regulations, such as the Food Safety Modernization Act (Laura, 2017), which requires increased attention toward the construction of a cold storage warehouse, are further expected to benefit the market. Issues associated with environmental concerns such as the emission of greenhouse gas are estimated to challenge the growth of the markets.

The recent growth of purchasing perishable food products online has created the need for innovative solutions to provide last-mile delivery, automated warehouses to manage inventories, and advanced temperature monitoring devices to maintain the safety of fresh food products. According to the National Statistics Bureau, in 2015 online purchases accounted for about 10.8% of China's retail sales. In online sales, vegetables and fruits accounted for a share of 55.2%, followed by meat and egg (at 17.4%). This growth has also encouraged around 4,000 companies to build their applications on smartphones, with specialization of fresh food sales in 2016. Emerging economies (such as China, India and Thailand) are investing in warehouse automation technologies (i.e., cloud technology, robots, conveyor belts and truck loading automation) to maintain the quality of temperature-sensitive products. According to the Global Cold Chain Alliance (GCCA), China has recorded a 41% growth in its refrigerated warehouse capacity, between 2014 and 2016, reaching about 107 million MT.

A cold chain is an interaction between a product, its destination and distribution channels. A product requires specific temperature and humidity conditions. These conditions dictate their successful transport to potential market destinations. Because of advances in cold chain logistics, it is possible to achieve long-distance distribution involving pre-cooling services, reefer trucks and warehousing facilities.

The cold chain is a science, a technology and a process. It requires the understanding of the chemical and biological processes linked with perishability, it relies on technology to ensure appropriate temperature conditions, and it is a series of tasks that must be performed to prepare, store, transport and monitor temperature-sensitive products (Rodrigue and Notteboom, 2016).

About 70% of the food consumed in the United States is handled by the cold chains, yet 25% of all food products transported in the cold chain are wasted each year due to breaches in integrity leading to fluctuations in temperature and product degradation. Supply chain integrity includes proper packaging, temperature protection and monitoring. Innovations in packaging, fruit and vegetable coatings, bioengineering and other techniques help reduce the deterioration of products.

1.3 IMPORTANCE OF COLD CHAIN AT DIFFERENT LEVELS OF THE SUPPLY CHAIN

1.3.1 AT THE FIELD LEVEL

Cooling the produce as soon as it is harvested and keeping it cool during storage and transport will slow pathogen growth and decrease the degradation of quality. To keep the produce in optimal condition, it is highly recommended to transfer the produce into a pre-cooling process such as forced air refrigerated area, vacuum cooling, hydrocooling, refrigerated trailer, refrigerated storage warehouse or applying a layer of ice to the produce. Early access to the cold chain is a key requirement for agribusinesses and farmers to take advantage of the growing demand for fresh produce in both domestic and international markets that require consistent quality, large volumes and high levels of food safety (Kitinoja et al., 2019). The long distances between harvest and distribution points, the lack of cold storage and poor transportation are the main drivers of postharvest losses (PHL) (Kader, 2005, 2006). Lack of cold chain infrastructure close enough to harvest points and the time lag in between harvest and first cold chain access causes high levels of over maturation and dehydration of perishable commodities, particularly leafy vegetables. Produce in informal markets usually does not enter into the cold chain (for which produce after harvest remains at 24°C–36°C per day). This significantly reduces the value and shelf life of the produce and, therefore, the amount of the produce that eventually reaches wholesale and retail markets. Pre-cooling, the rapid removal of field heat shortly after the harvest of a crop, is often omitted or not appropriately performed and this is the main cause of early desiccation. Very often small-scale farmers do not have access to pre-cooling solutions or believe standard cooling solutions can be applicable (Kitinoja, 2014; Kitinoja and Kader, 2015). The risk of desiccation is very high due to the long distances between the farms and the first cold storage available. This is then combined with logistical hurdles that increase delivery time and increase produce shrinkage often resulting in a loss in fresh weight and appearance. Mechanical damage accelerates produce dehydration which can claim up to 5% of volumes just between collection points and the peri-urban packing center and this, in turn, generates financial losses.

First-mile access to cooling and certification-ready facilities provides an opportunity for agribusinesses to reduce PHL and to store, aggregate and process produce sourced from several farmers. In rural areas, most farmers in the first mile of distribution lack the infrastructure required for developing the cold chain. Most farmers don't have access to the cold chain due to their inability to invest capital in infrastructure or due to the lack of cooling facilities close to them; thus, they are at a relative disadvantage in the supply chain (Kitinoja et al., 2019). Furthermore, these facilities need important energy and transportation infrastructures to be built, making it prohibitively expensive and not often possible in rural areas (Kitinoja and Barrett, 2015). The urgent need for solutions to create suitable environmental conditions to effectively reach and serve large numbers of small farmers and agribusiness in the first mile of distribution, especially in terms of physical access and affordability, is a way to stimulate innovation in the sector.

1.3.2 STORAGE

Cold storage facilities are crucial to minimizing PHL; however, losses occur at every step in the postharvest cycle, and therefore, cold storages cannot be considered as independent solutions to prevent postharvest spoilage but as one component that needs to be integrated with a cold chain network from the point of harvest to the point of purchase by the end consumer.

Temperature control in the shipment of products has continued to rise in relation to international trade. Several technologies are closely interacting sequentially to support the cold chain such as:

i. Devices and systems to monitor the condition of the cold chain, such as temperature and humidity, throughout all stages.
ii. Key technological innovations for storage are energy efficiency technologies that enable the facility to maintain a range of temperatures.

iii. Since a growing quantity of cold chain goods is shipped internationally, transport terminals such as ports are dedicating areas to cold chain logistics.
iv. A range of transport technologies are available and have been improved to transport cold chain products. Reefer vehicles and containers are among the most common technologies being used.

The success of industries that use the cold chain relies on knowing how to ship a product with temperature control adapted to the shipping circumstances. Cold chain operations have substantially improved in recent decades and the industry can respond to the requirements of a wide range of products. Different products require the maintenance of different temperature levels to ensure their integrity throughout the transport chain. The industry has responded with the setting of temperature standards that accommodate the majority of products. Staying within a temperature range is vital to the integrity of a shipment along the supply chain, and for perishables storage, it enables to ensure optimal shelf life. Any divergence can result in irrevocable and expensive damage as a product can simply lose market value or utility. For example, the shelf life of bananas stored below their threshold chilling temperature (13°C) can be reduced to only 5 days rather than 2 weeks if stored at 18°–20°C.

Different produce has different storage demands with regard to factors, such as optimum temperature, level of relative humidity, levels of ethylene production and sensitivity, and chilling sensitivity. Hence, the storage of single commodities is less complex than the storage of multiple commodities; however, the latter can still be the more viable option depending on external circumstances. Besides damages due to mismatches in temperature, the odors emitted by the produce can be transmitted. The ripening and decay processes can be accelerated through exposure to ethylene, which can result in changes in color, flavor and texture. Further, only high-quality produce should be "allowed" in the cold storage and produce should be sorted accordingly (FAO, 2009).

Tips for storage of high-quality horticultural produce:

- Store only high-quality produce, free of damage, decay and proper maturity (not overripe or under-mature).
- Know the storage requirements for the commodities and follow recommendations for proper temperature, relative humidity and ventilation.
- Avoid lower than recommended temperatures in storage because many commodities are susceptible to damage from freezing or chilling.
- Do not overload storage rooms or stack containers too high or too close together
- Provide adequate ventilation in the storage room.
- Keep storage rooms clean.
- Storage facilities should be protected from rodents by keeping the immediate outdoor area clean, and free from trash and weeds.
- Containers must be well ventilated and strong enough to withstand stacking. Do not stack containers beyond their stacking strength.
- Monitor temperature in the storage room by placing thermometers at different locations.
- Don't store onion or garlic in high-humidity environments.
- Avoid storing ethylene-sensitive commodities with those that produce ethylene.
- Avoid storing produce known for emitting strong odors (apples, garlic, onions, turnips, cabbages and potatoes) with odor-absorbing commodities.
- Inspect stored produce regularly for symptoms of injury, water loss, damage and disease.
- Remove damaged or diseased produce to prevent the spread of problems.

Examples of odor transfers that should be avoided:

- Apples/pears with celery, cabbage, carrot, potato or onion
- Citrus with strongly scented vegetables

- Green pepper will taint pineapple.
- Onion, nuts, citrus and potato should be stored separately (Kader, 2005).

Examples of ethylene-producing and ethylene-sensitive products:

- Ethylene-producing produce: apple, avocado, banana, pear, peach, plum, tomato
- Ethylene-sensitive produce: lettuce, cucumber, carrot, potato, sweet potato (Kader, 2006)

1.3.3 Transportation

Transporting fresh produce from point of origin (the field) to the point of use (retail store or food-service establishment) requires extensive planning. Horticultural commodities are susceptible to physical damage as well as microbial contamination during transport. Procedures and controls must be in place during the transportation process to assure the product arrives in good condition, safe and ready for consumption. Being able to ensure that a shipment will remain within a temperature range for an extended period is based on two important criteria: one is the type of container that is used and the other is the refrigeration method. Other important factors such as the duration of transit, the size of the shipment and the ambient or outside temperatures experienced are important in deciding what type of packaging is required. They can range from small insulated boxes that require dry ice or gel packs, rolling containers, to a 40-ft reefer that has its powered unit.

To minimize the effect of ripening and spoilage, it is important to keep the produce at the optimal storage temperature during transport. Transport agents and drivers must be aware of the optimal temperatures and be diligent in maintaining the proper temperature inside the refrigerated truck. Loading patterns and loading/unloading procedures will affect the temperature, so it is critical to manage the transport temperature to avoid damage to the product. Transportation logistics from the field to the retailer or foodservice establishment plays a major role in the longevity of the commodity. To assure freshness and optimum quality, produce deliveries must be made as quickly as possible and delays should be avoided.

The condition of the vehicle is critical when transporting fresh produce. The vehicles must be in good physical condition, dry and clean (washed and sanitized). They should be dedicated to carrying food products and must be pest-free. Refrigerated trucks or containers should be equipped with accurately calibrated thermometers and should be loaded to allow for adequate air circulation (Mohammed, 1993; Mohammed and Kitinoja, 2016).

A significant factor explaining the heterogeneity of produce temperature during land transportation is the heterogeneity of the airflow. The airflow is affected by factors such as the type of air delivery system (top-air delivery compared with bottom-air delivery) and the load patterns used in truck trailers, which will affect the amount of product warming or freezing that can occur. In top-air delivery trailers, all products should be placed on pallets or racks to provide adequate return air space under loads, while in bottom-air delivery trailers, special care should be taken to cover any vertical air channels left in a load from varying package sizes, shapes or numbers and to prevent short cycling of the circulating air and inadequate air circulation through and around the rest of the load (Brecht et al., 2016a, b, c). Maintaining the necessary air circulation channels to promote a uniform temperature is easier in straight loads of a single commodity. However, for mixed loads, which are common for short-distance transportation, the differences between the packages in terms of size, shape, design, required temperature, and the number will contribute to the heterogeneity of product temperature during land transportation. Whenever possible, only temperature-compatible commodities should be loaded in mixed loads. When that practice is not possible, a temperature suited to the most valuable or most perishable products should generally be used (Thompson et al., 2002b). Additional factors explaining the heterogeneity of product temperature during land transportation are different initial pallet temperatures as a result of nonuniform pre-cooling and greater sensitivity of pallets near the back of the truck to door opening (James et al., 2006; Pelletier et al., 2011; Nunes et al., 2014; de Micheaux et al., 2015).

Studies measuring time/temperature profiles of mixed loads of pumpkins and hot peppers during air transportation from Trinidad and Tobago to Canada indicated poor control of the temperature and relative humidity (Mohammed, 1993; Mohammed and Kitinoja, 2016). It is estimated that approximately only half of the transit time attributed to air transportation is spent in flight; during the other half, the perishable commodity is being transported to or from the airport, stored at the airport or loaded in or unloaded from the airplane (Mohammed, 1993). Both reports indicated that airport operations are divided into two categories: operations inside cargo terminals and operations on the tarmac. Inside cargo terminals, perishables are weighed, identified, labeled and placed into unit load devices (ULDs). Then, the ULDs are assigned to either a cargo flight or a passenger flight based on the volume and destination and are placed on the tarmac for cargo operations. After a flight, the ULDs are unloaded from aircraft cargo holds, brought to the cargo terminals to be inspected and cleared by customs and replaced again on regular pallets. High temperatures can be encountered, especially when the perishable products are on the tarmac before being loaded in the aircraft. On the tarmac, temperatures as extreme as $-50°C$ or $+50°C$ can be observed, depending on the season or location (Nunes et al., 2006; Pelletier, 2010; Mohammed and Kitinoja, 2016). Depending on airport operations, a difference of about $14°C$ inside ULDs was observed between the worst and best scenarios studied by Villeneuve et al. (2000). Bollen et al. (1998) measured the temperature inside pallets of asparagus shipped by air from Auckland to Tokyo and reported that the temperature of the asparagus increased from $4°C$ to $14°C$ within 30 minutes of ground operations in Auckland. Likewise, Mohammed (1993) reported increases of $10°C–12°C$ when hot peppers were shipped by air from Trinidad and Tobago to Toronto, Canada. Many factors, such as the availability of cooling and handling equipment at the airport, the type of ULDs used for transportation of perishables, flight delays, solar radiation on the tarmac and the geographical location of the airport, can result in temperature variations inside ULDs during airport operations (Villeneuve et al., 2000, 2001; Mohammed and Kitinoja, 2016). As a result of the possible increase in perishable produce temperature before or while the produce is being loaded in the aircraft, onsite pre-cooling may be required to ensure that produce safety is maintained (Laurin et al., 2003; Villeneuve et al., 2005). At certain major air hubs, such as Singapore and Dubai, some of the aforementioned effects are mitigated by airport cool storage capacity.

Sea transportation is much slower than transportation by air and may not always be appropriate for perishable commodities. Nevertheless, sea transportation may be more cost-efficient than air transportation and is an important mode of transportation for fruits, vegetables, dairy products, meats and fish products that are produced at a location far from the market and whose shelf life exceeds the transportation time. Sea transportation is generally performed in specialized vessels or refrigerated containers (Smale, 2004). Although transportation in specialized vessels remains significant for products such as bananas, refrigerated containers are the most frequent choice because they provide greater logistical flexibility and cost-efficiency for smaller shipments (Jedermann et al., 2014; Arduino et al., 2015). For both specialized vessels and refrigerated marine containers, the most typical airflow pattern is the delivery of cold air through T-floor gratings, and the air then flows vertically through the pallets before being returned to the cooling unit through the ceiling (Smale, 2004; Brecht et al., 2016a).

Amador et al. (2009) monitored the temperature inside a refrigerated marine container of pineapples shipped from Costa Rica to Florida, a trip that took 3 days. Defraeye et al. (2014) investigated sea transportation of citrus in refrigerated containers during a 21-day trip. Their findings showed that the temperature was lower (by as much as $3°C–4°C$) near the bottom of the pallets than near the top, a difference that is explained by the vertical airflow pattern inside refrigerated containers. In other investigations, Tanner and Amos (2003) monitored the temperature inside a refrigerated container and a specialized vessel during the shipment of kiwifruits from New Zealand to Belgium. Those authors reported a significant temporal and spatial heterogeneity of the temperature inside the container. Because of this heterogeneity, the temperature control system, which operated based on the measurement of the temperature at a single position within the container, was inefficient. As such, when the sensor measured a temperature above the set point of 0.5°C, a decrease in the delivery air temperature down to −5°C was observed, increasing the likelihood of freezing injuries for the pallets near the air delivery. Tanner and Amos (2003) also reported that the temperature was more uniform for transportation inside the hold of the vessel, although the number of kiwifruits outside the recommended range of temperature remained significant. The heterogeneity of the temperature can be attributed to several factors related to the operation and design of the container or the vessel as well as the properties of the product and packaging.

The guidelines for domestic transportation operations are:

a. Equipment maintenance is critical. Establish efficient defrost cycles for refrigeration units and insulated walls, floors and ceilings using thermostat devices. Manual defrost of a refrigeration unit should only be undertaken after unloading. Undertake regular calibration checks on refrigeration and sensor units. Check and maintain door seals regularly according to operating procedures.

b. Transport vehicles are not expected to significantly reduce product temperatures. Products should be loaded at or near their transport temperature. Loading warmer products at best may create heat transfer issues and at worst may create a microbiological hazard (if the transport vehicle's refrigerator, for example, is not able to bring the temperature down quickly enough to prevent spoilage).

c. Place a temperature data logger or thermometer at the warmest part of the refrigerated space (usually at the top of a wall nearest the door). If using a thermometer, carry out periodic temperature verification checks as requested by operator procedures and manually record observed temperatures. Record observed temperature after the product has been unloaded at each drop off point. (Wireless thermometers and data loggers are preferred as they do not require the refrigerated space to be opened to conduct temperature checks and can usually sound an alarm if a nominated temperature is exceeded.) Multiple data loggers help measure through load temperature variability, which can exceed 5°C, particularly in older assets.

d. If the product temperature is higher than the owner's specification, take corrective action to maintain product safety and quality. Do not assume that simply bringing a product's temperature back down is enough. Contact the manufacturer to identify whether further action needs to be taken.

When loading products, be conscious of the need for excellent air circulation. Do not overload the refrigerated compartment and never load above the compartment's load line.

1.3.4 Retail Stores and Domestic Storage

At several retail outlets, perishable produce is generally placed in a display cabinet or rotated between a display cabinet and a refrigerated storage room. Based on the data analyzed on a survey comprising 187 consumers conducted in Trinidad, it was concluded that 75%–83% of consumers at the point of purchase had confidence that perishable commodities were kept at the required temperature by supermarkets with chain stores (Mohammed and Craig, 2018). Nevertheless, it was noticeable that the time/temperature measurements indicated that storage in display cabinets was generally not the most efficient step in the cold chain, as the temperature was frequently elevated above the desired limit. An explanation for the poor refrigeration reported during the display at retail may be that some retailers are concerned more about the appeal of a product than about its preservation, for instance, overloading the front of display cabinets or placing the racks at the highest position (Mohammed and Craig, 2018; Villeneuve et al., 2002a).

Mohammed and Craig (2018) measured the temperature of ripe tomatoes, mangoes and cassava displayed on refrigerated cabinets in 45 supermarkets in Trinidad and Tobago, Guyana and St. Lucia and reported an average temperature of 12°C–13°C, significantly above the recommended temperature. Similarly, Villeneuve et al. (2002a) reported average temperatures of 8.6°C or 10.8°C for potatoes in display cabinets used by a Canadian retail chain, depending on whether the rack was placed at the lower or upper position, respectively. These contrasting results may be explained by several factors, a significant one being the heterogeneity of the temperature according to the position inside the display cabinet. Mohammed and Craig (2018) reported that the temperature of tomato and mango packed by produce managers toward the front of the refrigerated display cabinet was on average 2°C–3°C higher than the temperature of those placed toward the back. Similarly, Villeneuve et al. (2002a) reported that the temperature of potatoes on the side of a display cabinet was 3.9°C above the temperature of those near the center. These investigations showed that the temperature was generally warmer at the top of the display cabinets and that the placement of the fruits and vegetables was frequently inefficient given their recommended temperature. They further concluded that the causes of temperature heterogeneity inside display cabinets include the penetration of ambient air, the proximity to the lighting system, the defrost cycle and the heterogeneity of the airflow (Villeneuve et al., 2003; Nunes et al., 2009; Mohammed and Craig, 2018). Additional factors explaining the wide range of temperatures during the display at retail reported in the literature may include differences in the type, function, efficiency, setpoint temperature, turnover rate and door opening rate of display cabinets.

All studies on cold chains for perishable commodities that have measured time/temperature profiles from the moment the product is bought by the consumer until it is consumed indicate insufficient refrigeration practices. Mohammed and Craig (2018) in their study discussed the transportation of tomato, mango and cassava from harvest to 48 households in rural and urban locations in Trinidad and Tobago and Guyana and recorded that 88%–92% of these commodities attained a temperature above 9°C–10°C along the cold chain. Several studies have specifically investigated the temperature inside domestic refrigerators. Seven studies conducted in Europe or the United States and reviewed by James et al. (2008) indicated that temperatures inside domestic refrigerators were between 5.9°C and 7.0°C. In a similar investigation, Mohammed and Craig (2018) measured average temperatures inside household refrigerators to vary between 4°C and 8°C. The studies conducted on time/temperature measurements along the cold chain that have considered storage of perishable produce inside domestic refrigerators are in agreement with those that have focused specifically on this last step in the cold chain. Overall, the majority of studies points to an average temperature inside domestic refrigerators of between 6.0°C and 7.0°C, with the temperature being 1.0°C–2.0°C higher inside the door.

The higher-than-recommended temperature inside most domestic refrigerators could be attributed to various factors, including insufficient consumer awareness about the importance of proper temperature for food quality and safety. Even though the critical impact of consumer practices on the proper preservation of perishable commodities is well established, surveys show that most consumers are unaware of their important role in the cold chain and place most of the responsibility for maintaining food quality and safety in the industry. Other notable factors accounting for the high temperature inside domestic refrigerators include the frequent opening of the door, overloading or inadequate placement of the food, and inappropriate temperature settings. The high temperatures observed inside domestic refrigerators indicate that substantial improvements to consumer practices are required to improve perishable food preservation and limit food safety risks.

1.4   CURRENT STATUS OF DIFFERENT LEVELS OF THE COLD CHAIN

As discussed earlier, the cold chain is considered an essential tool to handle perishable produce and keep them in good condition, while connecting different markets. When maintained properly, the cold chain is a popular and relatively easy method to reduce food losses. The cold chain network and its components are well organized in high-income nations, but they are generally lacking in developing countries. Some nations, such as India and China, have experienced a remarkable increase in their cold storage capacity during recent years but still are not addressing needs in all the sectors. The majority of the cold capacity is directed toward the dairy and meat industries, while fresh fruits and vegetables are often neglected. The present status of different aspects of the cold chain in developing nations is highlighted below.

1.4.1 PRE-COOLING

Pre-cooling is the first and essential step of cold chain management where fruits and vegetables are treated with a low-temperature medium or cooling device. This process is important to maintain produce quality, lower decay rates as well as quantitative losses, extend shelf life and also reduce the energy consumption during storage (Thompson et al., 2002a). There are different options for pre-cooling such as hydro cooling, vacuum cooling, forced air cooling and hydro-air cooling. For smallholders, contact icing, evaporative cooling and portable forced air cooling can be easily used on the farm. The suitability of the options exclusively depends upon the produce being handled as well as the complexity of the value chain involved (Kitinoja and Thompson, 2010; Kitinoja, 2013). The significance of the pre-cooling operation is often undervalued which has limited its adoption. Technologies such as vacuum cooling involve high capital investment and maintenance and their usage is very limited in developing nations. Vacuum cooling is widely used in high-income countries for cooling leafy vegetables (Kitinoja and Thompson, 2010). Refrigerated room cooling (10°C–13°C) is comparatively cheaper than all other techniques, but the cooling is usually uneven and takes a long time to achieve proper pre-cooling. This method is best suited for relatively less perishable commodities such as onion, potato, sweet potatoes and apple fruit. Hydro cooling provides faster and uniform cooling, but the produce and the packaging material should be tolerant of wetting. It is used in some fruit, which receives wash, spray or dip after the harvest but holds high chances for the spread of pathogenic microorganisms. The water used for hydro cooling should be from a clean source and chlorinated to reduce microbial load (Thompson et al., 2002b). The package icing method also provides effective pre-cooling and maintains quality similar to that of hydro cooling or forced air cooling (Kochhar and Kumar, 2015). The small-scale stakeholders can choose suitable pre-cooling methods such as evaporative cooling systems, mechanical or other alternative sources of cooling, based upon the expenses, efficiency and type of commodity to be cooled (Kitinoja and Thompson, 2010).

1.4.2 COLD STORAGE

The global rise in demand for fresh fruits and vegetables as well as frozen commodities has increased the need for cold storage capacity. Cold storage serves a variety of needs in the food supply chain and ensures the year-round availability of seasonal foods, import and export logistics, and retail supply needs. The cold chain development in developing nations usually refers to cold storage only, with very less importance given to other aspects of the cold chain (Brondy, 2019). Over the past two decades, several public and private sector organizations have invested an enormous amount of money in the development of cold warehouses worldwide. According to the reports published by GCCA, the total capacity of refrigerated warehouses was about 616 million m³ in 2018, which is 2.7% more than the total capacity in 2016 and 11.4% more than that in 2014. Nearly 60% of world cold storage capacity is accounted for in India (150 million m³), the United States (131 million m³) and China (105 million m³) (GCCA, 2018). The average capacity range of the cold warehouses ranged from 15,000 to 25,000 m³ in developing nations, while it ranged over 100,000 m³ in high-income nations. The cold storage can be of different categories based upon the temperature range, i.e., frozen (<−18°C), low-chill (0°C–10°C) and mild-chill (10°C–20°C).

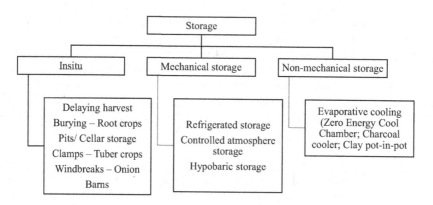

FIGURE 1.1 Different storage structures for horticultural produce.

Usually, fresh horticulture commodities fall in the category of low and mild chill, while many of the tropical fruits and vegetables develop chilling injury symptoms when stored below respective safe storage temperatures (Heyes, 2018). In developing nations, erratic electricity supply and fuel availability are the critical factors deciding the successful functioning of cold storage. In some locations, where a sustainable business case for investment is not developed, it may not be economical to maintain refrigerated cold storage. Establishing low-cost/nonmechanical cold storage structures to maintain low temperatures prove to be appropriate in such locations. These include zero energy cooling chambers, charcoal coolers, solar chillers, walk-in cold rooms, CoolBot™ equipped cold rooms, root cellars, etc. (Figure 1.1). These low-cost cold storage technologies can maintain the quality of produce and extend marketable life for a few extra days. These low-cost small-scale alternatives are becoming popular over refrigerated cold storage in low-income nations. For example, ColdHubs in Nigeria is a solar-powered walk-in cold storage installed in widely accessible places nearby marketplaces and farms. These units offer pay per crate per day to store their produce, which allows customers to extend the marketable life of their commodities for about 2–21 days (ColdHubs, 2020). Another example is CoolBot™ equipped cold rooms, which work using a normal domestic air conditioner with slight modifications enabling it to operate below the minimum set temperature (16°C). Installation and using CoolBot™ systems are much less expensive than commercial refrigeration systems and are being used in several regions of the world (Saran et al., 2013, www. storeitcold.com). Non-mechanical storage structures such as walk-in evaporative cooled warehouses can often achieve temperatures between 10°C and 18°C, which are adequate to extend shelf life in fresh tropical crops such as tomatoes and mangoes. But the results are best only in dry or semi-arid regions, with the relative humidity outside these structures lower than inside (Kitinoja, 2013). These low-cost structures are described in detail in the next chapters of this book.

1.4.3 REFRIGERATED TRANSPORTATION

Transportation of the produce from farm to wholesale/retail market forms an important link in the supply chain. A significant quantity of produce is lost during transport mainly due to rough handling, improper packaging, overloading, lack of good quality road infrastructure and lack of temperature management (Brondy, 2019). Temperature management plays a vital role during the long-distance transport, and therefore, the loads must be well stacked strategically to enable proper air circulation to carry away the heat generated from the produce as well as heat from the surrounding atmosphere (Kitinoja and Kader, 2015). A refrigerated transport vehicle mandatorily requires well-insulated walls and a proper cooling system to maintain the cool environment for the precooled commodities. But these factors involve high investment as well as regular maintenance and, therefore, understanding the return of investment for the specific business case is very important. There is a large gap between the required refrigerated transport and transport available in most of the developing countries. For example, in India, it is estimated that the difference between reefer vehicles required and the ones available is nearly 85% (NCCD, 2015). Appropriate packaging methods, maintaining the cleanliness of the transport truck and covering or providing the shade to the beds allowing proper air circulation, reduce considerable losses during transit. Transporting mixed loads in refrigerated trucks is common practice, but this could be a serious concern when optimum temperatures and ethylene sensitivity of the commodities transported are not compatible. High ethylene producers (like apples, pears, melons and bananas) can induce senescence and associated undesirable changes in color, aroma, taste and texture in ethylene sensitive commodities (like broccoli, cabbage and green leafy vegetables) (Kitinoja and Kader, 2015). Refrigerated rail transport is not very popular in the majority of the countries, but developing a well-maintained network of reefer railways would reduce the total number of road vehicles needed.

1.4.4 RETAIL MARKETS AND DOMESTIC STORAGE

Retail is the last link of the cold chain infrastructure, and the commodity transported into different locations does not often benefit from the whole cold chain until it is maintained at the recommended temperatures at retail. Yet, the cooling at the retail level is often neglected and, in most cases, the fresh fruits and vegetables are displayed at ambient temperatures, exposing them to temperatures as high as 30°C–35°C in tropical countries. This causes moisture loss, shriveling and softening of fruits and vegetables reducing their market price (Kitinoja and Kader, 2015). In the interiors of developing countries, the availability of refrigeration and electricity is very limited and, in such areas, basic temperature control methods such as shading the produce, sprinkling with water, covering with a moist cloth and using evaporative cooling structures could be used as alternatives. Care should be taken to use clean/potable water for this purpose to avoid any food safety issues. The supermarkets in some cities do have refrigerated display cases, but it is mostly limited to carbonated drinks, ice creams or some high-value commodities (Brondy, 2019). In some cases, some expensive commodities are given preference over other fresh fruits and vegetables, irrespective of the actual cooling requirements of a certain commodity.

The final step of the supply chain is the consumer, and the cold chain is continued as domestic refrigerators in the household. Home refrigerators significantly contribute to reducing domestic food waste. Brown et al. (2014) has explained in detail how temperature management in refrigerators could reduce financial and energy losses associated with food wastes. With increasing technologies and purchasing capacity, home refrigerators have become the most commonly used appliance in the majority of households. Though refrigerators were supposed to contribute to reducing food losses, some studies show that it also is responsible for food loss or waste when they are not used properly. For example, some tropical fruits and vegetables develop chilling injury symptoms and can lead to accelerated spoilage, when stored at temperatures lower than 10°C. Such commodities can be stored in the kitchen in a cool and shaded place. Consumer behavior of storing food for a long time in the refrigerator as well as misunderstood labels also lead to food waste. It is suggested to consume the product stored in the home refrigerators as soon as possible to prevent any wastage. Several indigenous storage methods such as earthen pot-in-pot, desert cooler, drip cooler, etc., are practiced in many parts of the world (Abiso et al., 2015).

1.5 REFRIGERANTS

The majority of mechanical cooling equipment functions using a vapor compression cycle in the presence of a working fluid called a refrigerant. There is a range of refrigerants available including water, ammonia, carbon dioxide, halocarbons, etc. Carbon dioxide, ammonia and sulfur dioxide were widely used, but they have been gradually replaced because of their toxicity, odor or the requirement of very high pressures. Anhydrous ammonia is being used in some large industrial installations even today. Ammonia has a remarkable performance in a wide range of refrigeration applications, but as it is a highly inflammable and toxic gas, the refrigeration system needs regular monitoring and servicing to prevent any hazard. In the 1930s–1940s, a class of compounds referred to as chlorofluorocarbons (CFCs) with one or two carbon atoms with a varying number of chlorine and fluorine atoms were very popular refrigerants. They were nontoxic to humans, noninflammable, noncorrosive with desirable thermodynamic properties. For about half a decade, CFCs especially R11 (trichlorofluoromethane, CCl_3F_2), R12 (dichlorodifluoromethane, CCl_2F_2) and R22 (chlorodifluoromethane, $CHClF_2$) were abundantly used for both domestic and industrial refrigeration. Later in the 1970s, it was found that CFCs had a negative impact on the environment as it oxidizes the ozone layer in the stratosphere.

The ozone-depleting potential (ODP) and global warming potential (GWP) are popular measures to calculate the environmental impact of refrigerants. The refrigerants can impact the environment either directly due to refrigerant leaking or due to the extra fossil-fuel resources spent on less-efficient refrigerants. Following the Montreal and Kyoto protocols, which were signed by almost all the countries, the CFCs are no more being used as refrigerant in developed nations and even in most developing nations. R134a (1,1,1,2-tetrafluoroethane, CH_2FCF_3) which has very low ODP was considered as a replacement for R12 because of the presence of fluorine it has several times higher GWP values than carbon dioxide. None of the existing refrigerants completely satisfy the criteria to be an ideal refrigerant. Carbon dioxide has relatively lower GWP and zero ODP, but it is not as efficient as ammonia or other refrigerants for it has a very low critical point. Ammonia though is excellent in environmental criteria having zero ODP and GWP, but safety could be a major concern. But with advances in the component design of refrigeration units, leakages have been a rare issue. Using nonflammable secondary refrigerants, which are cooled by the primary refrigerant, has also reduced the safety concerns of flammable refrigerants.

1.6 FOOD SAFETY

Food safety is an essential component in the supply chain to assure consumers that the food, when consumed according to its intended use, will not cause any harm. The foodborne hazards during postharvest handling and storage could be of three categories: biological, chemical, and physical. Biological hazards constitute bacteria, viruses, protozoans, fungi, helminthic parasites as well as insect pests and cause spoilage or some serious foodborne illness. Chemical hazards in fresh horticulture produce could be due to the agricultural chemicals used during production and postharvest handling or even by the toxins produced by certain bacteria and fungi. Physical hazards include foreign materials in the food items such as pieces of stones, metal, plastic, etc. Time-temperature abuse is one of the important factors causing contamination during fruit and vegetable handling and storage, which results in the production, growth and survival of pathogens (Piližota, 2014). Mishandling of the produce results in mechanical injuries or intrusion of foreign materials, which can further aggravate other hazards. It is important to maintain optimum ranges of relative humidity and temperatures throughout the storage period to maintain quality, safety and minimize losses. The low temperatures slow down microbial activity and growth, but some fresh fruits and vegetables develop chilling injury symptoms when stored below optimum temperatures (Kitinoja, 2013). The knowledge of the properties of certain fruit or vegetable crop is essential as the cold chain requirement varies with the crops at various stages of the supply chain.

In most cases, the facilities to store each commodity under individual optimum conditions are rarely available. The produce handlers usually have to make a choice of temperature and relative humidity based upon the value and demand of the commodity. The food safety of the commodity can be maintained by measures such as cleaning and washing the produce with sanitizing agents like chlorine, ozone and hydrogen peroxide after harvest; maintaining hygiene and proper cold holding during transport and storage; and appropriate packaging as well as by ensuring workers' hygiene at all stages of produce handling (Piližota, 2014). In recent times, several physical treatments (UV-C light, pulsed magnetic field, cold plasma, etc.) have been demonstrated to perform efficient surface decontamination of fresh fruits and vegetables during packaging and even at retail displays (Deng et al., 2020).

1.7 INNOVATIONS AND FUTURE SCOPE

Innovations in the cold chain are monitoring sensors and proper data analysis. Wireless sensors that detect the temperature in containers, trucks and refrigerated shelves that are battery-less and maintenance-free are being placed into existing IP, Wi-Fi or mobile networks that process data. The energy-harvesting principle—gaining energy from the surrounding environment—allows wireless modules to work without batteries and cables. In a cold chain monitoring system, temperature data from self-powered sensors are used to alert when temperatures approach thresholds and investigation or maintenance of the cooler is required. The information is sent to a central controller or even as a push notification to a smartphone. This enables the user to react in time before the cold chain is interrupted. Besides, these data, collected at thousands of containers, can be consolidated to find potential gaps in the temperature-controlled supply chain for an optimized logistic process.

Connecting the sensor system to IP provides ways to reach out and control the temperature from virtually anywhere on the globe. It allows centralized, or even outsourced computing resources (cloud-based computing), thus driving down infrastructure costs. Cloud-based computing resources enable the integration of external information; for example, local weather yielding an optimized cooling environment. This information is provided to the remote controller, enabling precise control. The same system that monitors temperature can also be used to ensure that only the actual needed amount of energy is provided, driving down energy waste and costs. Continuously collected data in the supply chain could improve food quality and significantly reduce loss.

In development are new refrigeration systems that do not use chemicals, fossil fuels or electricity. The clean cold chain is being researched and promoted by the University of Birmingham (UK) and tested in the United Kingdom by various companies that focus on engine manufacturing, energy storage and cryogenic power (https://www.birmingham.ac.uk/research/energy/research/-cold-economy/clean-cold-chains.aspx). One of the technologies being developed is the Dearman engine, which is being tested in reefers for providing cooling. The Dearman engine operates on liquid nitrogen or liquid air (highly compressed gases) and provides outputs of energy in the form of motion, with the by-products of cold and water vapor. The cold exhaust from the Dearman engine is being captured and used for cooling in reefer trailers.

Research and Development (R and D) for cold chain is required in the related areas of packaging (biofilms, biodegradable materials); produce protection (pest management, edible coatings for reducing water loss, etc.); and energy-efficient cooling, transport and cold storage systems. Many of the available technologies may require adaptation to meet the needs of developing regions.

1.8 CONCLUSIONS

Cold chain development is in various stages of awareness, application and adoption for fresh produce in developing regions. Improved access to needed tools and supplies, lower costs (perhaps via the development of local businesses related to cold chain management), training in more sustainable technologies and handling practices, and effective logistics and overall management will all be required for success in cold chain development.

REFERENCES

Abiso, E., Satheesh, N., and Hailu, A. (2015). Effect of storage methods and ripening stages on postharvest quality of tomato (*Lycopersicon esculentum* Mill) cv. Chali. *Annals of Food Science and Technology*, 6(1), 127–137.

Amador, C., Emond, J.P., and do Nascimento Nunes, M.C. (2009). Application of RFID technologies in the temperature mapping of the pineapple supply chain. *Sensing and Instrumentation for Food Quality and Safety*, 3(1), 26–33.

Arduino, G., Carrillo Murillo, D., and Parola, F. (2015). Refrigerated container versus bulk: Evidence from the banana cold chain. *Maritime Policy and Management*, 42(3), 228–245.

Bollen, A.F., Brash, D.W., and Bycroft, B.L. (1998). Air-freight cool chain improvements using insulation and supplemental cooling. *Applied Engineering in Agriculture, 14*(1), 49–53.

Brecht, P.E., Durm, D., and Rodowick, L. (2016a). Refrigerated Transportation Best Practices Guide. International Refrigerated Transportation Association.

Brecht, P.E., Durm, D., and Rodowick, L. (2016b). Summary and User Guide: FDA's Sanitary Transportation of Food Final Rule Advancing the Sanitary Transportation of Human and Animal Food. International Refrigerated Transportation Association.

Brecht, P.E., Durm, D., and Rodowick, L. (2016c) Sanitary Transportation of Food Compliance Matrix. International Refrigerated Transportation Association.

Brondy, A. (2019) Growing the cold chain in developing and emerging economies. *Agriculture for Development, 36*, 4–9.

Brown, T., Hipps, N.A., Easteal, S., Parry, A., and Evans, J.A. 2014. Reducing domestic food waste by lowering home refrigerator temperatures. *International Journal of Refrigeration, 40*, 246–253.

ColdHubs (2020) ColdHubs: Solar Powered Cold Storage for Developing Countries. http://www.coldhubs.com/. Accessed on 29 April 2020.

de Micheaux, T.L., Ducoulombier, M., Moureh, J., Sartre, V., and Bonjour, J. (2015). Experimental and numerical investigation of the infiltration heat load during the opening of a refrigerated truck body. *International Journal of Refrigeration, 54*, 170–189.

Defraeye, T., Lambrecht, R., Delele, M.A., Tsige, A.A., Opara, U.L., Cronjé, P., Verboven, P., and Nicolai, B. (2014). Forced-convective cooling of citrus fruit: Cooling conditions and energy consumption in relation to package design. *Journal of Food Engineering, 121*, 118–127.

Deng, L.Z., Mujumdar, A.S., Pan, Z., Vidyarthi, S.K., Xu, J., Zielinska, M., and Xiao, H.W. (2020). Emerging chemical and physical disinfection technologies of fruits and vegetables: A comprehensive review. *Critical Reviews in Food Science and Nutrition, 60*(15), 2481–2508.

FAO (2009) Horticultural Chain Management for Countries of Asia and the Pacific Region. [Online] Available from: http://www.fao.org/3/a-i0782e/.

GCCA (2018) 2018 GCCA Global Cold Storage Capacity Report. https://www.gcca.org/sites/default/files/2018%20GCCA%20Cold%20Storage%20Capacity%20Report%20final.pdf. Accessed on 27 April 2020.

Heyes, J.A. (2018). Chilling injury in tropical crops after harvest. *Annual Plant Reviews, 1*(1), 149–180. 10.1002/9781119312994.apr0605.

James, S.J., Evans, J., and James, C. (2008). A review of the performance of domestic refrigerators. *Journal of Food Engineering, 87*(1), 2–10.

James, S.J., James, C., and Evans, J.A. (2006). Modelling of food transportation systems–a review. *International Journal of Refrigeration, 29*(6), 947–957.

Jedermann, R., Praeger, U., Geyer, M., and Lang, W. (2014). Remote quality monitoring in the banana chain. *Philosophical Transactions of the Royal Society A: Mathematical, Physical and Engineering Sciences, 372*(2017), 20130303.

Kader, A.A. (2005). Increasing food availability by reducing postharvest losses of fresh produce. *Acta Horticulturae, 682*, 2169–2175.

Kader, A.A. (2006). The return on investment in postharvest technology for assuring quality and safety of horticultural crops. *Journal of Agricultural Investment, 4*, 45–52.

Kitinoja, L. (2013). Use of cold chains for reducing food losses in developing countries. PEF White Paper No.13-03.

Kitinoja, L. (2014). Exploring the potential for cold chain development in emerging and rapidly industrializing economies through liquid air refrigeration technologies. Liquid Air Energy Network, UK.

Kitinoja, L. and Barrett, D.M. (2015). Extension of small-scale postharvest horticulture technologies – A model training and services center. *Agriculture, 5*(3), 441–455

Kitinoja, L. and Kader, A.A. (2015). Small scale postharvest handling practices: A manual for horticultural crops (5th Edition). University of California, Davis, Postharvest Technology Research and Information Center. https://ucanr.edu/sites/Postharvest_Technology_Center_/files/231952.pdf. Accessed on 3 May 2020.

Kitinoja, L. and Thompson, J.F. (2010). Pre-cooling systems for small-scale producers. *Stewart Postharvest Review, 6*(2), 1–14. doi:10.2212/spr.2010.2.2.

Kitinoja, L., Tokala, V.Y. and Mohammed, M. (2019). Clean cold chain development and the critical role of extension education. *Agriculture for Development, 36*(3): 19–25.

Kochhar, V. and Kumar, S. (2015). Effect of different pre-cooling methods on the quality and shelf life of broccoli. *Journal of Food Processing Technology, 6*, 424. doi:10.4172/2157-7110.1000424.

Laura, M. (2017). Three things to expect during unannounced FDA inspections. Food Safety News.

Laurin, E., Nunes, M.C.N., and Émond, J.P. (2003). Forced-air cooling after air-shipment delays asparagus deterioration. *Journal of Food Quality, 26*(1), 43–54.

Market Insights Report. (2019). Global Cold Chain Monitoring Market (2019 to 2025). Industry analysis, trends, market size and forecasts. http://www.researchandmarkets.com/.

Mohammed, M. (1993). Postharvest constraints and solutions involved in export marketing of selected speciality and exotic vegetables from the West Indies. *Acta Horticulturae, 312*, 355–373.

Mohammed, M. and Craig, K. (2018). UN/FAO Food loss analysis: Causes and solutions. Case study on the tomato value chain in the Republic of Trinidad and Tobago. ISBN: 978-92-5-130586-7. 38 pp.

Mohammed, M. and Kitinoja, L. (2016). Gaps in the cold chain in the Caribbean. *Proceedings of the 3rd World Cold Chain Summit to Reduce Food Losses and Waste*. 1st–3rd December 2016, Singapore. 7 pp.

NCCD. (2015). All India Cold-chain Infrastructure Capacity (Assessment of Status and Gap), Delhi. (www.nccd.gov.in)

Nunes, M.C.N., Emond, J.P., and Brecht, J.K. 2006. Brief deviations from set point temperatures during normal airport handling operations negatively affect the quality of papaya (*Carica papaya*) fruit. Postharvest Biology and Technology, 41:328–340.

Nunes, M.C.N., Emond, J.P., Rauth, M., Dea, S., and Chau, K.V. 2009. Environmental conditions encountered during typical consumer retail display affect fruit and vegetable quality and waste. Postharvest Biology and Technology, 51:232–241.

Nunes, M.C.N., Nicometo, M., Emond, J.P., Melis, R.B and Uysal, I. (2014). ́Improvement in fresh fruit and vegetable logistics quality: Berry logistics field studies. *Philosophical transactions. Series A, Mathematical, Physical, and Engineering, 372*,20130307.

Pelletier, W., Brecht, J.K., do Nascimento Nunes, M.C., and Emond, J.P. (2011). Quality of strawberries shipped by truck from California to Florida as influenced by postharvest temperature management practices. *HortTechnology, 21*(4), 482–493.

Piližota, V. (2014). Fruits and vegetables (including herbs). In Y. Mortarjemi and H. Lelieveld (eds.). *Food Safety Management – A Practical Guide for the Food Industry* (pp. 213–249), Academic Press, Cambridge, MA. doi:10.1016/B978-0-12-381504-0.00009-3.

Rodrigue, J.P. and Notteboom, T. (2016). The Cold Chain and its Logistics. https://transportgeography.org/contents/applications/cold-chain-logistics/. Accessed on 27 April 2020.

Saran, S., Dubey, N., Mishra, V., Dwivedi, S.K. and Raman, N.L.M. (2013). Evaluation of cool bot cool room as a low-cost storage system for marginal farmers. *Progressive. Horticulture, 45*(1), 115–121.

Smale, N.J. 2004. Mathematical modelling of airflow in shipping systems: model development and testing [Doctoral dissertation]. Massey University, Palmerston North, New Zealand. Available from: http://encore.massey. ac.nz/iii/encore/record/C__Rb1816458? lang=eng.

Tanner, D.J. and Amos, N.D. (2003). Temperature variability during shipment of fresh produce. *Acta Horticulturae, 599*, 193–203.

Thompson, J.F., Brecht, P.E. and Hinsch, T. (2002a). Refrigerated Trailer Transport of Perishable Products. The University of California Div. of Agricultural and Natural Resources. Pub. No. 21614.

Thompson, J.F., Mitchell, F.G., Rumsey, T.R., Kasmire, R.F. and Crisosto, C.H. (2002b). Commercial cooling of fruits, vegetables, and flowers (No. 635.046 C734c 2002). California, US: University of California, Division of Agriculture and Natural Resources.

Villeneuve, S., Mercier, F., Pelletier, W., Ngadi, M.O. and Emond J.P. (2000). Effect of environmental conditions on air shipment of perishables during ground operations. *ASAE Annual International Meeting, Technical Papers: Engineering Solutions for a New Century, 2*, 599–609.

Villeneuve, S., Ngadi, M.O., and Emond, J.P. 2001. Heat transfer in air cargo unit load devices. Acta Horticulturae, 566:245–250.

Villeneuve, S., Emond, J.P., Mercier, F., and Nunes, M.C.N. 2002a. Analyse de la temperature de l'air dans un comptoir refrig ere. Rev. Gen. Froid, 1025:17–21.

Villeneuve, S., Emond, J.P., and Mercier, F. 2003. Comptoir refrigere avec distribution frontale de l'air. Rev. Gen. Froid, 1033:19–23.

Villeneuve, S., Bazinet, D., Mercier, F., Pelletier, W., and Emond, J.P. 2005. Systeme de refroidissement rapide fonctionnant a l'azote liquide pour les unit es de chargement d'avion. Rev. Gen. Froid, 1054:21–26.

2 Cooling Requirements of Selected Perishable Crops During Storage

Puran Bridgemohan
The University of Trinidad and Tobago

Majeed Mohammed and Vijay Yadav Tokala
The Postharvest Education Foundation

CONTENTS

2.1 INTRODUCTION

For most farmers, harvesting mature crops during or at the end of the growing or reproductive phase marks the culmination of the production process. However, this is not without its own peculiar set of challenges especially for the delivery of fresh, high-quality produce to markets. Overcoming biological and nonbiological challenges at each step in the postharvest handling system is critical for optimizing quality to secure the best bargaining position among stakeholders and to negotiate the highest market prices.

The lack of reliable and adequate cold chain facilities is one of the main causes of losses in perishable products, which are estimated to be about 40%–50% in the case of roots, tubers, fruit and vegetables (Gustavsson et al., 2011). The development and proper implementation of the cold chain facilitate postharvest loss reduction, helps meet market requirements for quality and safety and also improves food security initiatives. Low temperature is the main method used to extend the storage and market life of perishable commodities. Cold storage delays ripening and senescence processes by reducing the respiration rate (Brizzolara et al., 2020). Low temperature is effective

DOI: 10.1201/9781003056607-3

in decreasing the catalytic activities of different enzymes, including those involved in the different steps of respiration.

The postharvest operations of a given commodity include the appropriateness of the cold chain procedures with respect to temperatures and relative humidity (RH), in association with suitable packaging (Mangaraj et al., 2009). Nevertheless, each step in the cold chain has a significant impact on the final quality of the commodity, and temperature abuses that exceed the product tolerance level occurring at any point lead to food waste or raise safety concerns (Kitinoja et al., 2019).

In this chapter, the key aspects related to the suitable temperature regimes of selected perishable commodities along the cold chain as well as the potential improvements to the cold chain provided by real-time temperature monitoring of the various produce will be discussed. Studies conducted on the effects of the temperature of selected perishables along the cold chain are reviewed. From the analysis of previous studies, the current efficiency of the cold chain in maintaining the temperature of perishable commodities is established, and the most critical weaknesses that need to be corrected for better produce quality and safety are identified. Potential management systems to improve the cold chain based on the measurement of perishable food temperature are discussed, and challenges related to the implementation of such systems are identified. Finally, relevant prospective research projects for global and inclusive improvements to the cold chain are proposed.

2.2 PHYSIOLOGICAL CHANGES DURING COLD STORAGE

Proper storage conditions, such as temperature and humidity, are required to maximize storage life and maintain quality once the crop has been cooled to the optimum storage temperature. Generally, fresh fruits and vegetables need safe low temperatures and high RH for efficient storage with lower quality losses (Kitinoja et al., 2019). The higher the rate of respiration, the faster the produce deteriorates, while the lower temperatures slow down the respiration rates and thereby ripening and senescence processes, which prolongs the storage life of fruits and vegetables. Low temperatures also regulate the growth of pathogenic fungi which cause spoilage of fruits and vegetables in storage. High RH reduces physiological loss of weight by transpiration and maintains the nutritional quality as well as appearance. A significant challenge for an efficient cold chain is the different requirements of perishables (fruits and vegetables, fresh-cut fruits and vegetables, root crops and bulbs) to maximize shelf life and commercial potential. The choice of storage temperature for a specific commodity depends on several factors, including the commodity's mechanical properties and its sensitivity to chilling or freezing injuries. Injury from freezing temperatures can appear in fruit and vegetable tissues as loss of rigidity, softening and water soaking. Chilling injury (CI) is a physiological disorder that occurs to certain species when exposed to nonfreezing temperatures. CI can occur at temperatures from 0°C to 12.5°C (Saltveit, 2004). CI symptoms are varied and often do not develop until the produce has been returned to warmer temperatures.

Current market requirements for fresh fruits and vegetables are more demanding, having longer postharvest periods where high quality and food safety standards must be maintained. Because of this, other techniques such as controlled atmosphere storage (CAS) and modified atmosphere packaging (MAP) are used to enhance and augment cold storage. These, either actively or passively, alter the atmosphere composition surrounding and within the produce in order to influence cellular metabolism, causing a reduction in catabolism in climacteric fruit and vegetable and an inhibition of enzymatic reactions. Each commodity has its optimal CAS and MAP conditions which, together with controls on storage duration, RH and ethylene concentration, may influence shelf life and flavor life.

For CAS, the atmospheric composition is strictly monitored and adjusted in gas-tight rooms by control systems, whereas in MAP the changes in oxygen and carbon dioxide concentrations within the package are a function of factors such as the respiration rate of the produce as affected by cultivar, ripening stage, weight and temperature in combination with packaging film characteristics. Optimum CAS/MAP storage regimes for different fruit types have been mainly developed empirically based on their quality after storage. The main effects of low temperature and the CAS/MAP storage alone or in combination are associated with respiration, ethylene biosynthesis and its action and other metabolic processes, thereby decreasing the rates of change that occur during postharvest ripening, including color (chlorophylls, carotenoids and flavonoids), texture (softening as a result of cell wall disassembly and reduced cell turgor) and flavor (taste and aroma as a result of starch degradation, sugar–acid metabolism and synthesis of aromatic volatiles). These effects can apply regardless of whether the fruit is non-climacteric or climacteric, but for the latter, fruit types are important for the reduction of ethylene production. Ethylene has a key physiological role during the ripening process, a genetically regulated stage of development of climacteric fruits that is highly complex and coordinated by hormonal metabolism. In addition to physical methods, ethylene antagonist such as 1-methylcyclopropene (1-MCP), is used on specific fruit types. 1-MCP effects vary depending on the species, cultivar, maturity and ripening stage and factors such as 1-MCP concentration, treatment duration and temperature and posttreatment storage conditions. It is commercially applied to several commercially important fleshy fruits, such as apples and pears (Brizzolara et al., 2020).

2.2.1 Physiological Changes of Selected Commodities During Cold Storage

Storing fruits, vegetables and root crops under optimal conditions can extend shelf life and maintain quality. Deviations from the optimal can cause physiological disorders which negatively impact quality. The incorporation of the cold chain requirements of some crops is discussed.

2.2.1.1 Banana

Banana is a typical climacteric fruit, and important physicochemical changes take place during ripening. Since this fruit has a short green life, i.e., the elapsed time between harvest and the beginning of ethylene production, manipulation of environmental conditions, mainly the atmosphere and the temperature, is used to extend the storage time. Storage at low temperatures is a step in the cold chain, from the harvest to market, to extend the green life of fruit. Low temperatures temporarily inhibit ripening by maintaining low ethylene production, but most tropical fruits undergo physiological disorders and deterioration of quality when exposed to low temperatures. CI injury is an important disorder of bananas. Both green and ripe fruit are susceptible, with green fruit being slightly more sensitive than ripe fruit. CI results from exposing the fruit to temperatures below about 13°C for a few hours to a few days, depending on cultivar, maturity, condition of the fruit, temperature and duration of exposure. Symptoms include subepidermal discoloration visible as brown to black streaks in a longitudinal cut, a dull or greyish (smoky) cast on ripe fruit, failure to ripen, and in severe cases the peel turns dark brown or black, and even the flesh can turn brown and develop an off-taste. Chilled fruits are more sensitive to mechanical injury. Ripe fruit, if chilled, turn dull brown when later exposed to higher temperatures and are very susceptible to handling; the slightest pressure causes discoloration. Inflicted chill in green or ripe fruit may not become apparent until 18–24 hours after actual damage has occurred.

In bananas, the symptoms of CI appear to be cultivar-dependent and related to the genomic group. For example, the Nanicão banana cultivar from Brazil, a member of the AAA group, is commercially relevant but less tolerant to low temperatures than Prata, a cultivar of the AAB group (Der Agopian et al., 2011). According to Lichtemberg et al. (2001), the B genome in bananas contributes to cold resistance.

Some banana cultivars such as Green Nanicão may accumulate high levels of starch during low-temperature storage, which, otherwise, is most likely degraded during the ripening process and thus result in high amounts of soluble sugars in the fully ripe fruit (Peroni-Okita et al., 2013). However, the low cold resistance could be a disadvantage for the long-term storage of these banana cultivars as the fruits exposed to low temperatures for several days may accumulate a lower amount of sugars during ripening, although a marginal increase in the sucrose levels occurred during storage, which is reportedly a cryoprotective effect (Der Agopian et al., 2011). Although these changes could be partially responsible for cold tolerance, the net result was a decrease in fruit quality, as the per cent of soluble sugars in the ripe fruit was lower. Therefore, a better understanding of the effects of cold on the starch-to-sucrose metabolism of commercially relevant, cold-sensitive bananas is important in terms of food quality. To avoid CI, bananas are stored and shipped at a temperature of 13°C–14°C. The presence of ethylene increases the susceptibility of bananas to CI, while storage in MAP and CA reduces the incidence of CI.

2.2.1.2 Mango

Fully ripe mangoes are known for their aroma, peel color, good taste and nutritional value. Mango fruit is susceptible to various physiological disorders that influence fruit quality. Among the most important of these is CI. In general, storage temperatures below about 10°C–13°C but above freezing have been reported to injure mature-green mangoes (Saltveit, 2004). This problem limits the use of low storage temperature to manage postharvest ripening and seriously affects the ability of handlers to store or transport mangoes over long distances, because temperatures that are low enough to delay ripening, decay, and senescence may also be damaging to the fruit (Brecht et al., 2012).

The symptoms of CI described for mango fruit include greyish, scald-like discoloration on the skin, followed by pitting, uneven ripening and poor flavor and color development (Brecht et al., 2012). The latter two are especially important because the loss of flavor due to chilling may occur without the development of the other visual symptoms. The symptoms of CI are also often not apparent while the fruit are at a low temperature, but develop later when the fruit is brought to warmer temperatures for ripening or are displayed for sale. CI symptoms in mango fruit held at room temperature for 1–2 days after low-temperature storage were described as discolored and pitted areas on the surface, followed by irregular ripening with poor color and flavor and increased susceptibility to microbial spoilage. Ketsa et al. (2000) reported that the mangoes stored at 4°C for 3 weeks developed blackened lenticels and greyish patches on the peel after the transfer to ambient temperatures.

CI has other effects on mango fruit quality besides visual injury symptoms and flavor loss (Brecht et al., 2012). CI induced in mango fruit stored at 4°C accelerated the softening of the fruit after they were transferred to 20°C; humidification of the ambient atmosphere reduced the symptoms. Krishnamurthy and Joshi (1989) reported disruption of mesocarp cells and inhibition in carotene development after 4 weeks in fruit stored at 7°C. These fruits failed to ripen evenly after holding for up to 5 weeks at room temperature.

Tolerance of 'Keitt' mango fruit to CI was reported to increase after prestorage heat treatments (HTs) (McCollum et al., 1993). Brecht et al. (2012) reported that hot water quarantine treatment and other time-temperature combinations reduced the susceptibility of Tommy Atkins and Keitt mangoes to CI. Bender et al. (2000) also indicated that mangoes could be shipped for 2–3 weeks in controlled atmospheres at 8°C for tree-ripe fruit or 12°C for mature-green fruit without developing CI.

For the fruit in general, CI susceptibility decreases as the fruit develop, mature and ripen. Thus, immature mangoes are more susceptible to CI than mature fruit, and fruit that are mature but has not yet begun to ripen are more susceptible to CI than fruit that are undergoing ripening (Brecht et al., 2012). Mangoes that are exposed to chilling temperatures before they have begun ripening are never able to ripen normally. Most mango cultivars show injury below 10°C if the fruit has just reached full maturity, indicating that mango tolerance to CI increases during ripening (Mohammed and Brecht, 2002).

Variability among reports with regard to the lowest safe temperature to store mangoes without danger of CI may be due to differences in cultivar susceptibility as well as differences in the stage of fruit maturity or ripeness in different experiments. Knowing the critical combinations of time–temperature and the associated chilling threshold temperature(s) for the most important varieties of mangoes would provide basic information to decrease the incidences of CI and to deliver better quality, especially better tasting, mangoes to the consumer.

2.2.1.3 Papaya

Papaya fruit is an important tropical fruit with high economic value due to its rich nutritional constituents (carbohydrates, ascorbic acid, carotenoids and papain) (Jing et al., 2015). Nevertheless, as a typical climacteric fruit, papayas following harvest undergo a rapid ripening and softening process within 6–8 days when stored at ambient temperatures (Pan et al., 2019a). Similar to mango, the papaya fruit is also liable to CI when exposed to nonfreezing low temperatures below 12°C. Chilled papayas gradually exhibit the CI symptoms in form of pitting, scald and shriveling of the peel, hard lumps in the pulp around vascular bundles, water-soaked patches on the pulp, failure to ripen, loss of aroma and flavor and high susceptibility to postharvest decay. These symptoms ultimately lead to a severe decline in fruit quality and reduced marketability (Proulx et al., 2005).

HT combined with an edible coating could delay ripening, alleviate CI, increase fruit's resistance to fungal infection and, at the same time, maintain better storability quality of papaya fruits after prolonged storage under low temperature. Zhao and Dixon (2011) claimed that papaya fruits were progressively less susceptible to chilling stress as they ripened. Symptoms of CI occurred after 14 days at 5°C for mature green fruit and 21 days for 60% yellow fruit. Skin scald can be induced in colour-break fruit after chilling at 1°C for 24 hours. Singh and Rao (2005) concluded that MAP of individual papaya cv. Solo with LDPE extended the storage life to 30 days at 13°C without any CI symptoms and fruit ripened normally in 1 week at 20°C. MAP not only extended storage life and alleviated CI but also helped in the maintenance of the antioxidant potential of fruit by retaining acceptable levels of antioxidants like ascorbic acid and lycopene.

2.2.1.4 Eggplant

Eggplants must have a shiny fruit surface with color typical of the variety, a fresh unblemished calyx, and be free of any decay, discoloration or other defects. The freshness of the calyx is a very important quality parameter, and it is generally considered that calyx appearance declines more rapidly than does the quality of the fruit itself. Rapid precooling to 10°C immediately after harvest is necessary to retard discoloration, weight loss, drying of calyx and decay (Mohammed and Sealy, 1986). Eggplant is a perishable and chilling-sensitive tropical vegetable. Fruit are stored at 10°C–12°C with 90%–95% RH. Storage of eggplant is generally less than 14 days as visual and sensory qualities deteriorate rapidly. Decay is likely to increase after storage >2 weeks, especially after removal to typical retail conditions. Short-term storage or transit temperatures below this range are often used to reduce weight loss but result in CI after transfer to retail conditions. Eggplant fruit are chilling sensitive at temperatures below 10°C. CI is cumulative and may be initiated in the field before harvest. At 7.5°C chilling symptoms occurred after 12 days while at 5°C chilling symptoms occurred after 6 days (Wang, 2010). CI symptoms included surface pitting of the peel, browning of seeds, vascular bundles and calyx and possibly linked to lipid peroxidation. Accelerated decay by *Alternaria* spp. is common in chilling stressed fruit. CI and water loss can be reduced by storing eggplant in polyethylene bags or polymeric film overwraps but increased decay from *Botrytis* is a potential risk of this practice (Cantwell and Suslow, 2009).

2.2.1.5 Sweet Potato

Sweet potato roots are chilling sensitive and should be stored between 12.5°C and 15°C with high RH (>90%). Storage life of 6–10 months can be expected under these conditions, although sprouting may begin to occur after about 6 months depending on the cultivar. Temperatures above 15°C lead to rapid sprouting and weight loss. Careful handling during harvesting will minimize mechanical damage to the skin and reduce decay incidence during storage. Sweet potato roots freeze at −1.9°C and are very sensitive to CI at temperatures of <12°C. The severity of CI depends on the temperature and length of exposure below 12°C. Symptoms of CI include root shriveling, surface pitting, abnormal wound periderm formation, fungal decay, internal tissue browning and hardcore formation (Meyers, 2015). Islam et al. (2012) suggested that the chilling tolerant genotypes followed different mechanisms for tolerance. Thus, the chilling tolerance in sweet potatoes can be enhanced by breeding and selecting for chilling tolerance.

Pan et al. (2019b) investigated the effects of intermittent HT on CI and antioxidant capacity of sweet potato roots during cold storage at 5°C±0.5°C and 80%–85% RH. Roots were heat-treated in an air oven (45°C) for 3 hours continuously or intermittently. Intermittent treatment was achieved through the temperature recovery by room temperature after every 1 hour of continuous treatment. Their results indicated that both continuous and intermittent HTs maintained the root quality and improved cold damage resistance to varying degrees compared with the control group.

2.2.1.6 Avocado

Low-temperature storage (5°C–13°C) is predominantly used to extend the postharvest quality of avocados by suppressing the speed of cell metabolism and senescence. However, exposing avocados to temperatures below their critical threshold (10°C–15°C), but above freezing, can permanently cause irreversible damage to plant tissues, cells and organs leading to pulp spots and CI (Meyer and Terry, 2010). 'Pulp spot,' a low-temperature disorder, is commonly described as small dark spots in the flesh, and blackening of a region surrounding cut vascular bundles, which are either immediately visible when the fruit is cut, or which develop within a few minutes after cutting. Both pulp spot and CI disorders involve browning reactions implicating particularly the enzyme polyphenol oxidase (PPO) (Munhuweyi et al., 2020).

CI in avocado is cultivar dependent. In 'Hass' avocado fruit, CI symptoms occur 4 weeks at 6°C, based on maturity and growing conditions. External symptoms of avocado CI include skin pitting, scalding, water-soaked appearance, failure to ripen, blackening, off-flavor and decay. Internally, CI symptoms are associated with flesh browning (grey pulp, pulp spot and vascular browning) and increased susceptibility to pathogen attack.

This disorder in avocados can be controlled by applying physiological heat shock treatments. Accordingly, avocado fruits exposed to short-term high temperatures (>35°C) induced protective plant defense stress responses and provided repairs to the damaged membrane, organelles or metabolic pathways. Apart from alleviating CI symptoms, postharvest HTs can be applied for insect disinfestation, disease control and modifying fruit responses to cold (Lurie and Pedreschi, 2014). De Jesus Ornelas and Yahia (2004) reported that CI of 'Hass' fruit was effectively suppressed using dry (50% RH) and moist (90%–95% RH) forced air at 38°C for 6 hours before storage (5°C, 80%–85% RH up to 8 weeks) and fruit heated with dry air exhibited the best internal quality and the lowest CI incidence. However, the HTs induced higher weight loss and respiration rates. The integration of postharvest HTs into the commercial postharvest chains should, therefore, be applied with caution. Although they may reduce CI and control decay, they can also seriously damage fruit sensory quality, causing deterioration of flavor and aroma, and, therefore, consumer rejection. In a similar study, 'Hass' avocados were kept in CA (2% O_2 and 2% CO_2) at 5°C and 7°C to determine if CI could be prevented at the higher temperature. Both CA storage regimes, at 5°C or 7°C, resulted in better fruit quality than for control fruit kept in normal air at 5°C. However, after 4 weeks of storage, 7°C was less effective at retarding the progression of ripening in storage than CA at 5°C.

Exogenous polyamines, such as methyl jasmonate, can be used to control CI by simply enhancing plant defense mechanisms against the disorder. Polyamines induced synthesis of certain stress proteins (heat shock and pathogenesis-related (PR) proteins), maintained lipid oxidation stability and extended the shelf life of avocado fruit kept under low-temperature storage (Pathirana et al., 2011).

Cold damage during storage of 'Hass' avocado can also be controlled through preconditioning (pulp hydrocooling to 6°C) and waxing (Munhuweyi et al., 2020). The hydrocooling pretreatment alone cannot effectively prevent lenticellosis, but in combination with waxing creates a threefold reduction of internal damage and retarded fruit color break. The treatment-controlled rate of metabolism and ethylene production extended the commercialization period of the fruit kept at room temperature after storage at 3°C for 46 days (Mendieta et al., 2016).

2.2.1.7 Grapefruit

CI in grapefruit is usually represented on the peel that collapses and darkens to form pits although the occurrence of pitting is not targeted to the oil glands (Ritenour et al., 2003). However, less severe symptoms may show up as circular or arched areas of discoloration or scalding. Symptoms of CI are typically more pronounced after fruit are warmed to room temperature following exposure to the chilling temperature. CI symptoms generally require at least 3–6 weeks to develop at low (4.5°C) shipping and storage temperatures (Ritenour et al., 2003). Similar to other fruits, chilled grapefruits are also more susceptible to decay than are nonchilled fruits. Ritenour et al. (2003) also indicated that CI is often confused with another physiological disorder called postharvest pitting (PP) that is caused by low-oxygen concentrations (<9%) within waxed fruit and is visible as collapsed oil glands. PP requires only 2–4 days for symptom development after waxing and appears in fruit held at warm (>10°C) temperatures.

Depending on other predisposing factors, grapefruit storage and shipment below 10°C can cause severe CI. Studies show that CI is most severe when fruit are stored at temperatures from 3.3°C to 4.4°C compared with storage at higher or lower temperatures (Ritenour et al., 2003). Nevertheless, if grapefruits are kept at temperatures above 10°C, the potential for CI could be reduced, but it can also lead to the development of severe PP in waxed fruit. Thus, storage of waxed grapefruit at 7.2°C may often represent the best compromise to minimize the occurrence of both disorders. Preconditioning fruit for 7 days at 15.5°C can greatly reduce CI, but this may promote severe PP if fruit are preconditioned after the wax application. The conditions fruit experience during degreening can reduce grapefruit susceptibility to CI (Ritenour et al., 2003).

2.2.1.8 Watermelons

Watermelons can be stored up 14 days at 15°C with up to 21 days attainable at 7°C–10°C. For short-term storage or transit to distant markets (>7 days), 7.2°C and 85%–90% RH is recommended. Watermelons are, however, prone to CI at this temperature. Extended holding at this temperature will induce CI, rapidly evident after transfer to typical retail display temperatures. Many watermelons are still shipped without precooling or refrigeration during transit. These fruits must be utilized for prompt market sales as quality declines rapidly under these conditions. Symptoms of CI include pitting, decline in flesh color, loss of flavor, off-flavors and increased decay when returned to room temperatures.

2.3 TEMPERATURE REQUIREMENTS

Every horticultural produce has a specific safe range of cold storage temperatures, RH and potential storage duration. The storage temperatures below the optimum range but above freezing can cause CI (Table 2.1). CI is tissue damage in the commodity caused by exposure to temperatures lower than optimum storage temperatures for a long duration. Not only tropical and subtropical fruit and vegetables but also temperate crops exhibit CI symptoms during low-temperature storage. Many horticultural commodities are chilling sensitive and exhibit a range of symptoms (Table 2.2), which limits their postharvest life and results in major losses and waste.

2.4 STRATEGIES TO CONTROL PHYSIOLOGICAL DISORDERS DUE TO CHILLING INJURY

The cold chain management of fruit or vegetable is directly associated with the selection of crop type, cultivar, temperature, ethylene, storage duration and storage atmospheric composition and logistics in order to avoid accelerated senescence. Cold chain managers would need to focus on the critical temperature below which CI may occur (Table 2.2).

TABLE 2.1

Optimal Storage Conditions for Some Common Fruits and Vegetables

Product	Optimal Storage Temperature (°F)	(°C)	Optimal RH (%)	Ethylene Production	Sensitive to Ethylene	Approximate Storage Life
Apples	30–40	−1–4	90–95	High	Yes	1–12 months
Apricots	31–32	−1–0	90–95	High	Yes	1–3 weeks
Asparagus	32–35		95–100	No	Yes	2–3 weeks
Avocados, ripe	38–45	3–7	85–95	High	Yes	
Avocados, unripe	45–50	7–10	85–95	Low	Yes, Very	
Bananas, green	62–70	17–21	85–95	Low	Yes	
Bananas, ripe	56–60	13–16	85–95	Medium	No	
Beets	32–35	0–2	90–95	No	Yes	
Broccoli	32	0	95–100	No	Yes	10–14 days
Brussels Sprouts	32	0	90–95	No	Yes	3–5 weeks
Cabbage, Chinese	32	0	95–100	No	Yes	2–3 months
Cabbage	32	0	98–100	No	Yes	3–6 weeks
Cantaloupe	36–38	2–3	90–95	Medium	Yes	
Carrots	32	0	95–100	No	Yes	2 weeks
Cauliflower	32	0	95–98			3–4 weeks
Celery	32	0	98–100	No	Yes	2–3 months
Cherries	32–35	0–2	90–95	Very low	No	
Cherries, sour	32	0	90–95			3–7 days
Coconuts	55–60	13–16	80–85	No	No	
Cranberries	38–42	3–6	90–95	No	No	
Cucumbers	50–55		95	Very low	Yes	10–14 days
Eggplant	46–54		90–95	No	Yes	1 week
Figs	32–35	0–2	90–95	Low	No	
Garlic	32	0	65–70	No	No	6–7 months
Ginger	60–65	16–18	65–70	No	No	
Grapefruit	55–60	13–16	90–95	Very low	No	
Grapes	31–32		85	Very low	Yes	2–8 weeks
Guavas	45–50	7–10	90–95	Medium	Yes	
Kale	32		95–100			2–3 weeks
Kiwi, ripe	32–35	0–2	90–95	High	Yes	
Lemons	52–55	11–13	90–95	Very low	No	
Lettuce	32	0	98–100	No	Yes	2–3 weeks
Limes	48–55	9–13	90–95	Very low	No	
Lychees	40–45	4–7	90–95	Very low	No	
Mangoes	50–55	10–13	85–95	Medium	Yes	
Melons, Honey Dew	50–55	10–13	85–95	Medium	Yes	
Mushrooms	32	0	95	No	Yes	3–4 days
Nectarines	31–32		90–95	High	No	2–4 weeks
Okra	45–50		90–95	Very low	Yes	7–10 days
Onions	32–35	0–2	65–75	No	No	
Oranges	40–45	4–7	90–95	Very low	No	

(Continued)

TABLE 2.1 (*Continued*)
Optimal Storage Conditions for Some Common Fruits and Vegetables

Product	Optimal Storage Temperature (°F)	Optimal Storage Temperature (°C)	Optimal RH (%)	Ethylene Production	Sensitive to Ethylene	Approximate Storage Life
Papayas	50–55	10–13	85–95	Medium	Yes	
Peaches	31–32		90–95	High	Yes	2–4 weeks
Pears	29–31		90–95	High	Yes	2–7 months
Peas, green	32	0	95–98			1–2 weeks
Peppers, hot chili	32–50		60–70	No	Yes	6 months
Peppers, sweet	45–55	7–10	90–95	No	No	2–3 weeks
Persimmons	32–35	0–2	90–95	No	Yes, Very	
Pineapples	50–55	10–13	85–95	Very low	No	
Plums	31–32		90–95	High	Yes	2–5 weeks
Pomegranates	41–50	5–10	90–95	No	No	
Potatoes	45–50	7–10	90–95	No	Yes	
Pumpkins	50–55		65–70	No	Yes	2–3 months
Raspberries	31–32		90–95	Very low	No	2–3 days
Rhubarb	32	0	95–100	No	No	2–4 weeks
Strawberries	32	0	90–95	Very low	No	3–7 days
Sweet potatoes	55–60		85–90	No	Yes	4–7 months
Tangerines	40–45	4–7	90–95	Very low	No	
Tomatoes, mature green	55–70		90–95	Low	Yes	1–3 weeks
Tomatoes, ripe	55–70		90–95	Medium	No	4–7 days

TABLE 2.2
Symptoms of Chilling Injury of Selected Fruits and Vegetables

Commodity	Symptoms of Injury
Beans (string)	Rusty brown specks, spots or coalesced areas
Beans (snap)	Pitting and russeting
Cucumbers	Pitting, water-soaked spots, decay
Eggplants	Surface scald, *Alternaria* rot, blacken seeds
Cantaloupe	Pitting, surface decay
Honey dew	discoloration, pitting, surface decay, unripen
Watermelons	Pitting, objectionable flavor
Okra	Discoloration, water-soaked areas, pitting
Peppers (sweet and hot)	Sheet pitting, *Alternaria* rot on pods and calyxes, darkening of seed
Potatoes	Sweetening
Pumpkins and squashes	Decay, especially *Alternaria* rot
Sweet potatoes	Decay, pitting, internal discoloration; hard core when cooked
Tomatoes (ripe) (mature-green)	Water soaking and softening decay
	Poor color when ripe, *Alternaria* rot

The effectiveness of each treatment varies with the commodity, the maturity of the fruit and the dosage of the treatment (Li et al., 2016; Kitinoja and Thompson, 2010). Low-temperature conditioning and intermittent warming maintain high levels of phospholipids, increase the degree of unsaturation of fatty acids, increase the levels of spermidine and spermine and stimulate the activities of free radical scavenging enzymes. HT induces heat shock proteins (HSP), suppresses oxidative activity and maintains membrane stability. Application of compounds such as methyl jasmonate and methyl salicylate stimulates the synthesis of stress proteins, such as HSP, PR proteins and alternative oxidase (AOX), and these compounds, in turn, can activate lipoxygenase gene expression and induce the synthesis of abscisic acid and polyamines. The polyamines may act as free radical scavengers and membrane stabilizers. All of these processes can enhance the chilling tolerance of tissues and alleviate CI in tropical and subtropical fruits (Sharma et al., 2020; Zhang et al., 2019).

2.5 CONCLUSIONS

Horticultural crops are generally highly perishable and continue their physiological processes even after harvest throughout the supply chain. Therefore, maintaining the produce at low non-chilling temperatures extends the storage life and reduce postharvest losses by reducing the rates of biological processes and degradation. For each fruit or vegetable, the specific safe range of cold storage temperatures, humidity and potential storage period has been identified, with an appropriate method to improve storage efficiency. Maintaining an appropriate crop-specific storage environment is essential to achieve a successful supply chain, ensuring good quality of produce and fewer losses by the time it reaches consumer/end-user. At present day, various cooling technologies are available, which can satisfy the supply chain needs of different crops at various scales. Choosing appropriate technology and maintaining suitable conditions for the crops being handled are the key factors in effective cold chain management.

REFERENCES

Bender, R. J., Brecht, J. K., Sargent, S. A., and Huber, D. J. (2000). Low temperature controlled atmosphere storage for tree-ripe mangoes (*Mangifera indica* L.). *Acta Horticulturae, 509*, 447–458.

Brecht, J. K., Nunes, M. C.N., and Maul, F. (2012). Time-temperature Combinations that Induce Chilling Injury of Mangos. Center for Food Distribution and Retailing University of Florida.

Brizzolara, S., Manganaris, G. A., Fotopoulos, V., Watkins, C. B., and Tonutti, P. (2020). Primary metabolism in fresh fruits during storage. *Frontiers in Plant Science, 11*, 80.

Cantwell, M., and Suslow, T. V. (2009). Eggplant: Recommendations for maintaining postharvest quality. Department of Plant Sciences, University of California, Davis, CA 95616.

De Jesus Ornelas, P., and Yahia, E. M. (2004). Effects of pre-storage dry and humid hot air treatments on the quality, triglycerides and tocopherol contents in 'Hass' avocado fruit. *Journal of Food Quality, 27*(2), 115–126.

Der Agopian, R. G., Peroni-Okita, F. H. G., Soares, C. A., Mainardi, J. A., do Nascimento, J. R. O., Cordenunsi, B. R., Lajolo, F. M. and Purgatto, E. (2011). Low temperature induced changes in activity and protein levels of the enzymes associated to conversion of starch to sucrose in banana fruit. *Postharvest Biology and Technology, 62*(2), 133–140.

Gustavsson, J., Cederberg, C., Sonesson, U., Van Otterdijk, R., and Meybeck, A. (2011). Global food losses and food waste. UNFAO, Rome, Italy.

Islam, S., Izekor, E., and Garner, J. O. (2012). Lipid and fatty acid compositions of chilling tolerant sweet potato (*Ipomoea batatas* L.) genotypes. *American Journal of Plant Physiology, 7*(6), 252–260.

Jing, G., Li, T., Qu, H., Yun, Z., Jia, Y., Zheng, X., and Jiang, Y. (2015). Carotenoids and volatile profiles of yellow-and red-fleshed papaya fruit in relation to the expression of carotenoid cleavage dioxygenase genes. *Postharvest Biology and Technology, 109*, 114–119.

Ketsa, S., Wongs-aree, C., and Klein, J. D. (2000). Storage life and quality of 'Kluai Khai' banana fruit affected by modified atmosphere using bulk packaging. *Thai Journal of Agricultural Science, 33*(1/2), 37–44.

Kitinoja, L., and Thompson, J. F. (2010). Pre-cooling systems for small-scale producers. *Stewart Postharvest Review*, 6(2), 1–14.

Kitinoja, L., Tokala, V. Y., and Mohammed, M. (2019). Clean cold chain development and the critical role of extension education. *Agriculture for Development*, 36(3), 19–25.

Krishnamurthy, S., and Joshi, S. S. (1989). Studies on low temperature storage of Alphonso mango. *Journal of Food Science and Technology (Mysore)*, 26(4), 177–180.

Li, P., Yin, F., Song, L., and Zheng, X. (2016). Alleviation of chilling injury in tomato fruit by exogenous application of oxalic acid. *Food Chemistry*, 202, 125–132.

Lichtemberg, L. A., Malburg, J. L., and Hinz, R. H. (2001). Cold damage in bananas. *Revista Brasileira de Fruticultura (Brazil)*, 23, 568–572.

Lurie, S., and Pedreschi, R. (2014). Fundamental aspects of postharvest heat treatments. *Horticulture Research*, 1(1), 1–7.

Mangaraj, S., Goswami, T. K., and Mahajan, P. V. (2009). Applications of plastic films for modified atmosphere packaging of fruits and vegetables: A review. *Food Engineering Reviews*, 1(2), 133–158.

McCollum, T. G., D'Aquino, S., and McDonald, R. E. (1993). Heat treatment inhibits mango chilling injury. *HortScience*, 28(3), 197–198.

Mendieta, B., Olaeta, J. A., Pedreschi, R., and Undurraga, P. (2016). Reduction of cold damage during cold storage of Hass avocado by a combined use of pre-conditioning and waxing. *Scientia Horticulturae*, 200, 119–124.

Meyer, M. D., and Terry, L. A. (2010). Fatty acid and sugar composition of avocado, cv. Hass, in response to treatment with an ethylene scavenger or 1-methylcyclopropene to extend storage life. *Food Chemistry*, 121(4), 1203–1210.

Meyers, S. (2015). Avoid sweet potato chilling injury. Mississippi State University Extension Service. https://www.mississippi-crops.com/2015/02/18/avoid-sweetpotato-chilling-injury/

Mohammed, M., and Brecht, J. K. (2002). Reduction of chilling injury in 'Tommy Atkins' mangoes during ripening. *Scientia Horticulturae*, 95(4), 297–308.

Mohammed, M., and Sealy, L. (1986). Extending the shelf-life of melongene (Solanum melongena L.) using polymeric films. *Tropical Agriculture (Trinidad and Tobago)*, 63 (1), 36–40.

Munhuweyi, K., Mpai, S., and Sivakumar, D. (2020). Extension of avocado fruit postharvest quality using non-chemical treatments. *Agronomy*, 10(2), 212.

Pan, Y., Chen, L., Chen, X., Jia, X., Zhang, J., Ban, Z., and Li, X. (2019b). Postharvest intermittent heat treatment alleviates chilling injury in cold-stored sweet potato roots through the antioxidant metabolism regulation. *Journal of Food Processing and Preservation*, 43(12), e14274.

Pan, Y., Zhang, S., Yuan, M., Song, H., Wang, T., Zhang, W., and Zhang, Z. (2019a). Effect of glycine betaine on chilling injury in relation to energy metabolism in papaya fruit during cold storage. *Food Science and Nutrition*, 7(3), 1123–1130.

Pathirana, U. P., Sekozawa, Y., Sugaya, S., and Gemma, H. (2011). Effect of combined application of 1-MCP and low oxygen treatments on alleviation of chilling injury and lipid oxidation stability of avocado (*Persea americana* Mill.) under low temperature storage. *Fruits*, 66(3), 161–170.

Peroni-Okita, F. H., Cardoso, M. B., Agopian, R. G., Louro, R. P., Nascimento, J. R., Purgatto, E., Tavares, M. I., Lajolo, F. M., and Cordenunsi, B. R. (2013). The cold storage of green bananas affects the starch degradation during ripening at higher temperature. *Carbohydrate Polymers*, 96(1), 137–147.

Proulx, E., Cecilia, M., Nunes, N., Emond, J. P., and Brecht, J. K. (2005, December). Quality attributes limiting papaya postharvest life at chilling and non-chilling temperatures. *Proceedings of the Florida State Horticultural Society*, 118, 389–395.

Ritenour, M., Dou, H., and Mccollum, T. (2003). Chilling injury of grapefruit and its control. *Extension Publications*, 32611.

Saltveit, M. E. (2004). Effect of 1-methylcyclopropene on phenylpropanoid metabolism, the accumulation of phenolic compounds, and browning of whole and fresh-cut 'iceberg' lettuce. *Postharvest Biology and Technology*, 34(1), 75–80.

Sharma, S., Barman, K., Prasad, R. N., and Singh, J. (2020). Chilling stress during postharvest storage of fruits and vegetables. In: Rakshit A., Singh H., Singh A., Singh U. and Fraceto L. (eds.) *New Frontiers in Stress Management for Durable Agriculture* (pp. 75–99). Springer, Singapore.

Singh, S. P., and Rao, D. S. (2005). Effect of modified atmosphere packaging (MAP) on the alleviation of chilling injury and dietary antioxidants levels in 'Solo' papaya during low temperature storage. *European Journal of Horticultural Science*, 70(5), 246–252.

Wang, C. Y. (2010). Alleviation of chilling injury in tropical and subtropical fruits. *Acta Horticulturae*, 864, 267–273.

Zhang, M., Liu, W., Li, C., Shao, T., Jiang, X., Zhao, H., and Ai, W. (2019). Postharvest hot water dipping and hot water forced convection treatments alleviate chilling injury for zucchini fruit during cold storage. *Scientia Horticulturae, 249,* 219–227.

Zhao, Q., and Dixon, R. A. (2011). Transcriptional networks for lignin biosynthesis: More complex than we thought? *Trends in Plant Science, 16*(4), 227–233.

Section II

Cooling Systems

3 Traditional/Conventional Cooling Systems

Swarajya Laxmi Nayak, R.R. Sharma, and Shruti Sethi
ICAR-Indian Agricultural Research Institute

CONTENTS

3.1 INTRODUCTION

Fresh fruits and vegetables continue their ontogeny even after harvest, and the deterioration rate increases due to ripening, senescence and unfavorable environmental conditions (More et al., 2020). To avoid these huge postharvest losses and to substantially increase the returns to farmers, proper storage of fruits and vegetables is important. Being perishable, they need immediate postharvest intervention to curb the metabolic processes, reduce the microbial contamination and increase their shelf life (Falagán and Terry, 2018). This can be successfully achieved by storing them at a safe low temperature (0°C–15°C) and high relative humidity (RH) (65%–100%) based on the physiological characteristics of commodities. This helps preserve their freshness by restricting chemical, biochemical and physiological changes to a minimum (Basediya et al., 2013). The erratic supply of electricity with low voltage, especially in developing countries, has resulted in the development of several simple practices for cooling and enhancing storage efficiency.

For a long time known, farmers stored their produce in temporary bamboo structures constructed near the farmer's residential buildings. In the warm plains of India, fruits and vegetables are still stored in ventilated pits or cool dry rooms. Inside the hut, the fruits and vegetables are placed on the floor or over racks and covered with straw or plant leaves to avoid exposure to the atmosphere. This allows fruits and vegetables to be stored for a few days without much damage and farmers can sell the fresh produce in the local market in a staggered manner (Mobolade et al., 2019).

DOI: 10.1201/9781003056607-5

Mechanical refrigeration and other modern storage systems are expensive, energy-intensive and require an uninterrupted power supply. Such storage systems are not always readily available and cannot be installed quickly and easily. The above problems limit the development of on-farm cold storage of fresh horticultural produce in remote and inaccessible areas. Low-cost, low-energy and environment-friendly storage structures made from locally available materials are feasible options that are discussed in this chapter.

3.2 STORAGE METHODS

Several methods of storage are employed to keep the produce safe and enhance its longevity. Different modern storage systems such as refrigeration, hypobaric storage, modified atmospheric storage and controlled atmospheric storage have been adopted in developed countries. However, in developing countries like India, conventional storage methods are still practiced by the farmers as these are inexpensive and require no specialized infrastructure.

3.2.1 TRADITIONAL METHODS OF STORAGE

The modern cold storage units used for storing fresh produce are not widely accepted in low-income countries because of the high investment involved (Regina and Kumar, 2018). For this reason, farmers practice the use of traditional/conventional storage systems. Traditional storage systems are efficient and are extensively used in tropical and subtropical countries for prolonging the storage life of perishable commodities. They do not require any external energy source and can be installed on-farm. *In situ* or in-field storage, pits, clamps, windbreaks, cellars, barns, evaporative cooling, night ventilation, zero energy cool chamber (ZECC) and pot-in-pot are some conventional storage methods.

3.2.1.1 In Situ Storage

In situ storage involves mulching of harvested produce with a 6–8-inch layer of straw, hay or dry leaves. This prevents the harvested commodities from freezing and reduces disease/pest damage during on-farm storage. Vegetables such as beet (*Beta vulgaris*), carrot (*Daucus carota*), horseradish (*Armoracia rusticana*), turnip (*Brassica rapa subsp. rapa*), rutabaga (*Brassica napobrassica*), salsify (*Tragopogon porrifolius*) and parsnips (*Pastinaca sativa*) can be stored under the ground throughout the winter (Elansari et al., 2019). Potatoes can be stored *in situ* in hilly regions for up to 3–4 months (Sah and Kumar, 2008). The disadvantages of this system include quality defects, e.g., parsnips and horseradish can develop an undesirable bitter taste during prolonged storage, alternate freezing and thawing may damage the commodity and the land where the crop is grown remains occupied thereby preventing planting a new crop. Despite these disadvantages *in situ* storage is a common practice within the *Khasi* farmers of the Northeastern part of India (Sah and Kumar, 2008). Tomlins et al. (2007) reported that on-farm stored sweet potato (*Ipomoea batatus*) has a very good market quality (79%). Similarly, Abera and Haile (2015) reported a significant increase in fresh and dry weight by an average of 45% of *in situ* stored anchote (*Bixa orellana*) with an improved nutrient concentration, largely calcium and iron for up to 7 months.

3.2.1.2 Burying

Burying the produce in-ground is an age-old practice to store the root crops like potato (*Solanum tuberosum*) and ginger (*Zingiber officinale*). In India, potatoes are traditionally buried in the sand. Soil temperatures do not fluctuate much, and buried produce remains cooler than air temperatures (Ali et al., 2012). Sometimes root crops are packed in coir dust or sawdust to protect them from desiccation. Ezeocha and Ironkwe (2017) reported that potatoes buried underground showed higher starch content.

3.2.1.3 Pits or Trenches

These are on-farm storage structures that are dug at the edges of the field. Usually, they are located at the highest elevation in the field, to prevent water stagnation, especially in high rainfall regions. The pit or trench is lined with organic materials like straw or hay and produce is placed in it. The produce is then covered with a layer of organic materials followed by a layer of soil. Ventilation is maintained by making holes at the top to allow the passage of air to prevent rotting (Kale et al., 2016). Pits must be well-drained and protected from rodents. The most common way to make a pit is by immersing a barrel or galvanized can into the ground. A layer of sand is spread at the bottom of the pit or trench; the produce is placed in it, covered with a lid and with a sufficient amount of straw or mulch over the top to provide insulation. A layer of plastic can also be applied for further protection from moisture. Bricks can be used to prevent loss of the mulch layer due to wind or other disturbances.

Pits are widely used in Odisha and Himachal Pradesh states of India, for the storage of ginger and turmeric (*Curcuma longa*). Storage losses have been reported to be 10%–15% lower in ginger and 20%–30% lower in turmeric as compared to the normal storage practices (Babu et al., 2013). Sweet potato tubers stored in the pits showed reduced lignification and high carotenoid content during storage (Tumuhimbise et al., 2010). Yam (*Dioscorea alata*) tuber stored in an underground pit can be stored up to 56 days as compared to 35 days under ambient conditions (Umogbai, 2013). Ezeocha and Ironkwe (2017) recorded that potatoes stored under a pit covered with river sand and wood ash showed a lower percentage of sprouting, i.e., 6.1% and 3.5%, respectively, than those uncovered.

3.2.1.4 Clamps or Heaps

Clamps are convenient on-farm storage structures employed to preserve farm produce, namely potatoes, turnips, carrots, parsnip and other root crops. This traditional method is used for storing potatoes in certain parts of the world, such as Great Britain. An elongated conical heap of the produce is made on the ground within a marked dimension. The width of the clamp is usually 1–2.5 m and the height of the heap is about one-third of the width of the clamp (Elansari et al., 2019). The bottom of the mound is lined with straw, hay or leaves that act as an insulating material. At the top, straw is bent over the ridge such that rain tends to run off the structure. The straw thickness should be 15–25 cm when compressed. After 2 weeks, the clamp is covered with soil to a depth of 15–20 cm, but this may vary depending on the climate. The basic drawback of this method is that once the mound is open, it cannot easily be closed. A series of smaller mounds with a variety of vegetables in each can be a better method. Tortoe et al. (2010) successfully stored cured sweet potato for up to 28 days in a clamp system. Storing cassava (*Manihot esculenta* Cruntz.) cv. Kiboko and Kabete under clamps in double shade maintained a high carbohydrate content and had a lower loss in weight when compared to normal storage (Mdenye et al., 2016).

3.2.1.5 Windbreaks

This method has been used in the United Kingdom to store onions for up to 6 months. Windbreaks are constructed by wooden stakes into the ground in two parallel rows about 1 m apart (Figure 3.1). A wooden platform is built between the stakes about 30 cm from the ground. The hexagonal- or square-shaped iron mesh is fixed between the stakes and across both ends of the windbreak (Onifade et al., 2019). The windbreak should be located with its longer axis at the right angle to the prevailing wind.

3.2.1.6 Cellars

These are underground or partly underground rooms often constructed beneath the house. This imparts good insulation, provides a cooler environment in warm weather and protects from excessively low temperatures during cold climate. Cellars have been used to store potatoes, apples,

FIGURE 3.1 Windbreaks for storage of onion.

cabbages, onions as well as planting stock (El-Ramady et al., 2015). Harvested produce is spread out in thin layers on shelves to ensure adequate air circulation. The door of the cellar may be left open at night to ensure the low temperature inside the cellar. Apple saplings stored in an underground cellar showed significantly higher plant survival (92%–94%) than greenhouse-stored plants (37%–56%) (Angmo et al., 2018). An in-ground storage structure is used by the Khasi farmers of the North-East region of India for the storage of potatoes (Sah and Kumar, 2008).

3.2.1.7 Barns

Barns are on-farm storage structures constructed for storing, packing and processing produce. The roof of the building usually has a gambrel or hip shape to maximize the area. Although there are no precise scale or measure for the type or size of the building, the term barn is usually reserved for the largest or most important structure on any particular farm. Smaller or minor agricultural buildings are often labeled sheds or outbuildings and are normally used to house smaller implements or activities. Barns are mostly used to store apples, potatoes, tobacco leaves and yams (Aidoo, 1993). Amoah et al. (2011) successfully stored sweet potatoes in a barn that was evaporatively cooled for up to 12 weeks. About 60% of farmers in Ghana use yam barns for storage (Adeniji, 2019).

3.2.1.8 Night Ventilation

Night ventilation or night flushing is a passive cooling technique that uses the diurnal temperature difference in hot climates to store produce. This method is mainly used to store onions in bulk, or apples at higher altitudes. The storage room loaded with farm produce should be well insulated. A fan with a differential thermostat is installed into the storeroom which is switched on when the outside temperature at night becomes lower than the temperature within. The fan switches off when the temperature equalizes. Abubakar et al. (2019) recommended the improved naturally ventilated storage structure (INVSS) constructed using locally available materials such as sand, cement, wood, corn stalks, wire mesh and grasses to reduce storage losses in onions for up to 5 months.

3.2.1.9 Indoor Storage

The most convenient place to store fruits and vegetables is inside the home. Typically, crops like pumpkins, squash and sweet potatoes that store best at temperatures between 50°F and 60°F (10°C–15°C) and 60%–75% RH can be stored successfully inside. With proper ventilation, beets, carrots, parsnips, horseradish, turnips and winter radishes can also be stored. The basement of the storage structure is insulated by using straw or hay (Sah and Kumar, 2008). A vapor barrier of polyethylene film should be included to prevent condensation. An elevated window can be installed to allow warm interior air to escape while a lower window beneath the elevated one allows entry of cool air to maintain air circulation. White yam tubers stored in a box lost 9% of their weight, in comparison to 15% under normal storage conditions (Akpenpuun et al., 2018).

3.2.1.10 Evaporative Cooling

The evaporative cooling system is an on-farm storage structure employed for short-term storage of fruits and vegetables in hot and dry regions (Jha and Chopra, 2006) by reducing the temperature and increasing the RH of an enclosure (Odesola and Onyebuchi, 2009). Evaporative cooling is a physical phenomenon in which water evaporates based on latent heat of vaporization. The degree of cooling depends on the ambient humidity of the air, or more accurately the difference between the wet and dry bulb temperatures, and the efficiency of the evaporating surface. However, under very high RH, or when the wet and dry bulb temperatures are close, very little evaporation occurs, and cooling is inefficient. The working efficiency of an evaporative cooling system can be improved by using natural absorbent materials like jute, cotton and hessian waste (Olosunde et al., 2009). Evaporative cooling systems are environmentally friendly and are a promising cooling system especially for small farmers in rural areas (Dadhich et al., 2008).

There are two forms of evaporative cooling: passive/direct and active/indirect evaporative cooling. The difference is based on how the air movement through the moist materials (Dvizama, 2000). The passive form of evaporative cooling relies on the natural wind velocity to move air through the moist surface and effect evaporation while active systems use fans to provide air movement.

3.2.1.11 Zero Energy Cool Chamber

A zero-energy cooling system maintains the freshness of the commodity for a short-term period. Not only does it reduce the storage temperature, but it also increases the RH during storage which is essential for maintaining the quality of the commodities. Natural evaporation of water from a surface removes heat, creating a cooling effect. Water loss and desiccation of produce are lower at cooler temperatures. This storage method was originally developed in India by Dr. Susanta K. Roy and Dr. D.S. Khuridiya at the Indian Agricultural Research Institute, New Delhi, in the early 1980s to address postharvest losses, especially in rural areas where electricity is not available. In the same vein, ICAR-CIPHET also developed a low-cost evaporative cool room ($3 \times 3 \times 3$ m) to store the perishable commodities up to twice as long as ambient storage (Patil, 2010). This design is composed of a double-brick wall structure, supported by a base layer of brick and covered with a straw mat. This type of structure provides maximum benefit when such chambers are constructed in a hot and dry climate with RH less than 40% and a maximum daily temperature greater than 25°C. They should be located in a shady and well-ventilated area where a source for an ample amount of water is available.

ZECC is an affordable and eco-friendly storage system adopted by a wide range of farmers. The cost of ZECC depends on the size and cost of raw materials. Although ZECC can be constructed over a range of sizes, it is important to select a desirable size to avoid overbuilding and spending more money than needed. It can be made from locally available materials including bricks, sand, wood, dry grass, gunny bag, bamboo and twine. If the evaporative cooling chamber is not built in a well-shaded area, then a shed must be constructed to provide shade. However, the construction of the ZECC chamber is easy and cost-effective.

The usual dimension of ZECC is approximately 1.65×1.15 m, but the dimensions may vary according to the storage volume needed. The double-wall is about 0.68 m high with a cavity of 7.5 cm wide between the walls. The walls of ZECC are usually made from bricks that retain water for a prolonged period. The cavity between the two walls is filled with wet sand (Figure 3.2). The sprinkling of water twice or thrice daily is enough to maintain the moisture and temperature of the chamber, or an irrigation system can be installed to keep the sand wet. Subsequently, a frame made from wood/bamboo the same size as the foundation; covered with straw, dry grass or burlap sack; and secured with rope. The products are placed in plastic crates or baskets in the chamber. Some precautionary measures should be followed, such as the surface of the interior cooling space should

FIGURE 3.2 Construction of ZECC. (Roy and Pal, 1991.)

be cleaned regularly; the products should be segregated based on their ethylene sensitivity and stored separately to enhance the shelf life.

This chamber has been reported to increase the shelf life of many fruits and vegetables and the physiological loss in weight is reduced (Tables 3.1 and 3.2). Sharma et al. (2010) reported that Royal Delicious apples could be stored for 45 days in ZECC without adverse effects on fruit quality attributes. Islam and Morimoto (2012) reported that the shelf life of brinjals and tomatoes was increased up to 9 and 16 days, respectively, in ZECC when compared to storage at ambient conditions. The shelf life of potatoes increases up to 60 days as against 30 days in ambient storage, while tomato was safely stored for 14 days as against 7 days at ambient condition (Mishra et al., 2009). Mandarin fruits could retain their postharvest quality for up to 42 days (Bhardwaj and Sen, 2003). Jha (2008) stored potatoes, Kinnow mandarins and tomatoes for 50, 25 and 4 days, respectively, with only 10% loss in weight. Similarly, ginger treated with *Trichoderma harzianum* stored in ZECC has shown to have the highest percentage of germination (98.9%) as compared to conventional storage (Shadap et al., 2014).

TABLE 3.1
Storage of Fruits in Zero Energy Cool Chamber (where PLW is the physiological loss of weight)

Crop	ZECC Chamber		Ambient Temperature	
	Shelf Life (Days)	PLW (%)	Shelf Life (Days)	PLW (%)
Aonla (*Emblica officinalis* Gaertn)	18	1.7	9	8.7
Banana (*Musa acuminata*)	20	2.5	14	4.8
Grapefruit (*Citrus paradisi*)	70	10.2	27	4.9
Guava (*Psidium guajava*)	15	4.0	10	13.6
Kinnow mandarin (*Citrus reticulata*)	60	15.3	14	16.1
Lime (*Citrus aurantiifolia*)	25	6.0	11	25.0
Mango (*Mangifera indica*)	9	5.0	6	15.0
Sapota (*Manilkara zapota*)	14	9.5	10	20.9

Source: Anonymous (1985).

TABLE 3.2
Storage of Vegetables in the Zero-Energy Cool Chamber

Crop	ZECC Chamber		Ambient Temperature	
	Shelf Life (Days)	PLW (%)	Shelf Life (Days)	PLW (%)
Amaranth (*Amaranthus viridis*)	3	11.0	<1	49.8
Okra (*Abelmoschus esculentus*)	6	5	1	14
Pointed gourd (*Trichosanthes dioica*)	5	3.9	2	32.9
Carrot (*Daucus carota*)	12	9	5	29
Potato (*Solanum tuberosum*)	97	7.7	46	19
Mint (*Mentha arvensis*)	3	18.6	1	58.5
Turnip (*Brassica rapa* subsp. *rapa*)	10	3.4	5	16
Peas (*Pisum sativum*)	10	9.2	5	29.8
Cauliflower (*Brassica oleracea* var. *botrytis*)	12	3.4	7	16.9

Source: Anonymous (1985).

3.2.1.12   Pot-in-Pot Storage System

A Pot-in-pot type of storage system was a common practice in the Indus Valley Civilization during 3000 BC for cooling as well as storing water. These are similar to the present-day earthen pots (*ghara, matka, surahi*) used in India. In rural northern Nigeria in the 1990s, Mohamed Bah Abba used the pot-in-pot storage method. A pot-in-pot cooler is an affordable electricity-free refrigerator that uses the principle of latent heat of vaporization of water to maintain the low temperature inside the inner compartment. It consists of a porous outer clay pot lined with wet sand containing an inner nonporous glazed pot to prevent penetration of outside liquid, within which the food is placed. The evaporation of the outer liquid draws heat from the inner pot (Abhinav et al., 2018). The device can cool any substance and requires only a flow of relatively dry air and a source of water.

This device works efficiently in a hot and dry climate. It should be located in a shady and well-ventilated area; otherwise, clay pot coolers may not provide sufficient benefits to justify their use. The effectiveness of evaporative cooling varies with the temperature, humidity and airflow. The limitation of this system is that it can hold only a small number of items in large water containers (Odesola and Onyebuchi, 2009).

This system has been used for the preservation of carrots, tomatoes, okra, mangoes, guavas and citrus (Obura et al., 2015). Tomatoes, grapes and brinjals can be stored for up to 9 days without any deterioration in the pot-in-pot storage system (Murugan et al., 2011).

Table 3.3 provides a summary of the effects of the different methods of conventional storage systems used on the storage life of various perishable commodities.

TABLE 3.3
Storage of Perishable Commodities in Different Conventional Systems

Storage System	Commodity	Remark	References
ZECC	Strawberry (*Fragaria × ananassa*)	Mass loss (1.59%) was significantly reduced while shelf life increases more than 3 days compared to ambient conditions	Khalid et al. (2020)
ZECC	Brinjal (*Solanum melongena*) and Tomato (*Solanum lycopersicum*)	Shelf life increases up to 9 and 16 days, respectively	Islam and Morimoto (2012)
ZECC	Royal delicious apple (*Malus domestica*)	Stored up to 45 days without any adverse effect on quality parameters	Sharma et al. (2010)
ZECC	Capsicum, bitter gourd, cauliflower	Shelf life increases up to 5 days	Singh and Satapathy (2006)
ZECC	Mango (*Mangifera indica*)	Shelf life of mango was 9 days in ZECC as compared to 6 days under ambient storage conditions	Roy and Pal (1991)
ZECC	Kinnow mandarin (*Citrus reticulata*)	Fruits could be kept up to 40 days in cool chamber as against 15 days at room temperature	Pal et al. (1997)
ZECC	Brinjal, tomato, potato	Shelf life increases up to 9–15 days in brinjal and tomato; 90 days in potato	Verma (2014)
Porous evaporative cooling storage structure	Stem amaranth (*Amaranthus cruentus*)	Shelf life increased to 5 days as compared to 2 days in ambient conditions. Weight loss was found only 7.05% as compared to 28.62% as in room condition	Ishaque et al. (2019)
Evaporative cooling	Tomato (*Solanum lycopersicum*)	Retention of color, mass, firmness, respiration rate, TA and TSS up to 20 days	Nkolisa et al. (2018)

(*Continued*)

TABLE 3.3 (*Continued*)

Storage of Perishable Commodities in Different Conventional Systems

Storage System	Commodity	Remark	References
Solar-powered evaporative cooling	Tomatoes, mangoes, bananas and carrots	Stored up to 21, 14, 17 and 28 days, respectively, as against 6, 5, 5 and 8 days in ambient storage	Olosunde et al. (2016)
Barn accomplish with a fan	Yam (*Dioscorea rotunda*)	Rotting of tubers was only 1.85% as compared to 12.03% in tubers stored in the barn without fan during their 3 months of storage	Osunde and Orhevba (2009)
Silo with artificial cooling system	Onion (*Allium cepa*)	Internal and external chroma was higher up to 14 days of storage	Rêgo et al. (2019)
Ventilated wooden pack structure	Onion bulb (*Allium cepa*)	Minimum black mold was observed under wooden packed structure (3.23%), as compared to open ground (17.50%)	Soomro et al. (2016)
Clamp under double shade (CUDS)	Karembo and KME four varieties of cassava	Cassava cuttings stored under CUDS maintained their vigor up to 16 weeks	Mdenye et al. (2018)
Pit with two PVC vents	Cocoyam (*Colocasia esculenta*)	Sustained PLW to 11 % in 8 weeks as compared to 34% in control	Obetta et al. (2007)
Wooden cartoon box filled with rice husk	Turmeric (*Cucurma longa* L.)	Shrinkage and change in color of seed rhizomes were lesser in this storage condition compared to control	Nandini et al. (2014)
Pit storage with a layer of river sand	Sweet potato (*Ipomea batatas* L.)	Sweet potatoes can be stored up to 5 months without a major change in nutrient content	Dandago and Gungula (2011)
Heap	Kufri Bahar and Kufri Jyoti varieties of potato	Potatoes stored up to 90 days on-farm in heap and suitable for processing	Kumar et al. (2005)
Bamboo basket	Potato (*Solanum tuberosum*)	Storage up to 5–6 months	Kanwar and Sharma (2006)
Evaporative cooling basket	African star apple (*Chrysophyllum albidum*)	Fresh weight losses were lower in apples stored in ECB than in the refrigerator and ambient shade	Adindu et al. (2003)
Sack bag	Sweet orange (*Citrus sinensis*)	Fruits stored in basket and jute bag loss more weight (36.6% and 34.7%) than that stored in sack bag (20.8%)	Faasema et al. (2011)
Modified brick wall evaporative cooling chamber	Limes (*Citrus aurantifolia*)	Shelf life inside the ECC was extended by 5–20 days compared to ambient storage	Marikar and Wijerathnam (2010)
Heap	Tapioca, sweet potato, colocasia, yam, elephant foot yam	Tribes of West Garo hills of Meghalaya stored the mentioned commodities safely up to the next planting season	Das and Singh (2017)

3.3 CONCLUSIONS

A large quantity of perishables goes to waste every year due to a lack of storage facilities. If the environmental temperature is maintained sufficiently low in the supply chain, much of the wastage could be avoided. Proper storage of fruits and vegetables is necessary to ensure regulated distribution and extending the duration of their availability. Under low storage temperature, quality deterioration of freshly harvested commodities is slower as microbial growth and enzymatic or chemical reactions are retarded. The conventional storage system has potential in India and other developing countries where erratic power supply and inadequate facilities hinder the use of mechanical refrigerators. Most of the conventional storage systems work on the evaporative cooling system principle. It not only decreases the temperature but also maintains the RH which prevents desiccation of the perishable produce. Cooling through evaporation is an ancient and effective method of lowering the temperature. Conventional/traditional storage system requires no special skill to operate and, therefore, is most suitable for the rural application.

However, the preservation of perishable items is often limited in traditional methods because they cannot achieve temperatures close to refrigeration temperatures (4°C–7°C) (Obura et al., 2015). Although conventional methods are widely used by the rural population for storing their commodities, they are not practical for large-scale storage and their efficiency is limited by environmental conditions (Odesola and Onyebuchi, 2009).

REFERENCES

Abera, G., and Haile, D. (2015). Yield and nutrient concentration of Anchote (*Coccinia abyssinica* (Lam.) Cogn.) affected by harvesting dates and in-situ storage. *African Journal of Crop Science*, 3(5), 156–161.

Abhinav, M., Tiwari, S., and Sarkar, P. (2018). Eco-refrigerator - a sustainable approach towards the problem of food insecurity. *International Journal of Engineering Science*, 6(4), 94–98.

Abubakar, M. S., Maduako, J. N., and Ahmed, M. (2019). Effects of storage duration and bulb sizes on physiological losses of Agrifound light red onion bulbs (*Allium cepa* L.). *Agricultural Science and Technology*, 11(1), 90–97.

Adeniji, A. (2019). The methods of controlling yam tuber rot in storage: A review. *International Journal of Innovations in Biological and Chemical Sciences*, 12, 11–17.

Adindu, M. N., Williams, J. O., and Adiele, E. C. (2003). Preliminary storage study on African star apple (*Chrysophyllum albidum*). *Plant Foods for Human Nutrition*, 58(3), 1–9.9.

Aidoo, K. E. (1993). Post-harvest storage and preservation of tropical crops. *International Biodeterioration and Biodegradation*, 32(1–3), 161–173.

Akpenpuun, T., Akinyemi, B., and Oyesomi, T. (2018). A comparative study of using a wooden storage box and storage platform for white yam tuber storage. *Acta Agriculturae Slovenica*, 111(1), 17–23.

Ali, Z., Yadav, A., Stobdan, T., and Singh, S. B. (2012). Traditional methods for storage of vegetables in cold arid region of Ladakh, India. *Indian Journal of Traditional Knowledge*, 11(2), 351–353.

Amoah, R. S., Teye, E., Abano, E. E., and Tetteh, J. P. (2011). The storage performance of sweet potatoes with different pre-storage treatments in an evaporative cooling barn. *Asian Journal of Agricultural Research*, 5(2), 137–145.

Angmo, P., Chandel, J. S., Katiyar, A. K., Targais, K., Chaurasia, O. P., and Stobdan, T. (2018). Zero energy overwinter storage of apple nursery plants in trans-Himalayan Ladakh, India. *Defence Life Science Journal*, 3(2), 162–164.

Anonymous. (1985). Zero energy cool chamber. Research Bulletin No. 43. Indian Agricultural Research Institute New Delhi, 1–23.

Babu, N., Srivastava, S. K., and Agarwal, S. (2013). Traditional storage practices of spices and condiments in Odisha. *Indian Journal of Traditional Knowledge*, 12(3), 518–523.

Basediya, A. L., Samuel, D. V. K., and Beera, V. (2013). Evaporative cooling system for storage of fruits and vegetables - A review. *Journal of Food Science and Technology*, 50(3), 429–442.

Bhardwaj, R. L., and Sen, N. L. (2003). Zero energy cool-chamber storage of mandarin (*Citrus reticulata* Blanco) cv. 'Nagpur Santra'. *Journal of Food Science and Technology (Mysore)*, 40(6), 669–672.

Dadhich, S. M., Dadhich, H., and Verma, R. (2008). Comparative study on storage of fruits and vegetables in evaporative cool chamber and in ambient. *International Journal of Food Engineering*, 4(1), 1–11.

Dandago, M. A., and Gungula, D. T. (2011). Effects of various storage methods on the quality and nutritional composition of sweet potato (*Ipomea batatas* L.) in Yola Nigeria. *International Food Research Journal*, *18*(1), 271–278.

Das, T. K., and Singh, N. A. (2017). Indigenous tuber crops production system practiced by the tribe of West Garo hills district of Meghalaya. *Agricultural Extension Journal*, *1*, 1–5.

Dvizama, A. U. (2000). Performance evaluation of an active cooling system for the storage of fruits and vegetables. *PhD. Thesis*, University of Ibadan, Ibadan.

Elansari, A. M., Yahia, E. M., and Siddiqui, W. (2019). Storage systems. In E. M. Yahia (Ed.) *Postharvest Technology of Perishable Horticultural Commodities.* Woodhead Publishing, Cambridge, UK, pp. 401–437.

El-Ramady, H. R., Domokos-Szabolcsy, É., Abdalla, N. A., Taha, H. S., and Fári, M. (2015). Postharvest management of fruits and vegetables storage. In: E. Lichtfouse (Ed.) *Sustainable Agriculture Reviews.* Springer, Cham, 65–152.

Ezeocha, C. V., and Ironkwe, A. G. (2017). Effect of storage methods and period on the physiological and nutrient components of Livingstone potato (*Plectranthus esculentus*) in Abia State, Nigeria. *Open Agriculture*, *2*(1), 213–219.

Faasema, J., Abu, J. O., and Alakali, J. S. (2011). Effect of packaging and storage condition on the quality of sweet orange (*Citrus sinesis*). *Journal of Agricultural Technology*, *7*(3), 797–804.

Falagán, N., and Terry, L. A. (2018). Recent advances in controlled and modified atmosphere of fresh produce. *Johnson Matthey Technology Review*, *62*(1), 107–117.

Ishaque, F., Hossain, M. A., Sarker, M. A. R., Mia, M. Y., Dhrubo, A. S., Uddin, G. T., and Rahman, M. H. (2019). A study on low cost post harvest storage techniques to extend the shelf life of citrus fruits and vegetables. *Journal of Engineering Research and Reports*, 1–17.

Islam, M. P., and Morimoto, T. (2012). Zero energy cool chamber for extending the shelf-life of tomato and eggplant. *Japan Agricultural Research Quarterly: JARQ*, *46*(3), 257–267.

Jha, S. N. (2008). Development of a pilot scale evaporative cooled storage structure for fruits and vegetables for hot and dry region. *Journal of Food Science and Technology-Mysore*, *45*(2), 148–151.

Jha, S. N., and Chopra, S. (2006). Selection of bricks and cooling pad for construction of evaporatively cooled storage structure. *Journal of the Institution of Engineers (India)*, *87*, 25–28.

Kale, S. J., Nath, P., Jalgaonkar, K. R., and Mahawar, M. K. (2016). Low cost storage structures for fruits and vegetables handling in Indian conditions. *Indian Horticulture Journal*, *6*(3), 376–379.

Kanwar, P., and Sharma, N. (2006). Indigenous crop storage practices prevalent among rural people for food security in Himachal Pradesh. *Asian Agri History*, *10*(4), 281–92.

Khalid, S., Majeed, M., Ullah, M. I., Shahid, M., Riasat, A. R., Abbas, T., Aatif, H. M., and Farooq, A. (2020). Effect of storage conditions and packaging material on postharvest quality attributes of strawberry. *Journal of Horticulture and Postharvest Research*, *3*(2), 195–208.

Kumar, D., Paul, V., and Ezekiel, R. (2005).Chipping quality of potatoes stored in heaps and pits in subtropical plains of India. *Horticultural Science-UZPI (Czech Republic)*, *32*, 23–30.

Marikar, F. M. M. T., and Wijerathnam, R. W. (2010). Post-harvest storage of lime fruits (*Citrus aurantifolia*) following high humidity and low temperature in a modified brick wall cooler. *International Journal of Agricultural and Biological Engineering*, *3*(3), 80–86.

Mdenye, B. B., Kinama, J. M., Olubayo, F. M., Kivuva, B. M., and Muthomi, J. W. (2016). Effect of storage methods on carbohydrate and moisture of cassava planting materials. *Journal of Agricultural Science*, *8*(12), 100–111.

Mdenye, B. B., Kinama, J. M., Olubayo, F., Kivuva, B. M., and Muthomi, J. W. (2018). Effect of storage methods of cassava planting materials on establishment and early growth vigour. *International Journal of Agronomy and Agricultural Research*, *12*, 1–10.

Mishra, B. K., Jain, N. K., Kumar, S., Doharey, D. S., and Sharma, K. C. (2009). Shelf life studies on potato and tomato under evaporative cooled storage structure in Southern Rajasthan. *Journal of Agricultural Engineering*, *46*(3), 26–30.

Mobolade, A. J., Bunindro, N., Sahoo, D., and Rajashekar, Y. (2019). Traditional methods of food grains preservation and storage in Nigeria and India. *Annals of Agricultural Sciences*, *64*(2), 196–205.

More, A. S., Ranadheera, C. S., Fang, Z., Warner, R., and Ajlouni, S. (2020). Biomarkers associated with quality and safety of fresh-cut produce. *Food Bioscience*, *34*, 100524.

Murugan, A. M., Singh, R., and Vidhya, S. (2011). Evaporative cooling: A postharvest technology for fruits and vegetables preservation. *Indian Journal of Traditional Knowledge*, *10*, 375–379.

Nandini, K., Singh, M. S., Lhungdim, J., Nanita, H., Diana, S., and Dorendro, A. (2014). Effect of rice husk on storage of seed rhizome of turmeric (*Curcuma longa* L.). *Agricultural Science Digest-A Research Journal*, *34*(3), 199–202.

Nkolisa, N., Magwaza, L. S., Workneh, T. S., and Chimphango, A. (2018). Evaluating evaporative cooling system as an energy-free and cost-effective method for postharvest storage of tomatoes (*Solanum lycopersicum* L.) for smallholder farmers. *Scientia Horticulturae*, *241*, 131–143.

Obetta, S. E., Satimehin, A. A., and Ijabo, O. J. (2007). Evaluation of a ventilated underground storage for cocoyams (taro). *Agricultural Engineering International: CIGR Journal*, *9*, 1–17.

Obura, J. M., Banadda, N., Wanyama, J., and Kiggundu, N. (2015). A critical review of selected appropriate traditional evaporative cooling as postharvest technologies in Eastern Africa. *Agricultural Engineering International: CIGR Journal*, *17*(4): 327–336

Odesola, I. F., and Onyebuchi, O. (2009). A review of porous evaporative cooling for the preservation of fruits and vegetables. *Pacific Journal of Science and Technology*, *10*, 935–941.

Olosunde, W. A., Aremu, A. K., and Onwude, D. I. (2016). Development of a solar powered evaporative cooling storage system for tropical fruits and vegetables. *Journal of Food Processing and Preservation*, *40*(2), 279–290.

Olosunde, W. A., Igbeka, J. C., and Olurin, T. O. (2009). Performance evaluation of absorbent materials in evaporative cooling system for the storage of fruits and vegetables. *International Journal of Food Engineering*, *5*(3), 2.

Onifade, T. B., Uthman, F., and Iyalabani, K. A. (2019). Construction of storage structures for onion bulbs for food security. *Journal of Agricultural Engineering and Technology*, *24*(2), 35–43.

Osunde, Z. D., and Orhevba, B. A. (2009). Effects of storage conditions and storage period on nutritional and other qualities of stored yam (*Dioscorea* spp) tubers. *African Journal of Food, Agriculture, Nutrition and Development*, *9*(2), 678–690.

Pal, R. K., Roy, S. K., and Srivastava, S. (1997). Storage performance of kinnow mandarins in evaporative cool chamber and ambient condition. *Journal of Food Science and Technology (Mysore)*, *34*(3), 200–203.

Patil, R. T. (2010). Appropriate engineering and technology interventions in horticulture for enhanced profitability and reduction in postharvest losses. *Acta Horticulturae*, *877*, 1363–1370.

Regina, B., and Kumar, A. V. (2018). Review on the storage structure for longer shelf-life of vegetables. *Progressive Agriculture*, *18*(2), 145–161.

Rêgo, E. R., Ferreira, A. P. S., Pereira, D. M., Pereira, A. M., Pereira, O. L., and Finger, F. L. (2019). Artificially cooling of onion bulbs stored in brickwork-patterned vertical silos. *Horticultura Brasileira*, *37*(2), 234–238.

Roy, S. K., and Pal, R. K. (1991). A low cost zero energy cool chamber for short-term storage of mango. *Acta Horticulturae*, *291*, 519–524.

Sah, U., and Kumar, S. (2008). Traditional potato storage among Khasi tribes of Meghalaya Hills. *Asian Agri-History*, *12*(2), 129–137.

Shadap, A., Hegde, N. K., Lyngdoh, Y. A., and Rymbai, H. (2014). Effect of storage methods and seed rhizome treatment on field performance of Ginger var. Humnabad. *Indian Journal of Hill Farming*, *2*(1), 219–228.

Sharma, R. R., Pal, R. K., Singh, D., Samuel, D. V. K., Kar, A., and Asrey, R. (2010). Storage life and fruit quality of individually shrink-wrapped apples (*Malus domestica*) in zero energy cool chamber. *Indian Journal of Agricultural Sciences*, *80*(4), 338–341.

Singh, R. K. P., and Satapathy, K. K. (2006). Performance evaluation of zero energy cool chamber in hilly region. *Agricultural Engineering Today*, *30*(5 and 6), 47–56.

Soomro, S. A., Ibupoto, K. A., Soomro, N. M., and Jamali, L. A. (2016). Effect of storage methods on the quality of onion bulbs. *Pakistan Journal of Agriculture, Agricultural Engineering and Veterinary Sciences*, *32*(2), 221–228.

Tomlins, K., Ndunguru, G., Kimenya, F., Ngendello, T., Rwiza, E., Amour, R., Oirschot, Q. V., and Westby, A. (2007). On-farm evaluation of methods for storing fresh sweet potato roots in East Africa. *Tropical Science*, *47*(4), 197–210.

Tortoe, C., Obodai, M., and Amoa-Awua, W. (2010). Microbial deterioration of white variety sweet potato (*Ipomoea batatas*) under different storage structures. *International Journal of Plant Biology*, *1*(1), e1–e10.

Tumuhimbise, G. A., Namutebi, A., and Muyonga, J. H. (2010). Changes in microstructure, beta carotene content and in vitro bioaccessibility of orange-fleshed sweet potato roots stored under different conditions. *African Journal of Food, Agriculture, Nutrition and Development*, *10*(8).doi:10.4314/ajfand.v10i8.60888.

Umogbai, V. I. (2013). Design, construction and performance evaluation of an underground storage structure for yam tubers. *International Journal of Scientific and Research Publications*, *3*(5), 1–7.

Verma, A. (2014). Pre-cooling of fresh vegetables in low cost zero energy cool chamber at farmer's field. *Asian Journal of Horticulture*, *9*(1), 262–264.

4 Improved Small-Scale Cooling Systems

Lisa Kitinoja
The Postharvest Education Foundation

Charles Wilson
World Food Preservation Center LLC

CONTENTS

4.1 INTRODUCTION

A wide range of options and technologies exist for producing cold conditions for food handling, processing, storage, and transport. For precooling, operators can choose from simple farm-based methods, such as using water (hydrocooling) or ice, to more complex systems for forced air or vacuum cooling. For storage, options for food handlers exist that range from small walk-in cold rooms to large-scale commercial refrigerated warehouses. Small-scale cold rooms can be designed using traditional mechanical refrigeration systems, CoolBot™ equipped air-conditioner systems or evaporative cool chambers. During transport, cold temperatures can be provided via the use of ice, trailer-mounted refrigeration systems, evaporative coolers, or via passive cooling technologies such as insulated packages or the use of pallet covers during transport.

The suitability of these options depends on the food products being handled and the sophistication of the value chain. Winrock International (2009) and Kitinoja and Thompson (2010) have reviewed cooling practices utilized during precooling and cold storage for fresh horticultural crops and provided basic recommendations on cooling options and information regarding capital costs and energy use.

In general, the highest cost for mechanical refrigeration systems using electricity or diesel fuel occurs when temperatures are the hottest. Although costs may be high, the benefits of using cold chain technologies outweigh costs in regions where food losses due to lack of temperature management are the highest. Total construction and operating costs for refrigerated systems vary widely depending on the costs of local materials, labor, and electricity. Evaporative cooling (EC) systems are lower in cost but work well only in dry regions or during dry seasons. Postharvest losses can be greatly reduced with the use of cold storage, but the return on investment for any specific operation depends largely on the market value of the produce being cooled and stored and the use efficiency of the facility. Use efficiency includes the percentage of total capacity utilized and the number of days per year the facility is in operation.

4.2 TECHNIQUES FOR LOW-TEMPERATURE PRESERVATION

Precooling (initial cooling) refers to reducing the pulp temperature of fresh horticultural produce soon after harvest via various means of conduction or convection. Target temperatures depend on the produce type and its lowest safe temperature. Forced-air cooling (pulling or pushing cold air through packages of produce) and hydrocooling (immersing or spraying fresh produce with cold water) are commonly used precooling methods. Vacuum precoolers use a special high-cost system that mechanically lowers atmospheric pressure around the produce inside a sealed chamber that allows evaporation to reach temperatures as low as 1°C. Vacuum cooling is used mainly for cooling the pallets loaded with leafy green vegetables. **Chilling** involves the rapid lowering of the temperature of fresh or processed food products to less than 8°C via mechanical or nonmechanical means. Temperate crops typically have a target chilling temperature of 4°C or lower, while most tropical and subtropical crops are precooled to a target temperature of about 15°C. Tropical and subtropical crops can be damaged if precooling or storage temperatures are too low and will show a variety of chilling injury symptoms.

4.2.1 COMMERCIAL REFRIGERATION SYSTEMS (MECHANICAL)

4.2.1.1 Vapor/Recompressor Systems

Compressor refrigerators use electrically powered compressors to pressurize and heat gas that condenses back into a liquid through heat exchange with a coolant (usually air). Once the high-pressure gas has cooled and condenses into a liquid, it passes through a narrow orifice that creates a pressure drop, allowing the liquid to evaporate. The evaporation process absorbs heat, and the temperature of the refrigerant drops to its boiling point at the (now) low pressure. Hydrofluorocarbon (HFC) refrigerants or ammonia are used as refrigerant fluids.

4.2.1.2 Absorption/Adsorption Systems

Wang and Oliveira (2005) provide a full review and discussion of adsorption systems and their limited status and future potential. An absorption refrigerator uses a heat source (usually solar, kerosene, propane gas, or waste heat) to provide the energy needed to drive the cooling system. An absorption refrigerator changes gas back into a liquid using only heat and has few moving parts. Absorption refrigerators are popular alternatives to regular compressor refrigerators where electricity is unavailable, unreliable, or costly. Heat sources for adsorption systems can be of low temperatures such as solar power or waste heat from a processing operation, which can result in cost savings compared to fossil fuel-powered compression or absorption systems (Table 4.1).

The CoolBot™ is a recent development on the small-scale mechanical cooling technology front. It is used to equip cold rooms for the storage of chilled food products and fresh horticultural produce. A small cold room with a commercially installed refrigeration system costs about US$ 7000 for 3.5 kW (1 ton) of refrigeration capacity (Winrock, 2009). A small-scale option is to use a modified room air conditioner, a method originally developed by Boyette and Rohrbach (1993) to prevent ice buildup that restricts airflow and stops cooling. The control system of the window-style air conditioner unit is modified to allow the production of low air temperatures without the buildup of ice on the evaporator coil. Recently a US-based company developed an easily installed digital controller that prevents ice buildup while not requiring modification of the control system of the

TABLE 4.1

Examples of Mechanical Technologies Available for Low-Temperature Preservation

Cold Chain Step	Small Scale	Large Scale
Precooling systems	Portable forced air cooling systems	Vacuum cooling
	Coldwater sprays (hydrocooling)	Forced air cooling inside refrigerated cold rooms
		Hydrocooling via immersion or conveyor systems (recirculating water systems)
Cold Storage	Walk-in cold rooms	Refrigerated warehouses
	CoolBot™ equipped cold room	
Refrigerated transport	USDA Porta-cooler	Reefer vans
	Trailer equipped with CoolBot™	Refrigerated marine containers
		Refrigerated intermodal containers (for road, rail, and sea shipping)

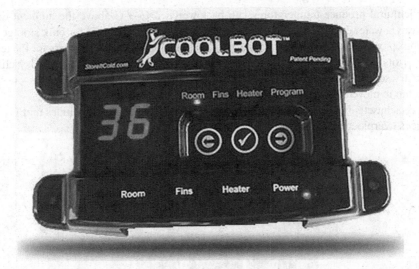

FIGURE 4.1 The CoolBot™ controller. (Photo source: The Postharvest Education Foundation.)

air conditioner (CoolBot™, Store It Cold, LLC, http://storeitcold.com). A 1-ton room air conditioner with a CoolBot™ control system currently costs about 80%–90% less than a commercial refrigeration system. This automated control system is designed so that any moisture condensation on the refrigeration coils is returned to the cold room air resulting in less product moisture loss than that by a commercial refrigeration system (Figure 4.1).

For refrigerated transport, small-scale producers and marketers can use portable coolers. Several types of portable coolers currently exist and have been tested extensively (Boyette, 1995; USDA, 1993). They can be self-constructed at a relatively low cost, and complete plans are available on the Internet on the ATTRA website (http://www.attra.ncat.org) and the North Carolina State University (NCSU) website (http://www.bae.ncsu.edu/programs/extension/publicat/postharv/ag-414-7/index. html). The USDA Porta-cooler can be carried on traditional small-scale transport vehicles, either pulled as a trailer or set into a pickup truck bed. The Porta-cooler consists of a small insulated box (3.5 m³), holding approximately 700 kg of produce, fitted with a room-sized air conditioner (2.9–3.5 kW) and a diesel-powered generator (2 kW). These units can be operated successfully at setpoint temperatures of 10°C or above with good results, making them most useful for transporting tropical and subtropical horticultural crops. At temperatures below 10°C, however, ice builds up on the coils, and the air conditioner will not work as designed. The CoolBot ™ control system described·above can be utilized to overcome this limitation. A full set of plans for the construction of an insulated trailer equipped with the CoolBot™ has recently been developed by scientists at NCSU and is available online to download free (http://plantsforhumanhealth.ncsu.edu/2012/08/20/pack-n-cool/).

4.2.2 NONMECHANICAL COOLING PRACTICES

4.2.2.1 Use of Shade

Keeping produce in the shade after harvest can decrease its temperature and slow deterioration. Rikard and Coursey (1979) were the first to demonstrate that shading the produce resulted in temperatures slightly higher than ambient air temperature (30°C–35°C), while produce exposed to the sun during the heat of the day had temperatures that rose to nearly 50°C. Since fresh produce experiences much higher rates of respiration and related deterioration at higher temperatures, reducing the temperature by 10°C can double or triple potential postharvest life.

4.2.2.2 Evaporative Cooling

Fresh horticultural produce temperatures can be lowered 2°C–3°C above the ambient dew point temperature via water evaporation. EC storage rooms are commonly used for bulk storage of tropical and subtropical crops (such as sweet potatoes) or as small-scale cool chambers for the temporary storage of fruits and vegetables in tropical climates. These systems work best in dry climates or during the dry season. Evaporative coolers can be passive or power-assisted (using a solar-powered or electric fan to move air through the storage chamber) (Figure 4.2).

The Massachusetts Institute of Technology D-Lab has published several guides on EC principles and practices (Verploegen et al., 2018, 2019).

FIGURE 4.2 Walk-in size charcoal cool room for small-scale cool storage in Semera, Ethiopia. (Photo credit: Lisa Kitinoja.)

Build your own charcoal cool room (Chemonics International).
http://www.postharvest.org/images/CharcoalcoolstoragePNACQ751.pdf.

FIGURE 4.3 Large-scale evaporative cooled storage facility for cured sweet potatoes, showing a row of large cooling units on the roof. (Photo credit: Robert Kasmire.)

A variety of designs for small-scale evaporative cooled storage chambers have been developed for fresh tropical and subtropical produce. Simple large-scale evaporative cool storage structures are used for storing huge quantities of sweet potatoes in the United States (Figure 4.3). Kitinoja and Thompson (2010) reviewed many designs currently available in Southeast Asia, India, and Africa, and most can be self-constructed using low-cost materials. The low-cost passive cooling chamber illustrated in Figure 4.4 is constructed from locally made kiln-fired clay bricks. The cavity between the walls is filled with clean sand, and the bricks and the sand are kept saturated with water. Fruits and vegetables are loaded inside, and the entire chamber is covered with a rush mat that is also kept moist. The storage capacity is 1 MT. During the hot summer months, this chamber can maintain an inside temperature between 15°C and 18°C below ambient temperature and relative humidity of about 95%.

The original developers of this technology at the Indian Agricultural Research Institute (IARI), New Delhi, have called it a "Zero-Energy Cool Chamber" (ZECC) because it uses no external energy (Roy and Khurdiya, 1982, 1985). A larger version of this chamber has been designed by IARI and constructed by the Central Institute of Post-harvest Engineering and Technology (CIPHET), Ludhiana, India, as a small cold room (6–8 MT capacity) and needs only the addition of a small water pump and a ventilation fan at the roofline (similar to the vent fans used in greenhouses). Since relatively large amounts of material are required to construct these cold storage chambers, they may be most practical when handling high-value products.

The construction cost for a small unit in India was US$ 200 (100 kg capacity) while the cost for a large walk-along unit was US$ 1000 (1 MT capacity). The cost of the larger commercial-sized unit is estimated to be US$ 8,000. Cooling results are best when the relative humidity conditions outside the ZECC are low, as during the dry season or in semi-arid regions.

Build your own ZECC (AVRDC)
12 min video (Swahili with English sub-titles) by Roseline Marealle
https://www.youtube.com/watch?v=ZUUVI8isvxU
Walk-in design for an evaporative cooled room (CIPHET)
8 min video https://www.youtube.com/watch?v=I0nOQFD7a3Q
Will EC work in your climate zone?
http://www.easycalculation.com/weather/dewpoint-wetbulb-calculator.php

Basediya et al. (2013) reviewed a wide range of EC technologies and provide additional guidance. In addition to these commonly used EC systems, other cooling systems are available when

FIGURE 4.4   Design for a 1 MT capacity ZECC (Kitinoja, 2010). (Diagram credit: Amity University, India.)

electricity is not available. Harvesting fresh produce early in the morning (except citrus crops because of fruit susceptibility to physical damage when turgid) ensures that produce is handled at lower temperatures than those that occur later in the day. The use of shade after harvesting keeps produce from warming in the sun before transport. Crushed or slurry ice is useful for rapid chilling or precooling of fish or vegetables that can tolerate water. Slurry ice is a solution of about 40% water, 60% ice, and 1% salt. Ice in large pieces or blocks can be used to cool water that can then be used in a shower or immersion-type hydrocooling systems. The cost of ice production can be very high compared to its cooling capacity (Kitinoja and Thompson, 2010), and melted ice can cause safety and sanitation problems during handling, storage, transport, and marketing.

Night air ventilation involves the opening of vents in the basement of an insulated storage structure during the cooler night hours and closing the facility during the daytime to keep the cool air inside. As a general rule, night ventilation effectively maintains a given product temperature when the outside air temperature is below the given product temperature for 5–7 hours per night (Kitinoja and Thompson, 2010). Natural underground cooling can be used in caves or root cellars. High altitude cooling can be used where ambient air temperatures are lower than average. Radiant cooling can be used in dry climates with clear night skies to lower the temperature of ambient air. When using a solar collector or a good radiating surface at night, air will cool as the collector surface radiates heat to the cold night sky. Temperatures inside the structure of 4°C less than night air temperature can be achieved (Thompson et al., 2002b).

Passive cooling (insulated packages or pallets covers) can be used during transport to keep precooled or chilled foods cold. The insulation acts to prevent rapid rewarming but has a limited range, and the distance or time that foods can be kept cool will depend on the outside air temperatures and desired product temperature upon delivery.

Solar-powered cooling systems that function via an ice bank or ice battery are in developmental stages (www.solarchill.org). Currently available solar chilling systems are very expensive and too small for commercial food handling or storage. Prototypes of this ice-based cool box are available via a United Nations program for storage of pharmaceuticals and vaccines. These boxes use a

TABLE 4.2

Examples of Nonmechanical Technologies Available for Low-Temperature Preservation

Cold Chain Step	Small Scale	Large Scale
Precooling systems	Portable evaporative forced air cooling systems	Slurry ice
	Use of shade after harvest	
	Use of crushed ice	
Cold Storage	Zero energy cool chambers (ZECC)	Evaporative cooled warehouses
	Evaporative cooled cool rooms (charcoal coolers)	Underground storage (caves)
		High altitude storage
	Underground storage (root cellars)	Radiant cooling
	Night air ventilation	
	High altitude storage	
	Radiant cooling	
	Solar chillers	
Refrigerated transport	Evaporative cooled insulated transport boxes or trailers	Insulated pallet covers of cooled produce

solar-powered 3×60 W photovoltaic array and ice as the energy storage medium (rather than acid batteries which tend to have a short life in hot climates and create environmental hazards if not recycled properly). Cost is estimated at US$ 1,500 for a unit that has a storage capacity of 50–100 L. These units would be best used for the temporary storage of highly perishable high-value foods such as fresh-cut fruits or vegetables, strawberries, cheeses, bean sprouts, or mushrooms (Table 4.2).

4.3 OTHER TECHNOLOGICAL ISSUES

A variety of design and logistic factors affect the effectiveness and efficiency of precooling, cold storage, and cold transport. The type of packages, use of packaging materials, and loading patterns can either facilitate or hinder cooling and cold storage technologies.

4.3.1 TYPE OF PACKAGES

The type of package used will have a direct effect on the effectiveness of precooling or chilling since in all cases the cooling medium needs to make contact with the food product to remove heat. Some foods are chilled without being packaged while others are packaged in plastic films, fiberboard cartons, or clamshell packages before cooling. Vented containers and vented packages allow fresh produce to cool faster than the same produce in closed containers or sealed packages. The recommended amount of venting for fresh produce being cooled using forced air cooling is 5% of the surface of the package. Plastic crates often have much more venting than 5%, so liners can be used to reduce the effective vent area and reduce the rate of water loss during precooling.

Aseptic packages and "shelf-stable" packages can greatly increase the shelf life of food products without refrigeration. These packages have excellent potential to preserve foods in areas where there is limited electric power for refrigeration. Facilities, equipment, and supplies needed for preparing aseptically packaged products are very costly and not obtained easily by small-scale producers, traders, processors, or marketers.

4.3.2 PACKING AND PACKAGING MATERIALS

Flushing packages with nitrogen gas can reduce oxidative deterioration. Nitrogen is used for packing some vegetable crops (shredded lettuce, leafy greens) and can further extend shelf life. However, if temperatures are not well maintained when using low oxygen or modified atmosphere packaging

(MAP), the lack of oxygen can lead to anaerobic conditions, more rapid deterioration, off-flavors, and potential health hazards.

Filters added to cartons of fresh produce (wraps, liners, shredded paper, etc.) will provide cushioning that can reduce bruising but will slow cooling and can interfere with airflow during cold storage and transport. Consumer packages with venting such as clamshells or punnets can decrease these problems while providing a ready-to-sell unit at the retail level. Each intended market will often have its standards for acceptable type, size, and quality of packages and packaging materials.

4.3.3 LOADING PATTERNS

Stacking packages directly on the floor or up against the walls interferes with cold air circulation during cold storage and transport. Loading a transport vehicle too high (too close to the ceiling) or stacking packages against the walls of the trailer can allow outside heat to more easily penetrate the load. Recommendations for loading patterns are available for a variety of cold storage facilities (Thompson et al., 2000b) and common sizes of road vehicles, marine, rail, and air containers (Thompson et al., 2000a, 2002a, 2004).

4.4 CURRENT PRACTICES VERSUS RECOMMENDED BEST PRACTICES

Globally, the limited use of cold and current cooling, cold storage, and cold chain management practices falls far short of recommended best practices, often leading to very high levels of food losses (Gustavsson et al., 2011; Kitinoja, 2010). There are many publications covering the topic of best practices in great detail (Kitinoja and Gorny, 1999; Thompson et al., 2002b; Vigneault et al., 2009; Kitinoja and Kader, 2015) and many websites provide current recommendations. The following are just a few websites where information on cooling and cold storage best practices can be found.

Web-based information portals:

Cold Chain News http://www.globalcoldchainnews.com/
Cold Chain Technologies http://www.coldchaintech.com
Refrigerated and Frozen Foods http://www.refrigeratedfrozenfood.com/
GCCA The Global Cold Chain Alliance http://www.gcca.org
UC Postharvest Technology Center http://postharvest.ucdavis.edu
The Postharvest Education Foundation http://www.postharvest.org
ColdChainInfo http://www.coldchaininfo.com/index.html
PurFresh Transport http://www.purfresh.com/app_transportation.htm

Many horticultural crops are harvested from the field or orchard and then handled at ambient temperatures until they reach the consumer. Traditional practices for handling horticultural crops depend upon quick handling and marketing to minimize deterioration and decay. But in practice, foods are often subject to handling delays, high temperatures, and long distribution chains as they move from producer to traders and then to markets.

Best practices for the period between harvest and packing/processing include precooling or chilling as soon as possible after harvest. Providing shade for fresh produce during packing can help reduce temperatures. Using simple on-farm practices such as shade or packing on ice can help reduce postharvest losses and increase shelf life.

Precooling by mechanical or nonmechanical means should be used for fresh produce whenever possible. Hydrocooling is best used for water-tolerant produce, while forced-air cooling can be used with nearly all kinds of fresh fruits and vegetables. Cold can be provided along the value chain by using a range of methods and approaches. Choices of mechanical refrigeration versus nonmechanical cooling are related to the level of sophistication of the value chain. Simple methods for providing cooling, such as shade, ice, or EC systems, may be cost-effective even when food products are being handled for a very short period between harvest and consumption, or if food products have low market value. More expensive cooling methods, such as forced air cooling, hydrocooling or refrigerated transport are economically feasible only when the product's market value is relatively high and when the facilities are used efficiently. This chapter has provided an overview of cooling and cold storage methods for small-scale users. Other chapters of this book will cover cold chain management logistics in more detail.

REFERENCES

Basediya, A. L., Samuel, D. V. K., and Beera, V. (2013). Evaporative cooling system for storage of fruits and vegetables-a review. *Journal of Food Science and Technology, 50*(3), 429–442.

Boyette, M. D. (1995). Cool and Ship: A low cost, portable forced air cooling unit. NCSU Extension Publication AG-414-7.https://content.ces.ncsu.edu/cool-and-ship-a-low-cost-portable-forced-air-cooling-unit.

Boyette, M. D., and Rohrbach, R. P. (1993). A low-cost, portable, forced-air pallet cooling system. *Applied Engineering in Agriculture, 9*(1), 97–104.

CoolBot™ (2021) Store It Cold, LLC. http://storeitcold.com. Accessed on 28 March 2021.

Gustavsson, J., Cederberg, C., Sonesson, U., Van Otterdijk, R., and Meybeck, A. (2011). Global Food Losses and Food Waste: Extent, Causes and Prevention. UN FAO: Rome. http://www.fao.org/fileadmin/user_upload/ags/publications/GFL_web.pdf.

Kitinoja, L. (2010). Identification of Appropriate Postharvest Technologies for Improving Market Access and Incomes for Small Horticultural Farmers in Sub-Saharan Africa and South Asia. WFLO Grant Final Report to the Bill and Melinda Gates Foundation, March 2010. 318 pp. http://ucce.ucdavis.edu/files/datastore/234-1847.pdf.

Kitinoja, L., and Gorny, J. R. (1999). Postharvest Technology for Small-Scale Produce Marketers: Economic Opportunities, Quality and Food Safety. Hort Series No. 21. University of California, Davis.

Kitinoja, L., and Kader, A. A. (2015). Small-Scale Postharvest Practices: A Manual for Horticultural Crops, 5th edition. Hort Series No. 8. University of California, Davis. 215 pp. http://ucanr.edu/sites/Postharvest_Technology_Center_/files/231952.pdf.

Kitinoja, L., and Thompson, J. F. (2010). Pre-cooling systems for small-scale producers. *Stewart Postharvest Review, 6*(2), 1–14.

Rikard, J. E., and Coursey, D. G. (1979). The value of shading perishable produce after harvest. *Appropriate Technology, 6*(2), 2 pp.

Roy, S. K., and Khurdiya, D. S. (1982). Keep vegetables fresh in summer (Vol. 27). *Indian Horticulture.* New Delhi: IARI.

Roy, S. K., and Khurdiya, D. S. (1985). Zero Energy Cool Chamber (Vol. 43). India Agricultural Research Institute: New Delhi, India: Research Bulletin.

Thompson, J. F., Brecht, P. E., and Hinsch, T. (2002a). Refrigerated trailer transport of perishable products. The University of California, Division of Agriculture and Natural Resources Publication no. 21614.

Thompson J. F., Mitchell F. G., and Kasmire R. F. (2002b). Cooling horticultural commodities. In Postharvest technology of horticultural crops, Third edition, Kader A.A. (Tech. Ed). The University of California, Division of Agriculture and Natural Resources, Publication 3311. 2002: 97–112.

Thompson, J. F., Brecht, P. E., Hinsh, T., and Kader, A. A. (2000a). Marine container transport of chilled perishable produce. Agriculture and Natural Resources, University of California, Davis, CA, USA. Publication no. 21595.

Thompson, J. F., Mitchell, F. G., Rumsey, T. R., Kasmire, R. F., and Crisosto, C. H. (2000b). Commercial cooling of fruits, vegetables and flowers. The University of California, Division of Agriculture and Natural Resources, Publication no. 21567.

Thompson, J. F., Bishop, C. F. H., and Brecht, P. E. (2004). Air transport of perishable products. Agriculture and Natural Resources, University of California, Davis (CA), Publication no. 21618.

USDA (1993). Transportation Tip: The PortaCooler. http://www.ams.usda.gov/tmd/MSB/PDFpubList/portacooler.pdf.

Verploegen, E., Ekka, R., and Gill, G. S. (2019). Evaporative Cooling for Improved Fruit and Vegetable Storage in Rwanda and Burkina Faso. Copyright © Massachusetts Institute of Technology. https://d-lab.mit.edu/resources/publications/evaporative-cooling-improved-vegetable-and-fruit-storage-rwanda-and-burkina.

Verploegen, E., Rinker, P., and Ognakossan, K. E. (2018). Evaporative Cooling Best Practices Guide. Copyright © Massachusetts Institute of Technology. http://d-lab.mit.edu/resources/publications/evaporative-cooling-best-practices-guide.

Vigneault, C., Thompson, J., Wu, S., Hui, K. C., and LeBlanc, D. I. (2009). Transportation of fresh horticultural produce. *Postharvest Technologies for Horticultural Crops*, 2(1), 1–24.

Wang, R. Z., and Oliveira, R.G. (2005). Adsorption refrigeration: an efficient way to make good use of waste heat and solar energy. International Sorption Heat Pump Conference, Denver CO. June 22–24, 2005.

Winrock International. (2009). Empowering agriculture: energy solutions for horticulture. USAID Office of Infrastructure and Engineering and the Office of Agriculture; 79p. Available online: http://ucce.ucdavis.edu/files/datastore/234-1386.pdf.

5 Sustainable Cold Chain Development

Toby Peters
University of Birmingham

Leyla Sayin
Toby Peters Consultancy

CONTENTS

5.1 INTRODUCTION

The world's population is expected to reach 9.7 billion by 2050 (United Nations, 2019), posing serious challenges for eradicating hunger, achieving food security and better nutrition. Even in 2019 before the COVID-19 pandemic, about 8.9% of the world's population or 690 million people were suffering from hunger, and without intervention, this number is expected to reach 840 million by 2030 (FAO, 2020).

The projections suggest that feeding 9.7 billion people would require increasing food production by 56% from the 2010 levels (WRI, 2019).* Closing this supply-demand gap, however, cannot be achieved sustainably by simply ramping up food production. Agriculture already occupies half of all habitable land, accounts for 70% of freshwater withdrawals and is responsible for around a quarter of greenhouse gas (GHG) emissions (Ritchie, 2019; World Bank, 2020; WRI, 2019). According to United Nations Food and Agriculture Organization (UNFAO) 1.3 billion tonnes or 30% of perishable food is lost or wasted annually, which is enough to feed approximately 2 billion people, resulting in an economic loss of about \$940 billion and an emission impact of 4.4 GT CO_2 in addition to wasting natural resources and agricultural inputs, and creating an unnecessary burden for the environment (FAO, 2011, 2014; Scialabba, 2015). Hence, rather than solely focusing on how to increase food production to meet the increasing food demand, we need to implement robust strategies into the food system for reducing food loss. In short, food saved is as important as food produced.

The majority of food wastage occurs postharvest in developing countries (food loss), mainly stemming from lack of robust cold chains alongside the lack of technical know-how on good agricultural practices as well as maintenance and operation of equipment, and limited market knowledge. According to UNFAO, 'the lack of sufficient and efficient cold chain infrastructure is a major contributor to food losses and waste in NENA (Near-East, North Africa), estimated to be 55%

---

* Measured in total calories.

DOI: 10.1201/9781003056607-7

of fruits and vegetables, 22% of meats, 30% of fish and seafood, and 20% of dairy'(FAO, 2015). UNFAO further estimates that 35% of fish caught globally are wasted and 'most of the losses are due to a lack of knowledge or equipment, such as refrigeration or ice-makers, needed to keep fish fresh'(FAO, 2018). Hence, a well-organised and sustainable cold chain supported by robust transport and energy infrastructure is key to translating supply into demand with minimum product losses to feed the surging population. With that in mind, refrigeration and cold chain infrastructure are essential to support small-scale and marginal farmers by providing effective and efficient market connectivity, linking demand with supply and enabling access to high-value markets. The important role of cold chains and market connectivity in matching increased supply to increased demand and bringing about sustainable development has been widely recognised (Peters, 2016; Willett et al., 2019; WRI, 2019). Cold chains are key to achieving the entire set of UN Sustainable Development Goals (SDGs), but, in particular, SDGs set ambitions to end hunger, achieve food security, end all forms of malnutrition, double the agricultural productivity and incomes of small-scale food producers and halve global food loss by 2030. However, little has been done to quantify what would be required to successfully meet these targets. Apart from the physical numbers and capacity of temperature-controlled road vehicles, cargo ships, multi-modal containers, domestic refrigerators, chilled display cabinets, cold storages, pack houses, ripening chambers, pre-coolers and a plethora of other cold chain-supporting equipment and infrastructure in the energy, transport, retail and food logistics sectors to effectively transport, store and distribute nearly 6 billion tonnes of food by 2050, adequate policies, business and finance models, cold chain management systems and training and skills development will be needed as well.

To put the challenge into perspective, currently, 70% of food in the United Kingdom passes through a cold chain at some point on the route at our table from the point of production, whereas by comparison in India barely 10% of the produce that could benefit from using the cold chain does so – with 1/20 of the population of India, the United Kingdom has 10 times more refrigerated vehicles (Peters et al., 2018; Timestrip, 2018). India is the world's second largest producer of vegetables and fruits and among the top ten in fish and meat but the bulk of these perishable goods face risk due to inadequate handling and logistical support. Reportedly, upwards of 25% of such produce is lost due to a lack of farm-gate preconditioning including pre-cooling, refrigerated vehicles and shipping containers and other supply chain bottlenecks (Peters et al., 2018). The social and economic implications of these inefficiencies across the supply chain are evident. Despite the large volumes of production, India has been ranked at 94 among 107 countries in the Global Hunger Index (GHI) 2020, and ranked 71 out of 113 countries in the Global Food Security Index (GFSI), and food losses result in a loss of around INR 920 billion, or USD 12.63 billion (GFSI, 2020; GHI, 2020; Rao, 2016).

Importantly, lack of cold chain and supporting infrastructure also restricts market expansion which in turn dissuades efforts to improve farm-level productivity. Any efforts to produce more, without concurrent logistics enablement, means higher supply at markets immediate to farms and a lower valuation as these markets are in surplus areas. Conversely, especially in the case of perishables, the markets located at a distance, evidence a disconnect with their supply, incurring high food loss and causing a demand-supply mismatch. This is not a win-win situation and not sustainable.

The addition of what will undoubtedly be a substantial amount of energy-consuming equipment and infrastructure to enable the movement of nearly 6 billion tonnes of food by 2050 will also have significant environmental impacts if achieved using today's fossil fuel-powered conventional technologies with high Global Warming Potential (GWP) refrigerants, posing major risks to achieving the net-zero target to deliver on the commitments to the Paris Agreement and Kigali Amendment to the Montreal Protocol (United Nations, 2015, 2016). Today, the food cold chain alone accounts for one-third of hydrofluorocarbon (HFC) emissions from refrigerants, or 1% of total GHG emissions, and these emissions are projected to rise significantly as new cold chain infrastructure comes online in developing countries (Carbon Trust, 2020).

A substantial proportion of GHG emissions from food that does not make it to market (around 1 GT CO_2e) comes from the lack or inefficiency of cold chains (GFCCC, 2015). However, we have to be mindful of the potential trade-off, while access to cold chains can offset some of the GHG emissions related to food loss, the expansion can result in significantly higher cooling emissions, exceeding the emission reduction benefits. The importance of sustainable and clean cold chain expansion has been highlighted by an academic study which concluded that the increased GHG emissions resulting from traditional cold chain operation in sub-Saharan Africa (SSA) would be about 10% greater than the food loss emissions avoided (Heard & Miller, 2019).

Alongside providing enough nutritious food to feed nearly 10 billion people with minimum environmental impact, we must at the same time also protect and indeed improve the livelihoods of the nearly half a billion small and marginal farmers who are the major stakeholders and are essential to today's global food system.

Today, the majority of the world's food is produced by smallholder farmers, contributing significantly to poverty reduction and food security (IFAD & UNEP, 2013). For example, according to the Africa Agriculture Status Report (AASR), 80% of Africa's 51 million farms are smaller than 2 hectares and deliver 70% of Africa's total food. Elsewhere, smallholder farmers provide around 80% of the food consumed in Asia, and this statistical pattern is repeated in developing economies around the globe (AGRA, 2018). Yet, at the same time, small-scale farmers are largely food insecure, often lack physical and/or economic access to cold chains and modern energy services threatening their contribution towards economic and social development, and leaving them vulnerable.

The deployment of well-functioning cold chains does not only reduce food loss and enhance food security but also allows for broader economic, social and environmental benefits to be reaped in the process. Cold chains enable farmers to earn more through not only quantity, but also the quality increase of the food produced, and facilitate their access to distant markets, such as consumption centres in distant cities and urban conurbations. Such capacity is empowering, as better access to markets might allow smallholders to produce higher-value crops, which will, in turn, allow farmers to generate higher incomes. Furthermore, access to cooling at the farm level can enable farmers to process their produce, including freezing, curing, drying, pasteurising and fermenting, and can provide significant economic benefits to farmers by extending the shelf life and increasing the value of the produce, and allowing the sale of the products off-season.

Apart from opening up new markets, the cold chain connectivity also provides opportunities to produce and sell food better suited to new growing conditions as they emerge, thereby helping them to adapt better to changing climates. Essentially, besides improving economic gains through safe and expanded market connectivity, cold chains also help with building capacity for the future resilience of farmers and the rural communities in which they are located (Peters et al., 2018)

The central philosophical challenge to be addressed is that feeding the world is largely perceived today as a large-scale industrial enterprise, with land consolidation into larger, mechanised production units being the default business model for economic efficiency. But, to achieve sustainability across all economic, social and environmental dimensions, we need radically innovative approaches to agricultural development which would economically empower the marginal and small farmer – and create rural employment and resilience, in alignment with an existing human backdrop and current ground reality (Peters et al., 2018).

The reality of business dictates that no matter how appropriate a shift in philosophy might be, cold chains will only be taken up by small and marginal farmers and associated supply chain players if they are affordable within the local economic context. As such, in developing economies, this is likely to require new 'pay as you use' type business and funding models, which eliminate the need for upfront investment in cold chain technologies and allow users to pay per unit of cooling they consume, driven through empowered Farmer Producer Organisations (FPOs), be they as farmer producer companies (FPCs) or co-operatives. As the equipment is owned and managed by the provider, these models also enable effective operation and maintenance of equipment, lowering the risk of unexpected breakdowns, downtime and efficiency loss, providing additional economic and environmental benefits. Equally key is to enable small and marginal farmers through FPOs and other knowledge transfer channels to understand how to avail services of the integrated components of storage and transport to gain the economic advantages available from cross-geography access, distance-price arbitrage, time-arbitrage and cross-seasonal trading.

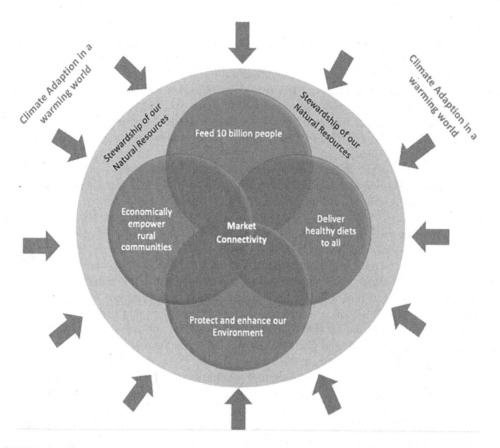

FIGURE 5.1   Climate adaptation in a warming world.

This is where new value is created, and management is as important as the physical infrastructure required to enable the flow of food from point of production to point of consumption (Peters et al., 2018). (Figure 5.1).

And where, in addition to FPCs and co-operatives, we incentivise commercial actors such as local entrepreneurs, medium-sized farmers and third-party logistics companies to become the local service providers, deploying and operating cold chain infrastructure on behalf of smallholder farmers and growers. For sustainability, we absolutely must ensure a fair and equitable flow of value back to the producers themselves.

In summary, sustainable cold chains sit at the intersection of the SDGs, the Paris Agreement and the Kigali Amendment. Application of *clean*[1] cooling in cold chains delivers on three core pillars for sustainable food: (i) enhance the income, economic wealth and financial security of farmers, growers and fishers; (ii) improve food quality, safety, nutritional content and value to consumers; and (iii) achieve this sustainably with minimum environmental and natural resource impact, (Peters et al., 2018).

The difficult question we need to answer is: 'How do you create the local and global, temperature-controlled "field to fork" connectivity to nutritiously feed 10 billion people sustainably from hundreds of millions of small-scale farmers whose livelihoods and well-being are often dependent

---

[1] Clean cooling is defined by Peters as "resilient cooling for all who need it without environmental damage and climate impact. It incorporates smart thinking to mitigate demand or active cooling where possible, minimised, and optimal use of natural resources, and a circular economy design that includes repurposing of waste heat and cold thermal symbiosis throughout the lifespan of the cooling system." (Peters & Chasserot, 2020)

on only 1–2 hectares, as well as ensure they are climate change adaptation ready and resilient, all without using fossil fuels?'

5.2   COLD STORAGE IS NOT A COLD CHAIN

Cold chain is a complex, multi-dimensional, temperature-controlled transport and storage system of perishable food from the harvest, through pre-conditioning, postharvest management to processing, distribution and consumption. However, approaches to date have been inherently top-down imposed, reductionist initiatives involving projects that test or demonstrate individual technologies and approaches. Such action disregards the essential cold connectivity of the cold chain that necessarily must include multiple static and mobile elements of cooling, and neglects interdependencies that exist between the economic decisions, available energy resources, technology choices, climate change mitigation and adaptation strategies, and social, cultural and political systems, and results in a sub-optimal outcome.

Establishing an efficient and swift cold chain requires the integration of many different elements across the food supply chain and the continuing management of those elements. However, approaches to cold chain have tended to focus on individual cold storage or food processing facilities rather than focusing on the long-term management and the maintenance of an integrated cold chain – not only a complete end-to-end cooling technology and supporting infrastructure, but the coordination between multiple stakeholders from farmer to retailer; robust operational management; continuous information flow from fork-to-farm and farm-to-fork; finance and business models to share value and create an economically sustainable system; and trained technical capacity to technology adoption, operation and maintenance seamless. In most developing countries, there is limited or even negligible cold storage equipment/refrigerated transport, from harvest to retail point, and where there is equipment, it is often old or obsolete (UNEP, 2019). Furthermore, the cold chain is often concentrated in the urban centres and transport terminals, and the use of a cold chain is particularly low in the first mile of the agricultural supply chain (Kitinoja, 2013; UNEP, 2019). Key challenges to sustainable cold chain development include:

- **Organisation:** There is a lack of collaboration among stakeholders for collective action and investments.
- **Financial capacity of marginal and small farmers:** Despite the potential upside given the capital cost small and marginal farmers do not have the individual financial capacity to invest in new equipment, and efforts to harness larger buying power via cooperatives to address cold chain shortages heretofore have been highly limited. For example, in SSA, 62% of small-scale fresh produce farmers cannot afford cooling technology (Power for All, 2021).
- **Fragmented nature of agricultural landholdings:** The majority of the farmers in developing countries have small landholdings, and farming practices are typically fragmented requiring multiple farm gate collection.
- **Lack of awareness:** Farmers lack awareness of simple techniques to take care of produce postharvest and training on the usage and maintenance of sophisticated cold chain equipment. Equally, there is sometimes a lack of awareness about the availability of much more efficient equipment, and there are no large-scale cold chain demonstrations to showcase its efficacy and impacts.
- **Lack of knowledge and skills:** In many countries, there are not sufficient numbers of trained engineers to maintain cold chain equipment in efficient working order, a problem compounded by new emission-reducing technologies that require an expansion in skills.
- **Lack of necessary infrastructure in rural areas:** Cold chains require robust transport and energy infrastructure to be in place to operate efficiently and effectively, ensuring

seamless market connectivity. In many communities in developing countries, electricity sources are unreliable or non-existent. Today, approximately 789 million people still do not have access to electricity in their homes or communities, and rural populations make up around 85% of this deficit (IEA et al., 2020). Due to high electricity tariffs coupled with lower operational efficiencies or lack of access to electricity, many rural cold chain services today typically rely on expensive fossil fuels with significant climate and environmental impact. Additionally, transport infrastructure requires equal attention. For example, one study in India reported that a typical driver would need to stop 49 times during their 2,400 km journey because of bad roads and also toll-gates that operate on different billing systems (Ray, 2010).

- **Risk of technology lock-in:** In the absence of affordable financing options and business models, the cold chain expansion could result in the technological lock-in of carbon-intensive technologies for the long term, adding to the climate challenge.
- **Lack of drivers for change and attention to cold chain components beyond large-scale storage:** Public sector's attention in developing economies remains focused on the provision of large-scale cold storage rather than encouraging the building of integrated cold chains for market connectivity.
- **One size does not fit all:** Models used in more industrialised countries may not always be successful in developing countries. Rather than pre-supposing solutions, there is a need to understand the portfolio of local cooling needs, resources, climate conditions as well as political, social and economic conditions to develop the optimum mix of innovative and fit-for-market solutions that cater for those needs sustainably.

System optimums across all dimensions can only be identified by taking a whole system but still localized perspective (understanding needs and requirements), identifying gaps and providing solutions. There is a need to integrate local and international trading companies, energy service companies, suppliers of cold storage facilities, assemblers and manufacturers and subsidized programs operating in the cold chain market.

A paradigm shift on how we approach the problem is necessary; it requires understanding the full value system and system-wide interventions which are multi-disciplinary that cut across the complete farm-to-fork temperature-controlled supply chain including at farm/first mile. There is a need to recognize the interdependencies that exist between the technological domain; energy storage and sources; manufacturing strategies; social, ecological and economic considerations; policy; governance; business models; finance; education and training. Given the many sub-sector focused actors, both in local and global supply chains, an integrated transition to energy-efficient, sustainable technology deployed in economically optimized, net-zero carbon cold chains is unlikely to emerge organically; this can risk climate targets as well as risk economic opportunity for food growers, fisheries and producers.

We, therefore, need to bring together the multi-disciplinary international and in-country expertise necessary to develop, demonstrate and provide the technical assistance for implementing the step-change pathways (culture and social, technology, manufacturing, skills, policy, business models, financing) for achieving (i) affordable (whole of life), (ii) greatest energy system resilience and (iii) lowest carbon emissions while (iv) meeting social and economic cold chain, postharvest management (PHM) and rural cooling needs (Peters, 2018).

5.3   WHERE DO WE START?

In rushing to deploy technologies and infrastructure, an underestimation of the scale of the cooling demand, the system complexities and its impact on energy consumption risk contributing to a lack of ambition in policies, infrastructures and technology developments as well as capacity building. It could ultimately have far-reaching social, economic and environmental consequences.

Efforts to identify pathways to sustainably meet the cooling and cold chain needs have been based on historical equipment trend analysis and projections typically driven by socio-economic variables, such as population growth and gross domestic product (GDP). The main problem with these approaches is that they fail to capture unmet needs, identify risk factors and deliver access to cooling for the benefit of all who needs it.

Equipment-based projections of cooling demand suffer from three significant weaknesses. First, data about unit stocks in each of the cooling categories is fragmentary: verified sales and disposal figures and second-hand transfers of equipment are not systematically collected or universally available, and unit stock data rarely includes expected product lifetimes. As a result, the equipment part is typically challenging to estimate and projections are unreliable. Second, equipment-based projections do not start by seeking to understand current and future household, business or community needs. Also, a focus on per capita equipment penetration rates pre-supposes a top-down technical solution to the cooling needs. This approach ignores the possibility of demand mitigation through the redesign of systems, the aggregation of demands, modal shifts, the use of waste or currently untapped resources, as well as support for existing cooling practices and behaviours (Peters et al., 2020).

Therefore, to meet the increasing cooling and cold chain needs without leaving anyone behind in alignment with the SDGs, the Paris Agreement on Climate Change and the Kigali Amendment to the Montreal, the first stage is a needs-based assessment of cooling which begins by going back to first principles and understanding at a macro level how much cooling would be required for the transport, storage and distribution required of nearly 6 billion tonnes of food by 2050, a substantial portion of which will be fresh and temperature-sensitive produce (Peters et al., 2019, 2020). Simultaneously at a micro level, we need to understand how people use cooling today and how they seek to maintain or enhance this level of cooling and how much cooling they need today, and in the future? Cooling needs assessment must fully account for people's perceptions of need at the same time; they must be service or practice-based, and they must take national, regional or community circumstances into account. Equally, needs are not constant but dynamic and change with other factors, which requires shifting away from 'passive' towards 'active' models of cooling assessment. They also need to consider projected future demand, the implications of increasingly hotter climates and the climate implications of cooling demand occurring at peak times and meeting those capacity requirements with sustainable energy.

From the gathered data, one can estimate the financial costs and environmental consequences of meeting all cooling and cold chain needs on a business-as-usual approach and understand the implications on our climate and developmental goals, targets and commitments. Equally, by combining needs with other resources and infrastructures such as energy, electricity and transport, this can then enable optimum and 'fit for the market' choices from demand mitigation to harnessing untapped thermal resources and traditional cooling provision technologies and renewable resources and traditional cooling provision technologies and renewable electricity for local conditions and context (environmental, social, economic, etc.) (Peters, 2018). Furthermore, we can understand the skills and capacity required to meet demand which would vary greatly across countries and communities.

Given that the cold chain is often weak or non-existent in developing countries, we have a unique opportunity to consider the energy systems and cold chains as a whole to achieve system efficiency and long-term sustainability. This objective can be done across all dimensions by identifying synergies between sectors and areas where energy and cooling demand can be aggregated and/or capacity can be shared. This is also important in the context of vaccination efforts to tackle the COVID-19 pandemic and future emergencies. Global distribution of vaccines manufactured for COVID-19 immunisation programmes will require the deployment of substantial additional cold chain capacity. Could this additional infrastructure and equipment be deployed in such a way that it not only facilitates the COVID-19 response and bridges existing gaps in routine immunisation programmes but also offers opportunities for re-purposing to help address the needs in other sectors, such as agriculture and food cold chain? Smart planning, along with system-level thinking, could contribute towards long-term sustainability and developmental goals, thereby leveraging COVID-19 investments that might otherwise result in redundant infrastructure and stranded assets.

The portfolio of cooling needs identified and quantified through the assessment will inform the design of cold chains that effectively minimise the demand for mechanical cooling. Hence, the energy system demand caters for the demand in a way that is holistically sustainable; harvests all available natural, renewable and waste energy resources; harnesses and leverages synergies between cooling/thermal sub-systems; and is supported by appropriate 'fit-for-market' cooling technologies. To this end, a needs-driven systems approach to cooling can be developed along a value chain with the following elements (Figure 5.2):

- <u>Making cold</u>: Harness unused or 'waste' resources such as cold-water bodies, 'wrong time' renewable energy (wind, solar), waste cold and heat, or ambient heat sinks.
- <u>Storing cold</u>: Store energy thermally, in physical mass or phase-change materials to make use of cyclical changes in ambient heat sinks and availability of energy at lower costs.
- <u>Moving cold</u>: Use new energy vectors and materials to move thermal energy, e.g. phase-change materials or liquid air or nitrogen can be used for cooling and generate mechanical energy.
- <u>Using cold</u>: Reduce cold loads by lowering cooling demand through passive techniques and approaches and behavioural changes, increasing equipment efficiency and substituting refrigerants with high GWP.
- <u>Managing cold</u>: Make cooling systems smart for real-time monitoring of cooling needs and performance (sensors), load adjustments (controls) and integrated system management and storage (communication). Deliver cooling only where and when needed while optimising system-wide impacts.

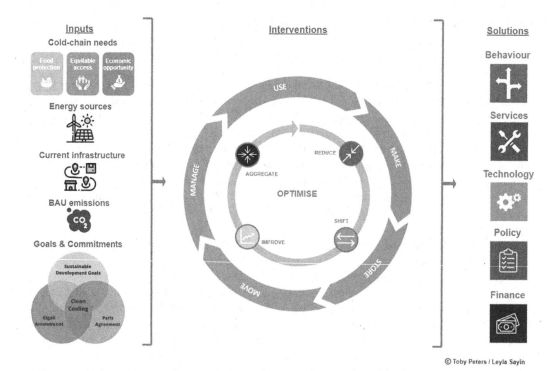

© Toby Peters / Leyla Sayin

FIGURE 5.2 A whole-system approach for designing clean and sustainable cold chains.

- <u>Financing cold</u>: Improve human decision-making and modify behaviour through smarter financing options (such as servitisation models), instant visibility of system status and consequences, integration of external costs/benefits (carbon pricing), markets for alternative energy vectors (e.g. waste heat) and other incentives.

Thereafter, to support both early successes and the deep systemic changes, the strategy is to use an 'avoid-shift-improve' approach, adding in the intervention of 'aggregate'. These four interventions, as illustrated in Figure 5.2, can simultaneously support the sustainable cold chain development through a range of solutions that significantly reduce cooling emissions via mitigation of mechanical cooling needs; move to highly energy-efficient active technologies that would avoid the use of high GWP refrigerants; optimise the use of all available renewable and waste thermal energy sources; and by harnessing and leveraging synergies between sectors and systems, create symbiotic yet resilient relationships.

- **Reduce**: Reducing the need for mechanical cooling through the use of passive cooling techniques and approaches, and behavioural changes.
- **Shift**: Transitioning to more sustainable technologies that avoid the use of high-GWP refrigerants; taking different approaches to cooling by integrating 'free', waste or currently untapped thermal energy resources to serve cooling needs more efficiently; and facilitating modal shifts that can substitute for cooling solutions, with much lower energy use and environmental impact.
- **Improve:** Using the 'best available' energy-efficient technologies, and maintaining the system efficiency with regular maintenance and monitoring.
- **Aggregate:** Aggregating different cooling needs into a single demand profile, integrating control systems to support 'thermal symbiosis', and bulk procurement of cooling or cooling equipment to maximise overall energy efficiency and minimise the economic and environmental cost of cooling. Across these three primary intervention classes, the overall efficiency and economics of the cooling system can be enhanced through aggregation, such as bringing different cooling needs together, integrating control systems, instigating bulk purchase arrangements and/or thermal symbiosis.

5.3.1 COMMUNITY COOLING HUBS

Cooling is needed in many sectors such as agriculture (food production, livestock, fisheries), post-harvest produce management (refrigeration, transport, food processing), health (vaccine storage, space cooling and comfort) and transport. Space cooling needs for humans and animals are needed not only for comfort but also because at extreme temperatures, susceptibility to diseases increases and work productivity decreases. The productivity of milch animals also reduces during the summer season leading to economic loss to the owner. Storage of animal and human vaccines which are highly temperature-sensitive requires cooling. In animal husbandry, artificial insemination is used for the production of animals at a large scale and cooling is needed for the preservation of semen.

Community cooling hubs (CCH) is envisaged as facilities that are highly accessible to villages or village clusters, managed by local entities, that meet diverse local cooling needs in an aggregated manner to enhance well-being while optimising energy and resource management and potentially bundling multiple revenues streams. By using cold chains as the foundation upon which to build, the CCH concept uses these as the anchor to develop an innovative, communitarian and integrated flexible systems approach to meet the broad portfolio of a rural community's cooling needs in developing markets including health and safe environments today and into the future (Figure 5.3). This will include the role of the CCH for both thermal energy services and, inter alia, emergency supports, i.e. pandemic vaccine delivery. Building on the principle of community ownership, the approach is to aggregate different cooling demands to optimize system efficient energy and resource management

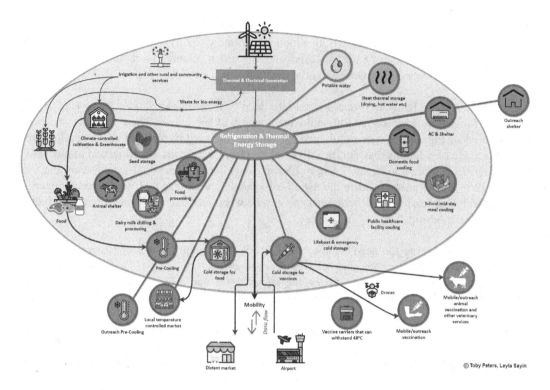

FIGURE 5.3 Community Cooling Hub model.

and bundle revenues in business models to maximize economic opportunity and inclusivity, to deliver substantial benefits to all from the domestic needs to the farmer to the food entrepreneur.

Taking a system approach, the first step in designing CCH is to undertake a comprehensive needs assessment to understand how the target communities use cooling today and the volume of cooling provision and cold chain required to meet the domestic, agricultural, health-related and commercial cooling needs. A pilot project has been conducted in three village clusters in Satara and Nashik districts of western Maharashtra, India, using an exploratory mixed-methods research design, and analyses show that multi-functional CCHs could contribute to different sectors, especially agriculture, dairy, buildings and healthcare (Debnath et al., 2021).

At its most basic level, such services could be a logical extension of the food packing activity into simple food processing and the provision of chilled space for food product manufacturing. For example, to turn milk into cheese or yoghurt or the production of pre-packed salads, or even ready-meals, to sell to nearby urban populations growing in affluence and looking for convenience, healthy eating and food provenance. Extending the concept further, the hub can offer personal refrigerated lockers accessible to village home-keepers.

At a more sophisticated level, the packhouse can be used to deliver the cooling for a community hall with a temperature safe environment, to serve as a crèche for the infants of the village or the elderly, or even to provide a schoolroom for classes on the hottest days of the year. Further to this, vaccines and medicines can also be safely stored at these hubs for local health care services. Ultimately, these community hubs could function as lifeboats in times of disaster, by storing an emergency supply of blood and medicines in response to the disaster threat assessment for the region.

Critically, these additional community supporting functions do not require a much larger energy load than that which is typical of a modern packhouse. However, through aggregation, such added services could be met sustainably and energy efficiently, including by the use of thermal storage to capture excess cold produced when the packhouse is not required to operate at full demand.

There will also be cases where the excess thermal energy can be rerouted to individual homes as an off-grid energy supply mechanism for cooling or hot water provision (using 'waste' heat from cooling equipment).

5.4 OVERVIEW OF RURAL COOLING NEEDS

See Table 5.1 for details.

TABLE 5.1
Overview of Rural Cooling Needs

Postharvest	
First to Last Mile Cold Chain including mobile cooling	Reduce food loss, market connectivity – feed the world; increase farmer's and fishers' (mostly are small and marginal) incomes; maintain the safety, quality and quantity of food
Food processing	Dairy – Milk processing, cheese, butter, yoghurt, ice cream production
	Create secondary economy/jobs within the rural or fishing community that require cooling/chilling/freezing in processing – enabled by secure, affordable, sustainable cold chain
Agriculture and horticulture and fishing/aquaculture	
Vaccines	Increase productivity of small-holder animals (goats, sheep, etc.) and also enable farmers to have insurance
Animal heat stress	Productivity (e.g. dairy cows) or even increased – think about 'cool sheds' for the housing of cattle, pigs, chickens, etc.
Storage of seeds, sperm, etc.	Maintain vitality of seeds, cryogenic storage essential for sperm
Mushroom cultivation, etc.	Some produce, e.g. mushrooms, require cold environments
'Coolhouses'	As global warming raises the temperature, some horticultural and agricultural produce will need temperature-controlled environments to grow in – a new concept of a 'coolhouse'.
Medical Services	
Vaccines and medicines	Reduce the loss of high-value products and mortality from preventable diseases – improve local health outcomes and life expectancy
Medical centre and services	Blood (significant for maternity – see Zipline drones) and other medical services, including temperature-controlled buildings.
	(Hospitals – MRI scanners, cryogenic treatments, etc., are not included)
Food	
Domestic refrigeration	Food loss and food poisoning also can store food so do not have to shop every day
Retail including food markets/stalls	Food loss, quality loss, food poisoning, to boost food's lifespan and an appearance by using unsafe preservatives (e.g. chemical preservatives including formalin)
School mid-day meal supply chain	Food loss and food poisoning
Workplace meal provision	Ditto above but for the workplace, mainly the farm or on the beach/quayside (fisherfolk etc.)
Safe/productive environments	
Risk of heat illness/hospitalization accelerated mortality for infirm, elderly, less able as well as those suffering from respiratory, cardiovascular and other chronic diseases (i.e. renal diseases)	
- at home	Old, young, pregnant, especially for sleep for all
- at work (indoor and outdoor)	Productivity and safety (e.g. accidents, fainting or heat collapse, heatstroke)
Community services – crèche	Safe environments
Community spaces	Community spaces where community socio-political business is enacted – i.e. 'town hall' meetings, governance, etc. – productivity
- at school	Productivity and improved educational outcomes.
Care-givers	Safety and productivity

(Continued)

TABLE 5.1 (*Continued*)
Overview of Rural Cooling Needs

Mobility and transport	
Public but also increasingly private	Longer-term issues of e-mobility and charging/range, safety (e.g. lack of adequate cooling in both public and private transport can have a severe health impact on vulnerable people such as the elderly, babies and young children)
Emergency – local life support systems	
'Cool lifeboats' at time of heat stress for old, young, etc., within a community	Mortality and reducing pressure on hospitals from heat stress and dehydration
Forward warehousing for supplies for emergency	Speed and cost of accessing essential temperature-sensitive supplies for natural disasters
Energy	
Cooling of clean power equipment (transformers, etc.)	
Disruption to power and transport	

5.5 IMPLEMENTATION – DESIGN AND TEST IN A VIRTUAL WORLD

The ultimate vision is to create an integrated clean and sustainable cooling system aligned to the goals of the Paris Agreement, Kigali Amendment and SDGs that:

- minimises the need for active cooling technologies using passive cooling techniques and approaches as well as behavioural changes,
- makes 'best use' of available natural, 'free' and waste thermal resources,
- supports the complete transition to highly energy-efficient cooling technologies,
- helps optimise energy use via integration of thermal energy storage systems that would enable cold to be used when and/or when needed,
- harnesses and leverages synergies between systems to create symbiotic yet resilient relationships,
- understands and designs for interdependencies across systems and broader infrastructure, and plans for unintended consequences, to ensure holistic sustainability and resilience capacity,
- supports the complete transition from carbon-intensive energy sources to renewables for remaining grid-based energy supply,
- enables aggregation of cooling needs via novel business models,
- is regularly monitored, optimised and adequately maintained,
- is supported by policy, regulation, appropriately structured finance and adequate skills training and development,
- enables safe decommissioning of system components for re-use, re-manufacture and re-cycling in a circular economy model, with no unanticipated impacts on overall system-of-systems sustainability,
- is fully inclusive, equitable, fair and ensures a just transition where socio-economic change is involved.

Developing a sustainable cooling and cold chain system is a multi-dimensional, multi-sectoral challenge, which requires the interdependencies that exist amongst the economic, environmental, energy, technological, social and political systems to be tackled and policies designed and implemented to address them. There is a need to strengthen government, industry and academic cooperation; incentivise accelerated innovation and market transformation via policy actions; support new technologies and business models via public and private finance; raise awareness of cooling issues among governments, businesses and end-users of cooling; and develop the skilled workforce required to facilitate deployment, operation and maintenance of new technologies as well as the decommissioning of existing systems and ultimately life-expired new technologies.

5.6 CONCLUSIONS

In the short term, it will continue to be the poor who will face the most significant challenges from lack of cold chain and it will remain a serious issue of equity. But food security is an integrated global issue. The United Kingdom is reliant on imports with 84% of fruits and 47% of vegetables were imported in 2019 (DEFRA, 2020). At the other end of the chain, Rwanda has a five-year strategy to double agri-exports by 2024–2025; this includes a nine-fold increase in high-value temperature-controlled horticulture exports. Global food demand is set to grow by 50% by 2050, driven by population and economic growth, and urbanisation. At the same time, climate change impacts all aspects of the food system from food production and availability, access, quality and safety to consumption patterns.

With one-third of food produced is lost or wasted globally, one effective strategy to closing the food supply-demand gap is saving the food which has already been produced, and cold chains sit at the nexus of the challenge. Cold chains are key to reducing post-harvest losses but also they improve the income, economic wealth and financial security of farmers, growers and fishers through effective market connectivity, and improve food quality, safety, nutritional content and value to consumers. Achieving this sustainably with minimum environmental and natural resource impact demands thinking at a systems level, that in turn requires a holistic and quantitative understanding of current and future needs alongside an integrated consideration of the available energy resources; the technical, socio-economic-political, cultural and environmental drivers and benefits; and the outputs and value chain steps required to meet those needs. All of this, while simultaneously ensuring that the demand for mechanical cooling is minimised, resources are allocated efficiently, lifecycle environmental impact is minimised, co-benefits are maximised at affordable costs and resilience is built.

REFERENCES

AGRA. (2018). Africa Agriculture Status Report 2018: Catalyzing government capacity to drive agricultural transformation. Nairobi, Kenya: Alliance for a Green Revolution in Africa. https://agra.org/wp-content/uploads/2018/10/AASR-2018.pdf.

Carbon Trust. (2020, September 23). Net-zero cold chains for food. https://www.carbontrust.com/resources/net-zero-cold-chains-for-food

Debnath, K. B., Wang, X., Peters, T., Menon, S., Awate, S., Patwardhan, G., Wadkar, N., Patankar, M., & Shendage, P. (2021). Rural Cooling Needs Assessment towards Designing Community Cooling Hubs: Case Studies from Maharashtra, India. *Sustainability*, 13(10), 5595. https://doi.org/10.3390/su13105595

DEFRA. (2020). Department for Environment, Food & Rural Affairs - Horticulture Statistics 2019. https://www.gov.uk/government/statistics/latest-horticulture-statistics.

FAO. (2011). Global Food Losses and Food Waste. http://www.fao.org/3/mb060e/mb060e00.htm

FAO. (2014). Food Wastage Footprint: Full-cost Accounting. Final Report. FAO. https://agris.fao.org/agris-search/search.do?recordID=XF2015001538

FAO (2015). Developing the Cold Chain for Agriculture in the Near East and North Africa (NENA). Policy Brief. Rome, Italy: FAO. http://www.fao.org/documents/card/en/c/395366e2-2fdb-4fac-9870-cf7ca26d4051/.

FAO (2018). The State of World Fisheries and Aquaculture 2018: Meeting the sustainable development goals. Rome, Italy: FAO (The State of World Fisheries and Aquaculture (SOFIA), 2018). Available at: http://www.fao.org/documents/card/en/c/I9540EN/.

FAO. (2020). The State of Food Security and Nutrition in the World 2020: Transforming food systems for affordable healthy diets. FAO, IFAD, UNICEF, WFP and WHO. https://doi.org/10.4060/ca9692en Also Available in: Arabic Russian French Spanish Chinese.

GFCCC. (2015). Assessing the Potential of the cold chain sector to reduce GHG emissions through food loss and waste reduction.

GFSI. (2020). Global Food Security Index (GFSI). http://foodsecurityindex.eiu.com/

GHI. (2020). Global Hunger Index Scores by 2020 GHI Rank. Global Hunger Index (GHI) - Peer-Reviewed Annual Publication Designed to Comprehensively Measure and Track Hunger at the Global, Regional, and Country Levels. https://www.globalhungerindex.org/ranking.html

Heard, B. R., and Miller, S. A. (2018). Potential changes in greenhouse gas emissions from refrigerated supply chain introduction in a developing food system. *Environmental Science & Technology, 53*(1), 251–260. https://doi.org/10.1021/acs.est.8b05322

IEA, IRENA, UNSD, World Bank, and WHO. (2020). Tracking SDG 7: The Energy Progress Report. World Bank.

Kitinoja, L. (2013). Use of Cold Chains for Reducing Food Losses in Developing Countries. The Postharvest Education Foundation.

Kumar, S., Sachar, S., Goenka, A., Kasamsetty, S., and George, G. (2018). Demand Analysis for Cooling by Sector in India in 2027. New Delhi: Alliance for an Energy Efficient Economy.

Peters, T. (2016). Clean Cold and the Global Goals. The University of Birmingham.

Peters, T. (2018). A Cool World: Defining the Energy Conundrum of Cooling for All. The University of Birmingham.

Peters, T. (2018). Clean and energy-efficient cooling can advance three internationally agreed goals simultaneously: The Paris Climate Agreement; the UN Sustainable Development Goals; and the Montreal Protocol's Kigali Amendment. Cooling for All – the 18th Sustainable Development Goal. https://www.ccacoalition.org/en/blog/cooling-all-%E2%80%93-18th-sustainable-development-goal

Peters, T., & Chasserot, M. (2020). Defining 'Clean Cooling'. Centre for Sustainable Cooling, shecco. https://issuu.com/shecco/docs/cleancooling

Peters, T., Bing, X., and Debhath, K. B. (2020). Cooling for All: Needs-based Assessment Country-scale cooling action plan methodology. https://www.sustainablecooling.org/wp-content/uploads/2020/06/Needs-Assessment-June-2020.pdf

Peters, T., Kohli, P., and Fox, T. (2019). The Cold-chain Conundrum. Birmingham Energy Institute – University of Birmingham. https://www.birmingham.ac.uk/Documents/college-eps/energy/Cold-Chain-Conundrum.pdf.

Power for All. (2021). Tech Spotlight: Solar-Powered Cold Storage. Power For All. https://www.powerforall.org/resources/images-graphics/tech-spotlight-solar-powered-cold-storage

Rao, R. K. (2016). Agri produce worth Rs 92K Crore wasted annually. https://rstv.nic.in/rs-92000-crore-worth-food-wasted-annually-minister.html

Ray. (2010). Supply Chain Management for Retailing. Tata McGraw-Hill Education.

Ritchie, H. (2019). Half of the world's habitable land is used for agriculture. Our World in Data. https://ourworldindata.org/global-land-for-agriculture

Scialabba, N. (2015). Food wastage footprint & Climate Change. FAO.

Timestrip (2018). 'The Fresh Cold Chain', 11 October. Available at: https://timestrip.com/the-fresh-cold-chain/.

United Nations. (2019). World Population Prospects 2019. https://population.un.org/wpp/

UNEP (2019). Sustainable Cold Chain and Food Loss Reduction - BRIEFING NOTE. UN Environment Programme. https://ozone.unep.org/system/files/documents/MOP31-Sustainable-HL_Briefing_Note.pdf.

United Nations. (2015). Paris Agreement. https://treaties.un.org/Pages/ViewDetails.aspx?src=TREATY&mtdsg_no=XXVII-7-d&chapter=27&clang=_en

United Nations. (2016). Amendment to the Montreal Protocol on Substances that Deplete the Ozone Layer. https://treaties.un.org/Pages/ViewDetails.aspx?src=IND&mtdsg_no=XXVII-2-f&chapter=27&clang=_en

Walpole, M., et al. (2013). Smallholders, food security, and the environment. International Fund for Agricultural Development.

Willett, W., Rockström, J., Loken, B., Springmann, M., Lang, T., Vermeulen, S., Garnett, T., Tilman, D., DeClerck, F., Wood, A., Jonell, M., Clark, M., Gordon, L. J., Fanzo, J., Hawkes, C., Zurayk, R., Rivera, J. A., Vries, W. D., Sibanda, L. M., … Murray, C. J. L. (2019). Food in the Anthropocene: The EAT-Lancet Commission on healthy diets from sustainable food systems. *The Lancet, 393*(10170), 447–492. https://doi.org/10.1016/S0140-6736(18)31788-4

World Bank (2020). Water in Agriculture [Text/HTML]. World Bank. https://www.worldbank.org/en/topic/water-in-agriculture

WRI (2019). Creating a Sustainable Food Future: A Menu of Solutions to Feed Nearly 10 Billion People by 2050. World Resources Institute.

6 The Container Mini Packhouse

Affordable and Effective Facility for Sorting, Packaging, and Storage of Fresh Produce for Small-/Medium-Scale Farmers

Ramadhani O. Majubwa, Theodosy J. Msogoya, and Hosea D. Mtui
Sokoine University of Agriculture

Eleni Pliakoni
Kansas State University

Steven A. Sargent
University of Florida

Angelos Deltsidis
University of Georgia

CONTENTS

DOI: 10.1201/9781003056607-8

6.1   INTRODUCTION

In the past three decades, the world has experienced an increase in engagement of developing countries in the production of horticultural crops for high value local and export markets (Zoss and Pletziger, 2007; Van den Broeck and Maertens, 2016). Among drivers for the engagement have been the availability of land, cheap labour, conducive climatic conditions, trade liberalization, market reforms, and change in trade policies (Gulati et al., 2007). Some developing countries in Africa, Asia, and South America are already engaged in off-season production and export of fresh horticultural produce to European and American markets, respectively (Jaffee, 1992; Hewett, 2012). Many other developing countries are also attracted to embark on fruit and vegetable production for export markets (Temu and Marwa, 2007; Gulati et al., 2007). Similarly, the demand for high-quality, safe fresh horticultural produce has been increasing in local and regional high-value markets such as tourist hotels, restaurants, and supermarkets (Davis, 2006; Van den Broeck and Maertens, 2016). Unfortunately, penetration to such markets requires compliance to stringent quality and safety standards that most small- and medium-scale farmers in developing countries are struggling to meet (Boselie et al., 2003; Zoss and Pletziger, 2007; Asfaw et al., 2010) due to inaccessibility and unaffordability of the necessary produce postharvest handling infrastructures including packing houses (Ahmad and Siddiqui, 2015).

Fruit and vegetable production enterprises for export in most developing countries start by involving small-scale farmers as suppliers to large exporters and later on production is taken over by large commercial farms with small-scale farmers serving as out-growers (Maertens et al., 2012). The shift is partly attributed to the low capacity of small-scale farmers to maintain quality and continuity of supply due to lack of produce handling, packing, and cold storage facilities (Davis, 2006; Asfaw et al., 2010; Jouanjean, 2013; Ahmad and Siddiqui, 2015). Studies indicate that, when proper facilities are available, small-scale farmers in groups can produce significant quantities of quality produce for local and export markets (Birthal et al., 2007; Ortmann and King, 2007; Narrod et al., 2009). Government and donor-funded projects in most developing countries have been investing in the construction of fruit and vegetable collection centres in key rural production areas. However, most of the already established and collection centres still under construction are lacking fresh produce handling and cold storage facilities (Kader, 2009; Kitinoja, 2013). Few of the existing collection centres equipped with mechanical cooling facilities tend to be non-functional by the end of the project due to lack of proper maintenance. Most value chain actors, including farmers and marketers, fail to operate and maintain such facilities because of high repair and operational costs that lower the return on investments (Kitinoja, 2013). Small- and medium-scale farmers in developing countries normally adopt technologies that are affordable in terms of construction and maintenance for the scale of business they operate (Kitinoja et al., 2011; Kitinoja, and Thompson, 2010). Sustainability of produce handling facilities in developing countries, therefore, need to consider proper sizing of farm operations, target markets and return on investment (cost/benefit) of the technology used (Kitinoja et al., 2011).

A packing house is a facility used for assembling and conducting postharvest handling operations and/or treatments that prepare fresh produce into quality categories required by the target consumer or markets (Yaptenco and Esguerra, 2012; Acedo and Weinberger, 2016; Ait-Oubahou et al., 2019). It provides shelter to fresh produce and facilitates postharvest handling operations including but not limited to; washing, postharvest treatments, cooling, sorting, grading, packing, and cold storage (Johnson and Hofman, 2009; NCCD, 2015; Asante et al., 2019). A typical packing house contains a shaded receiving area, packing line, produce packaging or carton stacking area, rapid-cooling area or equipment, refrigerated storage rooms, and loading dock for produce dispatch (Ait-Oubahou et al., 2019).

Poor and/or lack of handling infrastructures along the fresh produce value chain in developing countries accounts for a postharvest loss of 20%–50%, depending on the crop (Kitinoja, 2013). Among crops, postharvest loss also varies with produce handling stage along the value chain (Kitinoja and AlHassan, 2012). Postharvest losses of fresh produce in developing countries tend to be higher between harvest and storage or sale at the retail stage (Hodges et al., 2011; Gustavsson et al., 2011) due to poor temperature management and limited application of appropriate postharvest handling practices including sorting, grading, washing, packaging, cooling, and cold storage among farmers, handlers, and marketers (Kader, 2009; Kitinoja, 2013; Kitinoja and AlHassan, 2012).

Postharvest handling facilities such as packing houses and cold storage units are well distributed and utilized in developed countries, more so than in developing countries because of limited availability, unaffordability, and ineffective utilization. Availability and access to these infrastructures have been a challenge to most small- and medium-scale farming enterprises in developing countries due to the high establishment and running costs. The challenges lead to poor produce handling along the value chain and hence increase in postharvest losses and produce in compliance with quality and safety standards for access to local high value and export markets. To bridge this gap, Sokoine University of Agriculture (SUA) in collaboration with Kansas State University (KSU) and the University of Florida (UF) through the project "Capacity Building on Produce Postharvest Management in Tanzania" funded by the Feed the Future Innovation Lab for Horticulture at the University of California Davis, supported with the US Agency for International Development (USAID) funding, has designed and established an affordable mini packhouse facility equipped with cold rooms from used marine shipping containers.

6.2 ESTABLISHMENT OF A MINI PACKHOUSE FROM USED MARINE SHIPPING CONTAINERS

One of the limiting factors for availability, access, and use of packing house facilities in most developing countries has been related to high construction and maintenance costs. The container mini packhouse is considered as one of the affordable produce handling facilities that can be established in most developing countries. In Tanzania, the total establishment cost of the unit developed for this project, excluding the cost of electricity installation, was US$10,954 (Table 6.1). The cost of construction of a similar packhouse facility in other countries may vary with the cost of materials and labour charges. Also, the cost could be significantly reduced through the use of less expensive components, such as alternative door types in the receiving and packaging zones.

6.2.1 KEY CONSIDERATIONS IN DESIGNING A PACKHOUSE FACILITY

Packing house is a key infrastructural facility for produce preparation, handling, and management of a cold chain at the farmer level (Joshi et al., 2009). Being part of the cold chain system, packing houses are therefore, designed and constructed in a way that ensures smooth flow of produce, rapid cooling, and proper storage that minimizes exposure to any potential risks. The design should allow smooth conduct of operations such as precooling, cleaning or disinfecting, sorting and grading, packaging, cold storage and postharvest treatments such as waxing. Proper packhouse design and operations help to ensure produce quality, safety, and cost-effectiveness of the preceding handling stages (Heap, 2006; Arah et al., 2016). A proper packing house design should ensure that: (i) facility, equipment, and surfaces are erected using materials or in a manner that allows easy cleaning and does not keep pests or contaminants; (ii) adequate lighting is provided to facilitate operations such as cleaning, sanitation, and repair; (iii) packing, cooling, and all produce contact equipment/surfaces are kept away from the wall and well located for easy cleaning; (iv) cold storage areas are well insulated, drained, and graded to avoid water stagnation or infiltration leaks from outside; (v) air intake into the rooms is protected against potential sources of debris, dirt, animal wastes, or pests; and (vi) glass fixtures and light bulbs are of safety types or protected with screens or covers to prevent breakage. A well-designed packhouse facility should also be located in an area that is easily accessible, with a constant supply of potable water, a reliable source of

TABLE 6.1

Costs of Materials and Labour Charges for the Establishment of the Container Mini Packhouse

SN	Material Description	Cost (USD)
	A: Fabrication of the container mini packhouse facility	
1	Standard ISO marine shipping container of 6.06 m (20 ft) length (3 units)	2385.00
2	Painting materials	523.40
3	Terrazzo (8 tons terrazzo sand, aggregate, grinding stones, and polish)	944.30
4	Aluminium windows (4 pieces)	234.80
5	Insulated metal cold room doors (4 pieces)	1,588.00
6	Cement (40 bags of 50 kg each)	270.00
7	Rigid structurally insulated polyisocyanurate foam panels (24 pieces of 1 m width × 3 m length × 3 inches (76.2 mm) thickness)	1,436.50
8	Cement blocks (350 pieces)	185.00
9	Plumbing materials	319.00
10	Labour charges (cutting, welding, painting, insulation)	1,836.00
Subtotal 1		**9,722.00**
	B: Air conditioner installation	
1	1800 BTU, LG-Split room air conditioner (HS-C1865NQ4, Thailand)	661.00
2	CoolBot™ (E500290, CUL, US) control unit	350.00
3	Air conditioner installation labour	221.00
Sub total 2		**1232.00**
Grand total		**10,954.00**

Source: Own source.

power (electricity, solar, generator), and close to significant produce production or collection centres (Slama and Diffley, 2013; Ait-Oubahou et al., 2019).

6.2.2 Design of the Container Mini Packhouse

Organization of handling operations in the packhouse should ensure the quality and safety of final produce as required by the target destinations. That can only be achieved when the packhouse facility is designed to allow smooth and unidirectional flow of produce from receiving to the loading place/dock. It is, therefore, important to align the containers to allow connectivity from the receiving to packaging and cold storage. Such an organization will reduce the chances of produce recontamination during handling in the facility. The use of a prefabricated, sturdy container is considered beneficial over building a structure from scratch because it is faster to construct, low cost, and such containers can readily be accessible in most areas due to high expenses required for shipping companies to retrieve them. Occasionally, there is a possibility of finding already insulated containers which may reduce insulation costs.

The container mini packing house unit is composed of three used marine shipping containers. The containers are designed to form a U-shaped structure with each container zone designed for a specific function: receiving, packaging, and cooling zone. The mini packhouse entrance door leads to the receiving zone which is connected to the packaging zone through an insulated metal door of 1.25 m width by 2.40 m height made at an overlap between the receiving and packaging zone containers. The packaging zone has an exit door leading to a veranda between the receiving and cooling zone containers. The veranda gives access to the third container dedicated to cooling (cooling zone).

The three-container assembly of the mini packhouse is shielded from rainy and sunny conditions by an installed galvanized iron sheet roof shade.

The receiving zone is dedicated to produce cleaning and sorting. The zone has two washing stations each fitted with a tap and white-coloured ceramic sink of 0.96 m length (L) by 0.76 m width (W) by 0.52 m depth (D) devoted for produce cleaning and rinsing, respectively. Two movable tables of 0.82 m width each and 1.50 and 2.50 m length, respectively, are located adjacent to each washing station for holding washed produce. One 4.92 m long by 0.82 m wide table is positioned along the wall opposite the washing stations for produce sorting. The packing zone is equipped with two movable tables of 2.00 and 3.22 m length, respectively, and a width of 1.26 m each, used for produce grading and packing. All tables in the mini packhouse are made of angled metal frames and stainless-steel tops for easy cleaning and maintain hygiene. The cooling zone is divided into two cold rooms of 2.94 m length, 2.35 m width, and 2.35 m height (H) each, insulated with three inches (76 mm) thick cleanable closed-cell polyisocyanurate foam panels on walls and ceiling to prevent heat transmission and infiltration into the cooling area. Each cold room is fitted with an 18,000 BTU LG® split room air conditioner (HS-C1865NQ4, Thailand) coupled with a CoolBot™ (E500290, CUL, USA) temperature control device capable of reducing the room air temperature to as low as 3°C and relative humidity of about 80%–90%.

The design specification for the construction of a mini packhouse from used marine shipping containers requires strategic decision-making for the following parameters:

i. Size of the mini packing house facility and specific sections

The size of the mini packing house facility is determined by the type and size of containers to be used. Marine shipping containers come in varied sizes with different widths, lengths, and heights (Table 6.2). For a relatively larger packhouse facility, longer and taller containers such as "high cube" (2.44 m wide × 2.89 m height × 12.2 m length) or 12.2 m long standard containers can be used. Standard containers are those certified by the International Organization for Standardization (ISO) for use in marine shipping and hence can be found in most countries. Alternatively, smaller non-ISO standards (2.44 m long) or 6.06-m-long standard ISO containers can be used for small- and medium-sized mini packhouse facilities, respectively. Depending on the volume of produce to be handled, any of the packhouse areas can be made larger or smaller through the selection of the containers for that specific zone.

ii. Type and quality of insulation materials for the mini packhouse cooler

The efficiency of the packhouse cooler depends on the type and quality of insulation materials used. Many insulation materials are available; however, their insulation quality varies with the thermal resistance coefficient/value per unit of thickness and loss of

TABLE 6.2
Specifications of Marine Shipping Containers for Mini Packing House Construction

Container Type	Exterior Dimensions			Interior Dimensions		
	Width (m)	Height (m)	Length (m)	Width (m)	Height (m)	Length (m)
Non ISO Standard	2.44	2.59	2.44	2.11	2.35	–
	2.44	2.59	2.99	2.35	2.35	–
Standard ISO	2.44	2.59	6.06	2.35	2.35	5.88
	2.44	2.59	12.2	2.35	2.35	12.02
High Cube (Extra tall)	2.44	2.89	12.2	–	–	–

Source: Own source.

TABLE 6.3

Type and Quality of Rigid Board Insulation Materials and Air Conditioners

A: Insulation Materials

Rigid Board Insulations	Colour	Resistance (R) Value cm^{-1}	R-Value Loss Year^{-1} (%)
Polyurethane	Yellow	16	50
Extruded polystyrene (XPS)	Pink, Blue, Grey	13	20
Expanded polystyrene (XPS)	White	10	60–80
Polyisocyanurate	Off-white, pale yellow	17	–

B: Air Conditioner (AC) Brand, Capacity, and Room Size

Air Conditioner Brand	Cold Room Dimensions (m)	AC capacity (BTU)	Minimum Temp (°C)
Auto re-start AC with digital display is preferred;	1.20×1.20×2.40	8,000	3-9.9
	1.83×1.83×2.40	8,000	
Example:	1.83×2.40×2.40	10,000	
LG (any model), Danby (≥2011),	2.40×2.40×2.40	12,000	
Haier (any except HWE),	2.40×3.05×2.40	15,000	
Frigidaire (>2015), Samsung,	2.40×3.66×2.40	18,000	
and some models of General	3.05×4.27×2.40	24,000	
Electronic (GE).			

Source: Rivard et al. (2016) and CoolBot™ (2021).

resistance per year expressed in percentage (Table 6.3). The higher is the resistance value (R-Value) per thickness and the lower the percentage loss in R-Value, the better the insulation efficiency of the materials (Bogdan et al., 2005; Omer et al., 2007). The packhouse needs to be kept clean, hence the insulation material has to be water-resistant and easy to clean (Al-Homoud, 2005). Rigid structurally insulated panels (RSIPs) of varying thicknesses and other insulation forms are available and can be used depending on the desired quality of insulation (Al-Homoud, 2005; Aditya et al., 2017).

iii. Type and capacity of the air conditioner

A CoolBot™ cold room can be made using a window-type or split room air conditioner; however, not all types or brands work with CoolBot™. CoolBot™ works with most digital automatic re-start air conditioners including LG, Haier, Samsung, and General Electronic (Table 6.2). It must be noted that: "The mention of specific companies or trade names is not recommended to the exclusion of other similar companies or products." The ability of CoolBot™-coupled air conditioner to cool depends on three main factors: capacity of the air conditioner (BTU), the effectiveness of insulation materials, and size of the cold room (CoolBot™, 2021). Selection of air conditioner capacity to use in the cooler requires knowledge of the dimensions of the cold room. Each size of air conditioner has its maximum room volume for effective cooling. The larger the cold room unit, the higher the air conditioner capacity required (Table 6.3).

6.3   STEP-BY-STEP PROCEDURES FOR CONSTRUCTION OF THE CONTAINER MINI PACKING HOUSE

The key steps in the construction of the container mini packing house using marine shipping containers include:

i. Development of the mini packhouse floor plan

    ii. Establishment of the concrete foundation
   iii. Alignment and restoration of containers
   iv. Insulation of the cooling zone
    v. Installation of the air conditioner and CoolBot™ unit

6.3.1 Development of the Mini Packhouse Floor Plan

The packhouse floor plan is a sketch design that shows the orientation, measurements, and connectivity of zones and key fixtures in the mini packhouse (Figure 6.1). The plan is developed based on key considerations for packing house design. It defines the proper orientation of containers and hence the shape of the mini packhouse. The plan indicates the positioning, dimensions, and connectivity of the containers, concrete foundation, doors, and windows on the mini packhouse facility. It also indicates the size, partitioning of zones, and location of key fixtures like washing stations, air conditioners, and tables in the packing house. The plan illustrates the overlap of the two front containers to the middle rear container. The overlaps provide potential areas for door connections from front containers to the middle rear container. Three doors provisions are indicated, at the entrance

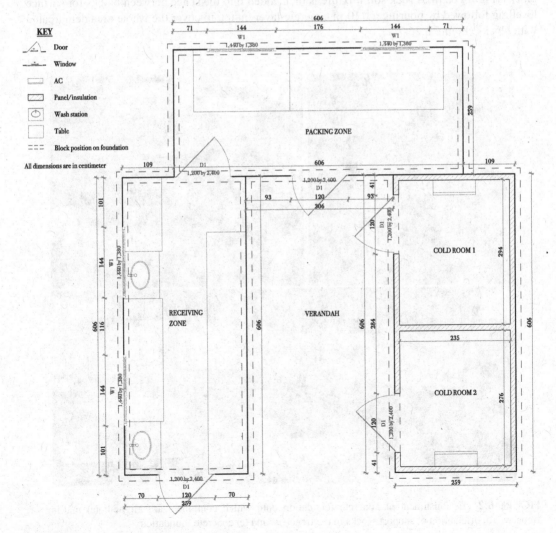

FIGURE 6.1 Packhouse design sketch plan.

to the receiving zone, between the receiving zone and packaging zone, and at the exit of the packaging zone to the outside space (veranda) between the three containers. Two more external door provisions, one to each cold room, are also indicated. Connectivity of the zones and door positioning are purposely made to guarantee the unidirectional flow of produce from receiving to cooling zones. The interior left wall and right wall of cold room 1 and cold room 2, respectively, are reserved for the installation of the split air-conditioner units.

6.3.2 ESTABLISHMENT OF A CONCRETE FOUNDATION

The concrete foundation is the base ground onto which the containers are aligned. It is established by transferring the container orientation measurements from the packing house design sketch plan onto the ground. Trenches of 0.60 m wide and 0.60 m deep are dug along the perimeter where the containers will be set (Figure 6.2a), and then a 2.5-inch (64 mm) layer of sand is added throughout the trench to allow levelling as blocks are laid in place. Blocks of 0.23 m (9 inches) wide, 0.46 m (18 inches) length, and 0.15 m (6 inches) height are oriented in the trench and fixed in place with mortar to support the outer base of the containers in the centre of the blocks (Figure 6.2b). A rock-soil mixture is then added into the space between blocks for surface levelling followed by pouring a 0.10-m concrete layer on top to cover the whole area demarcated with blocks (Figure 6.2c).

FIGURE 6.2 Establishment of concrete foundation onto which containers are aligned; (a) making of trenches, (b) orientation of support blocks in the trenches, and (c) concrete foundation.

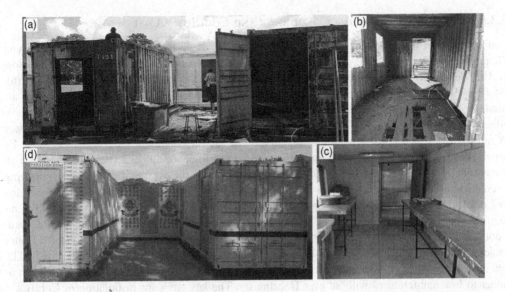

FIGURE 6.3 Alignment and restoration of used marine shipping containers; orientation and fabrication of containers (a), the initial look of the interior part of the containers (b), final look of the interior part (c), and final look of the exterior part of the containers (d).

6.3.3 ALIGNMENT AND RESTORATION OF CONTAINERS

Shipping containers are aligned using a crane aided with people pushing on sides for orientation onto the concrete foundation. The two front containers are placed first lengthwise onto the concrete base followed by the rear-middle container while maintaining an overlap of 1.5 m between each container (Figure 6.3). Three door openings of 1.20 m wide×2.40 m height each is made at the entrance to receiving zone (left), between receiving and packaging zones, and at the exit from the packaging zones, respectively. Two other similar doors, one to each cold room, are also installed. At each door opening, an insulated cold room door is fixed to reduce heat transfer between zones and prevent contamination during handling. Four window openings of 1.44 m wide×1.38 m height each are cut, two in the receiving area and two in the packaging zone and aluminium window panels fitted. Worn-out timbers on the container floor are replaced with 50 mm depth×250 m width treated hardwood timbers. Any rusted openings in the containers are welded and painted with oil-based paints. Finally, a layer of terrazzo concrete floor is installed in all three containers to enhance cleaning and drainage. The terrazzo floor has to be fixed after the installation of insulation panels. To avoid water infiltration between the floor and panels during cleaning, the terrazzo floor is made to overlap with the sidewall.

6.3.4 INSULATION OF THE COOLING ZONE

Based on the interior measurements of the containers, a total of 24 pieces of RSIPs of 1 m wide, 3 m height, and 76 mm (3 inches) thickness are used to insulate and partition the cooling zone container into two cold rooms of 2.35 m wide by 2.35 m height and 2.94 m length each. Panels are first sized based on interior container dimensions, aligned along the container walls and ceiling and then fixed using moisture curing, self-expanding aerosol polyurethane (Pu) foam (multipurpose Pu foam, 805 Akfix) dispensed through a straw adapter. For extra strength, aluminium angle liners are fixed between panels. Finally, spaces between the cold room door and panels on container walls are sealed using the Pu foam to prevent heat transfers.

6.3.5 Installation of the Air Conditioner and CoolBot™ Unit

CoolBot™ is an innovative device that can trick and override the air conditioner (Kitinoja and Thompson, 2010; Saran et al., 2013) in a well-insulated room and drop air temperature to as low as 3°C, depending upon the room size and capacity of the air conditioner (Rivard et al., 2016). CoolBot™ works with most auto re-start digital air conditioners including wall or split room air conditioning units as detailed in Section 2.2 (iii) and Table 6.3. Having decided on the air conditioner to use with the CoolBot™, the packhouse must be installed with a reliable source of power to run the units. Electrical wiring and installation of the air conditioner are preferably done by experts but not necessarily for the CoolBot™. The air conditioner unit has to be positioned in such a way that does not directly face the door outlet, to minimize the escape of cool air upon opening of cold room door (Figure 6.4a). Step-by-step installation of AC can be viewed on YouTube video by CoolBot™ (https://www.youtube.com/watch?v=4_OEyv8_sCo). Connection of a CoolBot™ unit requires knowledge on how to locate and connect the CoolBot™ sensors/cables to the AC unit. CoolBot™ has three sensors: CoolBot™ fin sensor (blue cable, black tip cover), CoolBot™ heater cable (black cable, red tip cover), and CoolBot™ room sensor (blue cable, black tip cover). The heating element (heater cable) and programmed micro-controllers (sensors) direct your air conditioner to low temperatures without ever freezing up. The key steps for installation of CoolBot™ can be viewed on YouTube (https://www.youtube.com/watch?v=WocdCnMjxA0) and involve the following:

Step 1: Mount the CoolBot™ unit at about 0.15–0.30 m right or left side adjacent to the AC unit (Figure 6.4a) but never below it to avoid wetting, and then plug the cables/sensors into corresponding labelled ports at the bottom of CoolBot™.

Step 2: Remove air filters at the front of the AC unit and locate AC's temperature sensor, which is normally attached to the front grill of the AC.

Step 3: Place the CoolBot™ heater cable next to the AC temperature sensor and wrap them together tightly with a 50 mm square piece of aluminium foil. Additional ties can be added using insulation tape below the foil to keep the wrap together. The foil connection mustn't be kept in front of the AC. In case there is a secondary sensor, as it has been for some AC brands, place and fix in place the secondary sensor on top of the wrap of primary temperature sensor and CoolBot™ heater cable using a piece of electrical insulation tape but do not over-insulate.

Step 4: Make a slit on the fins in front of the AC between the bottom and second cooling pipes using a pen or pencil and insert a quarter of the CoolBot™ fin sensor tip. As you do that make sure it does not touch the cooling pipes. Secure the fin sensor in the fins by pushing them gently.

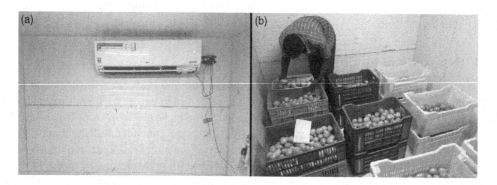

FIGURE 6.4 Air conditioner and CoolBot™ installation in the container mini packhouse cold room (a) and produce storage (b).

Step 5: When using the CoolBot™ Pro version and have a Wi-Fi connection, mount the CoolBot™ jumper outside the cooler to a suitable and well-protected area against rain, snow, and sunshine depending on the length of the data cable and then connect the jumper to CoolBot™ using the provided data cable.

Step 6: Turn on the AC and set it to cool mode at the lowest safe temperature setting and highest fan speed.

Step 7: Plug in the CoolBot™ or CoolBot™ Pro power supply and set the temperature onto the CoolBot™ unit by pressing the checkmark button. When the current set temperature blinks, use the right or left arrows to set to the desired cold room temperature. Save the set-point temperature by pressing the checkmark. Other settings can be changed by following settings given in the CoolBot™ manual which is normally provided with the unit during purchase. The cold room will then be ready for storing the produce (Figure 6.4b).

6.4 COOLING EFFECTIVENESS OF COOLBOT™ COLD ROOMS

The performance of a cold room is measured by its capacity to maintain the set point temperature at the lowest possible energy consumption (Vignali et al., 2017). The power consumption of most refrigeration units is determined by the compressor running time which varies with the amount of heat energy to be removed (refrigeration load) from the cold room to maintain the desired storage temperature (Khan and Afroz, 2015). Performance of refrigeration units including CoolBot™ cold room (CB-CR) is affected by five main factors: (i) building heat transmission, (ii) infiltration heat gain, (iii) product load, (iv) heat gain due to evaporator fans, and (v) miscellaneous heat gains (Adre and Hellickson, 1989). *Building heat transmission* refers to the heat gain through walls, roof, windows, doors, and floor of the storage room. Heat transmission varies with type and thickness of insulation, the surface area of the building, construction design, air temperature difference between outside and inside, and solar radiation incident to exterior surfaces of the cold room. *Infiltration heat* refers to heat gained due to penetration of outside air through cracks around the cold room walls or doors. The cooling efficiency of a packhouse cold room can, therefore, be improved by minimizing the frequency of cold room door opening, installation of rubber gaskets around the door frames, and fixing hanging plastic door strips to reduce the escape of cold air and infusion of hot air (Slama and Diffley, 2013). *Heat gain due to evaporator fans* refers to the heat generated by motors of evaporator fans in the cold room. As evaporator fans run to facilitate the movement of heat from stored produce to evaporator coils, it also generates heat. Such heat is equivalent to the electrical power used to drive motors of the fan (including efficiency losses). *Produce load* is the sum of the heat generated by the produce from the inside of the cold room (Adre and Hellickson, 1989). Horticultural crops brought in from the field must be cooled to the set-point temperature; this field heat represents the greatest cooling load for the AC. Also, since fresh crops are still living, they continue to undergo respiration during which carbohydrates (sugars) are broken down into carbon dioxide, water, and energy in form of ATP and sensible heat (called the heat of respiration) (Osorio and Fernie, 2014; Kader, 2002). The amount of heat of respiration varies with the type and quantity of produce stored. The produce heat load requiring removal during cooling can be computed according to Adre and Hellickson (1989) following *Equation 1*.

$$q = m \times C_p \times (t_1 - t_2) \tag{6.1}$$

Where q=produce heat removal, kJ; m=mass of produce, kg; C_p=Specific heat of produce, kJ kg^{-1}°C^{-1}; $t_1 - t_2$=initial and final temperature, respectively, °C.

Miscellaneous heat gains refer to the heat generated by lights, equipment, and people. For proper produce cooling, it is, therefore, recommended to install refrigeration units with slightly higher cooling capacity than actual refrigeration load to offset such heat gains.

6.5 APPLICATION AND PERFORMANCE OF THE CONTAINER MINI PACKHOUSE

The container mini packhouse can be used for a number of produce handling operations including washing, sorting, grading, packaging, precooling, and cold storage. Apart from these operations, a model container mini packing house can be used for demonstration and training of agriculture extension workers, farmers, marketers, and students on best postharvest handling operations for maintaining produce quality and reduction of postharvest losses.

6.6 COMPARISON BETWEEN DIFFERENT COLD STORAGE TECHNOLOGIES

In one of the storage trials conducted at SUA, four low-cost storage conditions were compared on their potential to maintain produce quality and shelf life; zero energy cooling chamber (ZECC), ambient storage, and CB-CR storage at 13°C±1°C or 16°C±1°C. Tomato fruits of the variety "Assila" (Seminis Vegetable Seeds, Inc.) were harvested at three maturity stages, i.e. mature green, breaker, and light red (USDA chart-1975) and stored at each of the conditions. Those stored in the CB-CR at 13°C±1°C or 16°C±1°C in the mini packing house had delayed colour change and more marketable fruits after 12 days of storage than those stored in the ZECCs and ambient conditions (Table 6.4). Similarly, CB-CR storage at 13°C±1°C or CB-CR at 16°C±1°C retained more than 50% marketable mature green fruits by day 42 of storage (Figure 6.5).

The economic evaluation of the four storage conditions revealed a lower cost-benefit ratio on mature green tomato fruits store in CB-CR storage at 16°C±1°C (0.45) or 13°C±1°C (0.58) than in

TABLE 6.4

Effect of Maturity Stage and Storage Conditions on Tomato Marketability Following 12 Days of Storage

	Marketable Fruits (%)			
Maturity Stage	13°C±1°C	16°C±1°C	ZECC	Ambient
Mature green	99.89 cD	97.78 bC	71.00 cB	57.00 cA
Breaker/turning	90.33 bB	84.56 abB	50.33 aA	43.00 aA
Light red	71.56 aBC	74.11 aC	49.67 bAB	39.33 aA

Means bearing the same lower cases within a column or the same upper cases in a row are not significantly different based on Turkey HSD at $P \leq 0.05$.

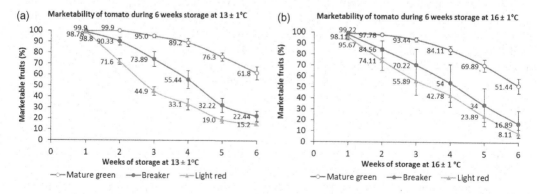

FIGURE 6.5 Storability of mature green, breaker and light red tomato fruits in CB-CR at (a) 13°C±1°C or CB-CR at (b) 16°C±1°C storage conditions for 6 weeks.

ZECC (1.64) and ambient storage (2.27). The evaluation was based on a total produce load of 256 kg of tomato fruits, although the maximum carrying capacity of the room is 4,104 kg equivalent to 144 crates of 300 fruits (28.5 kg) each. It is important to note that the effectiveness of most postharvest handling technology can vary with crop, season, and intensity of use during the season (Kitinoja et al., 2011; Kitinoja, 2013).

6.7 SUMMARY AND CONCLUSIONS

The quality of fresh horticultural crops at the consumer level is a function of pre-harvest crop management, harvest, and postharvest handling practices. However, greater loss in value is expected when produce deterioration occurs towards the last stages of handling. Application of appropriate postharvest handling practices and technologies is necessary for compliance with key quality and safety standards demanded by local high-value and export markets. Failure to adopt appropriate postharvest handling technologies and practices by most small- and medium-scale farmers in developing countries is attributed to the high cost of establishment and maintenance in relation to the size of enterprises they operate.

The container mini packing house is considered to be the appropriate scale of produce handling facility to fulfil the needs of small- and medium-scale farmers in developing countries due to the relatively lower cost of construction and maintenance compared to industrial facilities. It uses a low-cost walk-in cooler equipped with a standard residential air conditioner unit fitted with a CoolBot™ device which is also easy and cheap to construct and repair. Such a facility can offer significantly higher returns on the investment when intensively used with the right crop(s) and seasonal timing. Value chain actors can realize even higher profits when integrating the use of the facility with other successful harvest and postharvest handling practices, such as optimal harvest maturities and appropriate packaging methods and materials. Storage timing in such a facility should, therefore, consider produce price changes over the season. Effective use of the facility can enhance produce compliance to quality and safety standards and allow value chain actors to pass over glutted markets (with low prices) to reach high-priced markets.

ACKNOWLEDGEMENTS

Thanks go to the USAID/UC Davis Horticulture Innovation Lab, who sponsored the project Capacity Building on Produce Postharvest Management in Tanzania at the Sokoine University of Agriculture through the Kansas State University (prime project leader). The authors would also like to thank Mr. Shafiki Sharoim for his enthusiasm in working on the re-fabrication of the containers and installation. We dedicate this chapter to our valued colleague, Theodosy Msogoya, who passed away before publication occurred.

REFERENCES

Acedo A., and Weinberger, K. (2016). Vegetables postharvest: Simple techniques for increased income and market. Cambodia: AVRDC – *The World Vegetable Center, Taiwan and GTZ-Regional Economic Development Program*. Cambodia (KH). Hal, 37.

Aditya, L., Mahlia, T. M. I., Rismanchi, B., Ng, H. M., Hasan, M. H., Metselaar, H. S. C., and Aditiya, H. B. (2017). A review on insulation materials for energy conservation in buildings. *Renewable and Sustainable Energy Reviews*, 73, 1352–1365.

Adre, N., and Hellickson, M. L. (1989). Simulation of the transient refrigeration load in a cold storage for apples and pears. *Transactions of the ASAE, 32*(3), 1038–1048.

Ahmad, M. S., and Siddiqui, M. W. (2015). Factors affecting postharvest quality of fresh fruits. In Ahmad, M. S., and Siddiqui, M. W. (Eds.) *Postharvest Quality Assurance of Fruits- Practical Approaches for Developing Countries*. Springer, Cham, pp. 7–32.

Ait-Oubahou, A., Brecht, J. K., and Yahia, E. M. (2019). Packing operations. In *Postharvest Technology of Perishable Horticultural Commodities*. Woodhead Publishing, Cambridge, UK, pp. 311–351.

Al-Homoud, M. S. (2005). Performance characteristics and practical applications of common building thermal insulation materials. *Building and Environment*, 40(3), 353–366.

Arah, I. K., Ahorbo, G. K., Anku, E. K., Kumah, E. K., and Amaglo, H. (2016). Postharvest handling practices and treatment methods for tomato handlers in developing countries: A mini review. *Advances in Agriculture*. doi:10.1155/2016/6436945.

Asante, A. A., Yin, M., Abbew, C. K., and Nyumuteye, D. (2019). Optimizing the operational process of a cold chain fruit pack house. *African Journal of Engineering Research*, 7(3), 64–73.

Asfaw, S., Mithöfer, D., and Waibel, H. (2010). What impact are EU supermarket standards having on developing countries' export of high-value horticultural products? Evidence from Kenya. *Journal of International Food and Agribusiness Marketing*, 22(3–4), 252–276.

Birthal, P. S., Jha, A. K., and Singh, H. (2007). Linking farmers to markets for high-value agricultural commodities. *Agricultural Economics Research Review*, 20, 425–439.

Bogdan, M., Hoerter, J., and Moore Jr., F. O. (2005). Meeting the insulation requirements of the building envelope with polyurethane and polyisocyanurate foam. *Journal of Cellular Plastics*, 41(1), 41–56.

Boselie, D., Henson, S., and Weatherspoon, D. (2003). Supermarket procurement practices in developing countries: Redefining the roles of the public and private sectors. *American Journal of Agricultural Economics*, 85(5), 1155–1161.

CoolBot™ (2021). Store It Cold, LLC. http://storeitcold.com.

Davis, J. R. (2006). How can the poor benefit from the growing markets for high value agricultural products? *Available at SSRN 944027*.

Gulati, A., Minot, N., Delgado, C., and Bora, S. (2007). Growth in high-value agriculture in Asia and the emergence of vertical links with farmers. In: Swinnen, J. F. M. (Ed.) *Global Supply Chains, Standards and the Poor*. CAB International, New York, pp. 91–108.

Gustavsson, J., Cederberg, C., Sonesson, U., van Otterdijk, R., and Meybeck, A. (2011). Global Food Losses and Food Waste: Extent Causes and Prevention, Rome, Food and Agriculture Organization (FAO) of the United Nations.

Heap, R. D. (2006). Cold chain performance issues now and in the future. *Bulletin of the IIR*, 4. 1–13. https://www.crtech.co.uk/pages/environmental-testing/COLD_CHAIN_PERFORMANCE_ISSUES.pdf.

Hewett, E. W. (2012). High-value horticulture in developing countries: Barriers and opportunities. *Plant Sciences Reviews*, 229.

Hodges, R. J., Buzby, J. C., and Bennett, B. (2011). Postharvest losses and waste in developed and less developed countries: Opportunities to improve resource use. *The Journal of Agricultural Science*, 149(S1), 37.

Jaffee, S. (1992). *Exporting High-Value Food Commodities: Success Stories from Developing Countries*. The World Bank, Washington, DC, USA. 117 pp.

Johnson, G. I., and Hofman, P. J. (2009). Postharvest technology and quarantine treatments. In Litz, R. E. (ed.) *The Mango* (2nd Edition). CAB International, New York, 529–606 pp.

Joshi, R., Banwet, D. K., and Shankar, R. (2009). Indian cold chain: Modeling the inhibitors. *British Food Journal*, 111(11), 1260–1283.

Jouanjean, M. A. (2013). Targeting infrastructure development to foster agricultural trade and market integration in developing countries: An analytical review. *London: Overseas Development Institute*, 1–26.

Kader, A. A. (2002). Postharvest biology and technology: An overview. In: Kader, A. (Ed.) *Postharvest Technology of Horticultural Crops*. (3rd Edition). University of California, Agriculture and Natural Resources, Los Angeles, p. 3311.

Kader, A. A. (2009). Handling of horticultural perishables in developing versus developed countries. In *VI International Postharvest Symposium*, 877 (pp. 121–126).

Khan, M. I. H., and Afroz, H. M. (2015). Effect of phase change material on compressor on-off cycling of a household refrigerator. *Science and Technology for the Built Environment*, 21(4), 462–468.

Kitinoja, L. (2013). Use of cold chains for reducing food losses in developing countries. *Population*, 6(1.23), 5–60.

Kitinoja, L., and AlHassan, H. Y. (2012). Identification of appropriate postharvest technologies for small scale horticultural farmers and marketers in sub-Saharan Africa and South Asia-Part 1. Postharvest losses and quality assessments. *In International Symposium on XXVIII International Horticultural Congress on Science and Horticulture for People (IHC2010)*, 934 (pp. 31–40).

Kitinoja, L., and Thompson, J. F. (2010). Pre-cooling systems for small-scale producers. *Stewart Postharvest Review*, 6(2), 1–14.

Kitinoja, L., Saran, S., Roy, S. K., and Kader, A. A. (2011). Postharvest technology for developing countries: Challenges and opportunities in research, outreach and advocacy. *Journal of the Science of Food and Agriculture*, *91*(4), 597–603.

Maertens, M., Minten, B., and Swinnen, J. (2012). Modern food supply chains and development: Evidence from horticulture export sectors in Sub-Saharan Africa. *Development Policy Review*, *30*(4), 473–497.

Narrod, C., Roy, D., Okello, J., Avendaño, B., Rich, K., and Thorat, A. (2009). Public–private partnerships and collective action in high value fruit and vegetable supply chains. *Food Policy*, *34*(1), 8–15.

NCCD (2015). All India Cold-chain Infrastructure Capacity (Assessment of Status and Gap). Delhi: National Centre for Cold-chain Development (NCCD). https://iifiir.org/en/fridoc/4676.

Omer, S. A., Riffat, S. B., and Qiu, G. (2007). Thermal insulations for hot water cylinders: A review and a conceptual evaluation. *Building Services Engineering Research and Technology*, *28*(3), 275–293.

Ortmann, G. F., and King, R. P. (2007). Agricultural cooperatives II: Can they facilitate access of small-scale farmers in South Africa to input and product markets? *Agrekon*, *46*(2), 219–244.

Osorio S., and Fernie, A. R. (2014). Fruit ripening: Primary Metabolism. In: Nath, P., Bouzayen, M., Matto, A. K., and Pech, J. C. (Eds.) *Fruit Ripening, Physiology, Signalling and Genomics* (Eds.) CABI International, London, UK.

Rivard C., Oxley, K., Chiebao, H., Gragg, S., and Pliakoni, E. (2016). The KoolKat: A demonstrational mobile cooling unit to support the development of small and/or urban farms. ASHS conference 24252, Atlanta, GA. 9th August 2016.

Saran, S., Dubey, N., Mishra, V., Dwivedi, S. K., and Raman, N. L. M. (2013). Evaluation of coolbot cool room as a low cost storage system for marginal farmers. *Progressive Horticulture*, *45*(1), 115–121.

Slama J., and Diffley A. (2010). *Wholesale Success: A Farmer's Guide to Food Safety, Selling, Postharvest Handling, and Packing Produce*. (4th Edition). Family Farmed Org., 315pp.

Temu, A. E., and Marwa, N. W. (2007). Changes in the governance of global value chains of fresh fruits and vegetables: Opportunities and challenges for producers in Sub-Saharan Africa. Geneva: South Centre. FAO, AGRIS. https://agris.fao.org/agris-search/search.do?recordID=GB2013203388.

Van den Broeck, G., and Maertens, M. (2016). Horticultural exports and food security in developing countries. *Global Food Security*, *10*, 11–20.

Vignali, R. M., Borghesan, F., Piroddi, L., Strelec, M., and Prandini, M. (2017). Energy management of a building cooling system with thermal storage: An approximate dynamic programming solution. *IEEE Transactions on Automation Science and Engineering*, *14*(2), 619–633.

Yaptenco, K., and Esguerra, E. (2012). Good Practice in the Design, Management and Operation of a Fresh Produce Packing-House. Food and Agriculture Organization of the United Nations/FAO Regional Office for Asia and the Pacific, Bangkok. Rap Publication.

Zoss, M., and Pletziger, S. (2007). Linking African vegetable smallholders to high-value markets: Potentials and constraints in smallholders' integration into GLOBALGAP-certified and/or domestic African high-value supply-chains. In: *Conference on International Agricultural Research for Development*, The University of Göttingen, (pp. 9–11).

7 Clean Cold Chain Technologies

Deirdre Holcroft
Holcroft Postharvest Consulting
and
The Postharvest Education Foundation

Wynand Groenewald
Future Green Now

Vijay Yadav Tokala
The Postharvest Education Foundation

CONTENTS

7.1   INTRODUCTION

Access to affordable and reliable cooling at every step of the supply chain (precooling, packing, processing, storage, transport, distribution, retail) is key to reducing losses, maintaining produce quality, and extending the availability of perishable foods (Kitinoja et al., 2019). Currently, cold chain infrastructure relies on non-renewable fossil fuels which contribute to carbon dioxide (CO_2) emissions. It is estimated that traditional refrigeration and air conditioning units contribute more than 10% of global CO_2 emissions, three times more than that of shipping and aviation combined (Teverson et al., 2015). The awareness of the impact of current energy sources on global climate change is encouraging the adoption of cleaner technologies. The initiatives outlined in the Kyoto Protocol and Paris Agreement have affected the refrigeration industry. In addition to the prohibition of refrigerants that adversely affect the environment, there is an interest in cooling technologies that reduce greenhouse gas emissions and climate change by conserving natural resources (Peters, 2019).

Developing countries require refrigeration technologies that are accessible, affordable, scalable, and often off the grid, especially in those countries with an unreliable electrical supply. This chapter reviews clean energy sources and storage systems used directly for the cold chain, with an emphasis on currently available technologies or those close to commercialization. Clean energy sources that feed into the electrical grid were not part of this review but are discussed in the IPCC (2011) report. In addition, this chapter covers alternative refrigerants that contribute to a cleaner cold chain.

7.2 CLEAN ENERGY SOURCES

7.2.1 Solar

Until relatively recently, solar-powered cold chains were expensive and limited to small containers of vaccines and medicines. In the last decade, there has been considerable progress in the adoption of solar or hybrid-solar cold chain technology for precooling (i.e., removing field heat from perishables), cold storage (i.e., cold rooms at or above freezing), transportation, and retail.

Solar cooling can be achieved by photovoltaic (PV) powered compression chillers or by solar thermal collectors. Since the investment costs of PV-powered systems are lower than for solar thermal collectors (Lazzarin, 2020), this chapter will focus on the former. The capital cost of solar-powered cold rooms has been reduced as the price of PV panels decreased during the last decade. This has been coupled with improvements in the efficiency of the panels (EnergySage, 2020). The conversion efficacy, i.e., the panel's ability to convert sunlight into usable electricity, is about 15%–22%, and panels usually last about 20–25 years (Vourvoulias, 2020).

Likewise, the energy storage system which is required to maintain power during nights and cloudy periods can vary widely in cost and efficiency. Storing energy using phase change materials (PCMs) is cheaper than lithium battery storage, while vanadium batteries, which have a longer life, are double the price of lithium batteries (Table 7.1). Hybrid solar-powered cold rooms connected to the electrical grid, or with fuel-powered backups which can be activated manually or automatically, are increasingly common.

The high capital cost of solar-powered cold storage is balanced by the lower operating costs. Singh et al. (2016) found that a 6–8 metric tonne (t) capacity, solar-powered cold storage facility with a battery backup in India cost USD 29,770 and that solar power saved 16,500 kWh of electricity per year valued at USD 1,712. (The figures are based on the average exchange rate for 2016 of USD 1= INR 67.18).

Cold or cooling as a service (CaaS), where the user pays for what they need rather than investing in the infrastructure, has become a reality. CaaS is essentially the separation of the ownership of the asset and the user. Typically, this involves the purchase of the equipment for cooling, cold storage, or cold transportation by a third party with the user paying a set price based on the volume of produce cooled or stored and duration in cold storage. The CaaS business model is a practical solution to overcoming the financial barrier to cooling in developing countries (Fox, 2019). CaaS has been implemented by at least three groups that use solar or hybrid cooling systems in sub-Saharan Africa, namely, ColdHubs (Nigeria; Chapter 13), Freshbox (Kenya), and InspiraFarms (Table 7.1).

Solar-powered cold technology suppliers for developing countries are summarized in Table 7.1. Most of these systems are developed for cold storage at farms, collection centres, markets, distribution centres, and retail outlets. These cold rooms vary widely from marine containers (20 or 40 ft.) to purpose-built solar cold rooms of varying sizes and range in price from USD 500 to 5,000 per m² (Table 7.1). Custom-built, larger facilities can be even cheaper at about USD 300 per m² (Bruni, 2020, personal communication). In comparison, the capital cost of conventional cold storage ranges in price from USD 1,615 to 1,830 per m² in the United States (The Hollingsworth Company, 2020).

Precooling systems using both vacuum and forced air systems have been developed using solar power. Adjustments to the cooling capacity, airflow, humidity control, and pressure of InspiraFarms' cold rooms resulted in the ability to precool produce and are being used commercially with avocados (80,000 kg per day), berries, and green beans (12,000 kg per day) (Bruni, 2020, personal communication).

FrigoMobile, a mobile cart fitted with a 180-L capacity cold box, was designed for small-scale retail in developing countries. While this is ideally suited to fish and meats, it could also be used to retail perishable fruits and vegetables (FrigoMobile, 2020; Table 7.1).

TABLE 7.1
Commercially Available Solar Cooling, Cold Storage, and Retail Options for Fresh Produce with Details on the Respective System Including a Cost per Unit Area

	Manufacturer/ Supplier, Country	Model	Details	Power Storage and Backup Systems	Area (m²)	Volume (m³)	Capacity (tonnes)	Capacity (Pallets, Crates)	USD per m² Unless Stated	Website
1	Aldelano, USA	Solar ColdBox (10ft)	Refrigerator: −17.8°C to 12.8°C Freezer: −23.3°C to 4.4°C	AGM maintenance-free batteries, generator backup options	5	9		5 pallets	6,000	https://solarcoldbox.com/
2	Aldelano, USA	Solar ColdBox (20ft)	Refrigerator: −17.8°C to 12.8°C Freezer: −23.3°C to 4.4°C	AGM maintenance-free batteries, generator backup options	9	19		10 pallets	5,111	https://solarcoldbox.com/
3	Aldelano, USA	Solar ColdBox (40ft)	Refrigerator: −17.8°C to 12.8°C Freezer: −23.3°C to 4.4°C	AGM maintenance-free batteries, generator backup options	20	50		20 pallets	3,200	https://solarcoldbox.com/
4	Aldelano, USA	BackPack	Solar-powered equipment for existing facility $42,000	AGM maintenance-free batteries, generator backup options	–	–	–	–	42,000 /unit	https://solarcoldbox.com/
5	ColdHubs, Nigeria		Daily flat fee per crate		9	20	3	150 crates	3,000	www.coldhubs.com
6	Coldinnov, France	Freecold	Variable in size	AGM or OPzV battery	Variable	10–30	0.2–0.6		from 1,337	https://www.coldinnov.com/en/freecold-photovoltaic-solar-refrigeration/solar-powered-cold-room/
7	Coldinnov, France	FrigoMobile	180L solar-powered street vending cart	PCM Ecotainer	–	0.18	–		1,282–2,061 /unit	https://www.coldinnov.com/en/freecold-photovoltaic-solar-refrigeration/-refrigerators-freezers/

(Continued)

TABLE 7.1 (Continued)

Commercially Available Solar Cooling, Cold Storage, and Retail Options for Fresh Produce with Details on the Respective System Including a Cost per Unit Area

	Manufacturer/ Supplier, Country	Model	Details	Power Storage and Backup Systems	Area (m²)	Volume (m³)	Capacity (tonnes)	Capacity (Pallets, Crates)	USD per m² Unless Stated	Website
8	DanSolar, Denmark		Solar panels have 30-year life	Lithium batteries	9		10		2,066	http://dansolar.dk/en/international/-cold-storage-container/
9	DanSolar, Denmark		Solar panels + battery have 30-year life	Vanadium batteries: 30-year life; non-polluting; function at 50°C	9		10		4,132	http://dansolar.dk/en/international/-cold-storage-container/
10	DGridEnergy, USA	Solar Cool Cub 188 Single Chamber	100% solar	12×325 W panels and 48V 1,100 Ah deep-cycle battery bank	8	19		10 pallets		https://www.dgridenergy.com/products/solar-cool-cube/
11	DGridEnergy, USA	Solar Cool Cub 188 Single Chamber	Hybrid solar/grid	6–12 panels, batteries 48V, 400Ah battery + auto-starting generator	8	19		10 pallets		https://www.dgridenergy.com/products/solar-cool-cube/
12	DGridEnergy, USA	Solar Cool Cub 426 Single/Dual Chamber	100% solar; option dual zones	24×325 W panels and 48V 2,000 Ah battery bank	19	43				https://www.dgridenergy.com/products/solar-cool-cube/
13	DGridEnergy, USA	Solar Cool Cub 426 Single/Dual Chamber	Hybrid solar/grid; option dual zones	12–24 panels, batteries 48V, 800 Ah battery bank + auto-starting generator	19	43				https://www.dgridenergy.com/products/solar-cool-cube/

(Continued)

TABLE 7.1 (Continued)

Commercially Available Solar Cooling, Cold Storage, and Retail Options for Fresh Produce with Details on the Respective System Including a Cost per Unit Area

	Manufacturer/ Supplier, Country	Model	Details	Power Storage and Backup Systems	Area (m²)	Volume (m³)	Capacity (tonnes)	Capacity (Pallets, Crates)	USD per m² Unless Stated	Website
14	EcoZen, India	Ecofrost 12.5×8×10ft	Solar; 4°C–10°C range; purchase, lease, or community model	Thermal energy with low-cost backup <30hours (no battery)	9	28	5			http://ecozensolutions.com/innovation/-micro-cold-storage
15	EcoZen, India	Ecofrost 16×8×8ft	Solar; 4°C–10°C range; purchase, lease, or community model	Thermal energy with low-cost backup <30hours (no battery)	12	29				http://ecozensolutions.com/innovation/-micro-cold-storage
16	EcoZen, India	Ecofrost 15×10×10ft	Solar; 4°C–10°C range; purchase, lease, or community model	Thermal energy with low-cost backup <30hours (no battery)	14	43			1,321	http://ecozensolutions.com/innovation/-micro-cold-storage
17	Ecolife, Uganda	Eco cold room	Insulated with recycled PET bottles; roof ventilation; 5°C–10°C range	Thermal energy storage		20	4–5		–	https://www.ecolifefoods.org/products
18	FreshBox, Kenya		Low-cost subscription model (USD 0.50/crate/day)				2			https://www.freshbox.co.ke/
19	InspiraFarms, UK/Italy	CS30	Replace or supplement grid; Precooling capacity of 8kW	Thermal energy storage (60 kWh) with 48h backup (no battery)	30	90	14	15 pallets	1,300	http://www.inspirafarms.com/products/
20	InspiraFarms, UK/Italy	FP120	Cold room+75m³ processing room; replace or supplement grid; Precooling capacity of 8kW	Thermal energy storage (60 kWh) with 48 hours backup (no battery)	30	90	15	16 pallets	542	http://www.inspirafarms.com/products/

(Continued)

TABLE 7.1 (Continued)

Commercially Available Solar Cooling, Cold Storage, and Retail Options for Fresh Produce with Details on the Respective System Including a Cost per Unit Area

	Manufacturer/ Supplier, Country	Model	Details	Power Storage and Backup Systems	Area (m²)	Volume (m³)	Capacity (tonnes)	Capacity (Pallets, Crates)	USD per m² Unless Stated	Website
21	Inviro Choice, UK	Vacuum cooler	Solar; improved efficiency (70% better); vacuum cooler with energy storage							https://www.vacuumcooling.eu/vacuum-cooling-benefits/
22	Solar Freeze, Kenya	Solar Freeze							4,800	http://www.solarfreeze.co.ke/solutions/?1
23	Off-Grid Factory, The Netherlands	The Village Fridge	Solar system for communal cold storage		6	15.6				https://www.theoffgridfactory.com/en/village-fridge/
24	Vink Solar, The Netherlands	20 ft	−0.5°C to 5°C range	Thermal energy storage	10	22	6	10 pallets	4,950	https://solar-coldroom.com
25	Vink Solar, The Netherlands	40 ft	−0.5°C to 5°C range	Thermal energy storage	23	49	12	20 pallets	2,913	https://solar-coldroom.com

Singh et al. (2016) calculated the return on investment of a 6–8 t solar cold room which was used to extend the storage life of mangoes and bell peppers from 4 to 5 days to 15 and 21 days, respectively. With operating costs of about USD 0.03/kg/week, their pay-back period was 9.4 years and the system lasted 15 years, assuming that the batteries would be replaced after 7 years.

InspiraFarms worked with a Zimbabwean exporter to design and deliver a packhouse that enables forced air cooling, climate-controlled packing, final cooling, and dispatch of 10 tons of berries and green beans per day. This facility has a pre-cooling capacity of 1,800 kg per hour. The implementation of pre-cooling reduced losses by 50%, i.e., shrinkage decreased from 6.5% to 3.0% and increased shelf life by 50%, as well as enabling the use of sea-freight and access to premium markets in the Far East. These results were achieved during the logistical disruptions and extended transit times that resulted from the global Covid-19 pandemic and have been attributed to InspiraFarms' technology as well as the excellent production and operational capabilities of the client.

7.2.2 Liquefied Natural Gas

Liquefied natural gas (LNG), predominantly methane, is formed by cooling the gas to −160°C. The gas is transported in the liquid phase. When the LNG is exposed to ambient conditions, it reverts to the gaseous phase, removing energy from the surroundings (the latent heat of vaporization) and expanding in volume 700 times. The waste cold from the phase change back to gas is a potential source of cooling but is usually limited to the industries clustered around the LNG terminals (Kitinoja, 2014; Peters, 2016).

However, this energy could be used to liquefy air or nitrogen (boiling points of −194°C and −196°C, respectively), which can be stored in unpressurized, insulated tanks, transported, and used in conjunction with the Dearman engine. This piston engine uses the volume expansion during the phase change to produce both power and cooling. In Europe, the 'waste cold' from the LNG imports of seven EU countries in 2014 could have cooled 210,000 transport refrigeration units (TRU) (Peters, 2016). The use of TRU reduces CO_2 emissions by 80% compared to conventional diesel-powered reefers (Clean Cold Power, 2021).

A case study using liquid air engine technology to cool tomatoes during transportation was compared to the current practice of transporting in a truck with no cooling. The improved practice included using smaller capacity boxes (20 kg) and smaller loads than the current practice. This reduced the losses in tomatoes from 50% to 5% and resulted in higher returns (Table 7.2; Kitinoja, 2014).

7.2.3 Biogas

Biogas, which is generated by anaerobic digestion of organic matter, is composed of methane (50%–75%), CO_2 (30%–45%), water (1%–5%), and small amounts of other gases. Biogas can be produced from waste materials from farms, households, or food processing facilities. Biogas can be stored and used when needed making it a sustainable source of fuel (Gomez, 2013). The success of these systems depends upon the continuous availability of biomass and the efficacy of the digesters.

According to a recent UNFAO report, biogas can be used directly for heating and cooking, or for producing electricity. Biogas has been used for cooling milk in Pakistan, where Winrock International installed four biogas generators in 2012–2013 (Sharma and Samar, 2015), and in Kenya and Tanzania (Flammini et al., 2018). These small-scale chillers are not as efficient as mechanical compression chillers and can be difficult to repair and maintain, but they can be alternatives in areas without other options (Flammini et al., 2018). SimGas, a Dutch company, is commercially manufacturing biogas-powered milk chiller units of up to 10 L capacity, with the capacity to cool seven times faster than a domestic refrigerator (SimGas, 2020). Some modifications in the units might make them suitable for short-term storage of fresh horticultural produce. The high capital investment has prevented the widespread adoption of this technology (Sharma and Samar, 2015).

TABLE 7.2

Case Study Details for Transport of Tomatoes for 10 hours over Distances of 400–650 km from Arusha, Ngarenanyuki, Lushoto, or Iringa to Dar es Salaam, Tanzania

Details	Current Practice (No Cooling)	Improved Practice (Cooling During Transportation)
Truck	Typical open truck	Reefer truck (20 ft)
Metric tonnes (t)/load	8 t	6 t
Packaging	200×40 kg wooden boxes	300×20 kg wooden boxes
Losses	50%	5%
Market price/kg	$0.47/kg	$0.63/kg
Cost of extra boxes	$0	$200
Cost of cooling (from 30°C to 18°C)	Not used	$259/load ($43 per t×6 t)
Maximum market value at wholesale	$3750/load; $469/t	$3562/load; $593/t
Value – costs/load	$3750	$3562–200 (boxes) -259 (cooling)=$3103
Value/metric tonne	$469	$517

Adapted from Kitinoja (2014).

7.3   CLEAN ENERGY VECTORS OR STORAGE SYSTEMS

PCMs are substances that absorb or release heat as they change between solid, liquid, or gaseous states by using the latent heat of fusion or vaporization. These materials are very efficient at storing energy with the most useful phase transition being between solid and liquid phases (Zalba et al., 2003). Good PCM should have melting point range in the operating temperature of that product, high specific heat to retain energy for longer durations, as well as high thermal conductivity in both solid and liquid phases. The material should undergo a minimal change in volume with no degradation during the phase change. Also, these materials should be easily available, cost-effective, recyclable, and have no health or environmental hazards. These criteria can be met by a wide range of compounds or mixtures (Regin et al., 2008).

PCM can be used for storing energy, for example, from solar panels, or be used directly for cooling during storage, transportation, or retail marketing. However, in the latter case, the energy source is not always renewable (Oró et al., 2012). PCM walled containers are already being used for short-distance transport of hot or cold foods. Croda and Cryopak are examples of manufacturers of PCM-based packing options for perishables and temperature-sensitive materials. (Croda, 2020; Cryopak, 2020).

PCM is used to replace the heat sink system in the thermoelectric refrigeration systems to improve the cooling capacity (Riffat et al., 2001). Wang et al. (2007) demonstrated that using PCM in this manner can result in 8% energy savings. Using PCM as the precondenser in a refrigeration system can enhance the coefficient of performance by 6%. Combining solar photovoltaic (SPV) systems with PCM has been used in solar cold rooms (Table 7.1), but it can also be optimized to create mobile cold storage units with PCM walls (Sinha and Tripathi, 2014).

Liu et al. (2012) developed a refrigerated transport system using PCM in a storage unit in the vehicle which was cooled by a land-based refrigeration unit before the trip. The PCM used in this transport was an inorganic salt-water solution which was cheaper than many of the commercially available PCM and could maintain refrigerated truck temperatures as low as -18°C. Paraffin-based PCM has also been tested in refrigerated transport (Ahmed et al., 2010).

The University of Birmingham and China Railway Rolling Stock Corporation Limited (CRRC) have designed and constructed a refrigerated truck-to-train container using PCMs that could maintain 5°C–12°C for up to 120 hours. This technology has been evaluated commercially over 35,000 km of road and 1,000 km of rail in different climatic regions (University of Birmingham, 2019).

One of the disadvantages of using PCM in the cold chain is the inability to control the temperature regime as its efficacy is affected by environmental conditions. There is room for developing PCMs with the capacity to maintain constant cooling temperatures for longer periods which would allow their adoption in cold storage and transportation in rural areas where electricity is erratic or mechanical cooling is not available.

The CryoFridge by Valeo uses liquid nitrogen for cooling during transportation. The liquid nitrogen is released from a storage tank in the vehicle through a special evaporator. This is a quiet and clean cooling technology that has no emissions, low maintenance, and is simple to operate. It does, of course, require a supply of liquid nitrogen which limits its application. This system is being used in South Africa (Valeo, 2021).

7.4 CLEANER REFRIGERANTS

Refrigerants are the substances used within a refrigeration cycle that absorb or reject heat energy as required to effect cooling or heating. The properties of the refrigerant determine the efficiency of a refrigeration system and the operational parameters. During the industrial revolution, the demand for refrigeration increased and has continued to do so ever since. There was a need to develop refrigerants with low-pressure requirements to make refrigeration simple and cost-competitive. An unintended consequence of these refrigerants is that they contribute to global warming.

In the 1990s the depletion of the ozone layer was attributed to the use of chlorofluorocarbons (CFC) in refrigerants, aerosols, and other household products. Therefore, refrigerants with ozone-depleting potential (ODP) were banned under the Kyoto Protocol and replaced by hydrochlorofluorocarbon (HCFC) and hydrofluorocarbon (HFC) refrigerants. While these formulations did not deplete ozone, they increased global warming, i.e., have a high global warming potential (GWP).

The GWP of refrigerant gas is the contribution to global warming resulting from the emission of one unit of that gas relative to one unit of the reference gas, CO_2, which is assigned a value of one. GWPs are classified into six groups, i.e., 1=very low (<10), 2=low (<150), 3=medium (<750), 4=high (>750), 5=very high (>1,500), and 6=extremely high (>2,500). Table 7.3 shows the GWP of commonly used refrigerants over 20 and 100 years. The industry typically uses the 100-year metric; however, it is more accurate to use the 20-year metric as this corresponds to the lifespan of these refrigerants and more accurately indicates the effect refrigerants have on the environment (Clodic et al., 2013; Maté and Kanter, 2010).

The total equivalent warming impact (TEWI) of a refrigeration system is the sum of direct and indirect emissions. Direct emissions are an indication of the impact on the environment due to refrigerant leakage and the inevitable loss of a portion of the refrigerant at the end of life. The GWP

TABLE 7.3
The Global Warming Potential (GWP) of Selected Refrigerants over 20 and 100 Years

Refrigerant	Atmospheric Lifetime (years)	GWP (20 years)	GWP (100 years)
HFC-134a	14	3,830	1,430
HFC-404a	34.2	6,010	3,922
HFC-410a	–	4,340	20,88
HCFC-22	12	5,160	1,810
Propane		<3	<3
NH$_3$		0	0
CO$_2$		1	1

Adapted from Maté and Kanter (2010).

of a refrigerant is the main determinant of direct emissions. Indirect emissions are the CO_2 emissions of the power supply source and the amount of power used by a refrigeration system. Therefore, both energy-efficient systems and those using sustainable power sources will have reduced indirect emissions.

Arias (2005) found that for a direct expansion supermarket refrigeration system using R404a as a refrigerant, the TEWI was as follow:

$$\text{TEWI}(\text{R}404\text{a}) = \text{direct emission} + \text{indirect emission}$$
$$= 1,222,500\,\text{kg}\,CO_2\text{e} + 2,840,700\,\text{kg}\,CO_2\text{e}$$
$$= 4,063,200\,\text{kg}\,CO_2\text{e}$$

A 30% reduction of the TEWI is possible by replacing the R404a refrigerant with a natural alternative.

Although refrigeration systems are closed-loop systems, leaks are inevitable releasing the refrigerants into the atmosphere. The leakage of 1 kg of R404a, an HFC blend, has the same impact as releasing 6,010 kg of CO_2. The average annual leakage rates in refrigeration systems in the United Kingdom demonstrates the extent of this problem (Table 7.4). Leakage rates are higher in mobile than in stationary systems, and these rates tend to be higher in developing countries as refrigeration systems are replaced less frequently.

The first step towards a cold chain with a lower impact on the environment is the selection of the refrigerant. Natural refrigerants including air, water, hydrocarbons (HC), CO_2, and ammonia (NH_3) have no negative impacts when released into the atmosphere and have lower GWP (Table 7.3). These natural refrigerants are being used by the refrigeration industry and their use is increasing. NH_3 is a well-known and efficient refrigerant that has been used for over 100 years. Air is an efficient alternative for ultra-low temperature (ULT) applications. Water is used as a refrigerant in chiller applications for space cooling and heating applications. HC are found in stand-alone refrigeration systems and household fridges. CO_2 was introduced as a refrigerant about 20 years ago and is a safe and efficient refrigerant in the commercial and industrial sector with global adoption (Table 7.5).

There is a natural option for every heating and cooling application, and products, knowledge, and training are available worldwide to support the adoption of natural refrigerants. Natural refrigeration systems are available worldwide making these technologies accessible in developed and developing countries.

7.5 PUBLIC AND PRIVATE SUPPORT

The success of a clean cold chain in developing countries depends on technology transfer, financial support, and/or policy changes. Various public and private organizations are currently contributing

TABLE 7.4

Leakage Rates (%), Refrigerant Charge, and Estimated Number of Refrigeration Systems for the UK Stationary Refrigeration Sector

Sector	Leakage Rate (%)	Typical Refrigerant Charge (kg)	Estimated Number of Systems
Retails cabinets	<1	<3	4,000,000
Small commercial	<1	3–30	300,000
Supermarket	20–30	30–300	50,000
Industrial	15–20	>300	50,000

Adapted from Koronaki et al. (2012).

TABLE 7.5

Number of Transcritical Carbon Dioxide Installations (i.e., Installations Using CO_2 as a Refrigerant) Around the World as of October 2020 (Shecco, 2020)

Country/Region	CO_2 Installations
Europe	29,000
Japan	5,000
The United States	650
Canada	340
South Africa	>220
New Zealand	100
Australia	95
South America	75
Central America	20
Indonesia	13
Russia	9
Mexico	6
China	3
Taiwan	2
India	1
Jordan	1
Malaysia	1

---

to the development of a clean cold chain. The Basel Agency for Sustainable Energy and the Kigali Cooling Efficiency Program (K-CEP), with the endorsement of the Global Innovation Lab for Climate Finance, launched the CaaS Alliance in November 2019 (www.caas-initiative.org) to promote a pay-per-service model for more efficient cooling systems and to support the implementation of this model in developing countries. The Solar Cooling Initiative of the International Solar Alliance (https://isolaralliance.org/) aims to spread the use of solar and solar-hybrid energy in agriculture.

7.6 CONCLUSIONS

The development of a cleaner cold chain is needed throughout the world but in countries with expensive energy and unreliable access, there is an even greater demand. Using sustainable energy sources and cleaner refrigerants and improving the energy efficiency of refrigeration are all important steps in achieving a cleaner cold chain.

The success of the clean cold chain will depend on selecting the most viable options available in each region, the collaboration between different sectors, creative financing, and sound policy. In addition, the role of extension in the success of the cold chain is critical, especially with regard to the development of appropriate training and effective implementation of various outreach methods (Kitinoja et al., 2019).

REFERENCES

Ahmed, M., Meade, O. and Medina, M.A. (2010). Reducing heat transfer across the insulated walls of refrigerated truck trailers by the application of phase change materials. *Energy Conversion and Management*, *51*, 383–392.

Arias, J. (2005). *Energy Usage in Supermarkets - Modelling and Field Measurements*. PhD Thesis. Department of Energy Technology, Institute of Technology, Sweden.

Clean Cold Power. (2021). Cool technology. https://cleancoldpower.com/cool-technology/. Accessed on 30 January 2021.

Clodic, D., Pan, X., Devin, E., Michineau, T. and Barrault, S. (2013). Alternatives to high GWP in refrigeration and air-conditioning applications. Final Report. https://iifiir.org/en/fridoc/alternatives-to-high-gwp-hfcs-in-refrigeration-and-air-conditioning-4538.

Croda. (2020). CrodaTherm™ Bio-based phase change materials - Food Delivery and Supply Chains. https://www.crodatherm.com/en-gb/products-and-applications/crodatherm-wax/food-and-refrigeration/food-delivery-and-supply-chains. Accessed on 8 November 2020.

Cryopak. (2020). The perfect degree of protection - Packaging and Refrigerants. https://www.cryopak.com/packaging-and-refrigerants/. Accessed on 8 November 2020.

EnergySage. (2020). How has solar panel cost and efficiency changed over time? Energy Sage. https://news.energysage.com/solar-panel-efficiency-cost-over-time/. Accessed on 18 August 2020.

Flammini, A., Bracco, S., Sims, R., Cooke, J. and Elia, A. (2018). Costs and benefits of clean energy technologies in the milk, vegetable and rice value chains. Food and Agriculture Organization of the United Nations. http://www.fao.org/3/i8017en/I8017EN.pdf. Accessed on 8 November 2020.

Fox, T. (2019). Community cooling hubs: A route to sustainable economic development. *Agriculture for Development, 36*, 47–50.

FrigoMobile. (2020). Solar-powered refrigerators and freezers – Freecold FrigoMobile. https://www.coldinnov.com/en/freecold-photovoltaic-solar-refrigeration/refrigerators-freezers/. Accessed on 5 November 2020.

Gomez, C.D.C. (2013). Biogas as an energy option: An overview. In A. Wellinger, J.D. Murphy, and D. Baxter (Eds.) *The Biogas Handbook*, 1–16. Woodhead Publishing, Sawston.

The Hollingsworth Company. (2020). Cold storage as a business model https://hollingsworthcos.com/2019/04/cold-storage-as-a-business-model/. Accessed on 8 November 2020.

IPCC. (2011). *Renewable Energy Sources and Climate Change Mitigation. IPCC (International Panel on Climate Change) Special Report*, Cambridge University Press. http://srren.ipcc-wg3.de/report. Accessed 8 November 2020.

Kitinoja, L. (2014). Exploring the potential for cold chain development in emerging and rapidly industrializing economies through liquid air refrigeration technologies. Liquid Air Energy Network, UK.

Kitinoja, L., Tokala, V.Y. and Mohammed, M. (2019). Clean cold-chain development and the critical role of extension education. *Agriculture for Development, 36*, 19–25.

Koronaki, I.P., Cowan, D., Maidment, G., Beerman, K., Schreurs, M., Kaar, K., Chaer, I., Gontarz, G., Christodoulaki, R.I. and Cazauran, X. (2012). Refrigerant emissions and leakage prevention across Europe - Results from the RealSkillsEurope project. *Energy, 45*, 71–80.

Lazzarin, R. (2020). Solar cooling. 40th Informatory note on refrigeration technologies. International Institute of Refrigeration. https://iifiir.org/en/fridoc/solar-cooling-2020-40-lt-sup-gt-th-lt-sup-gt-informatory-note-on-143007.

Liu, M., Saman, W. and Bruno, F. (2012). Development of a novel refrigeration system for refrigerated trucks incorporating phase change material. *Applied Energy, 92*, 336–342.

Maté, J. and Kanter, D. (2010). The benefits of basing policies on the 20-year GWP of HFCs. https://ozone.unep.org/system/files/documents/Benefits%20of%20Basing%20Policies%20on%20%2020%20GWP%20of%20HFCs.pdf.

Oró, E., de Gracia, A., Castell, A., Farid, M.M. and Cabeza, L.F. (2012). Review on phase change materials (PCMs) for cold thermal energy storage applications. *Applied Energy, 99*, 513–533.

Peters, T. (2016). The best use of waste cold from LNG re-gasification - Liquid air as an energy vector. https://setis.ec.europa.eu/system/files/integrated_set-plan/bham_input_action6_0.pdf. Accessed 30 January 2021.

Peters, T. (2019). Editorial. The clean cold chain in developing countries. *Agriculture for Development, 36*, 2–3.

Regin, A.F., Solanki, S.C. and Saini, J.S. (2008). Heat transfer characteristics of thermal energy storage systems using PCM capsules: A review. *Renewable and Sustainable Energy Reviews, 12*, 2438–58.

Riffat, S.B., Omer, S.A. and Ma, X. (2001). A novel thermoelectric refrigeration system employing heat pipes and a phase change material: An experimental investigation. *Renewable Energy, 23*, 313–23.

Sharma, D. and Samar, K.K. (2015). Bio chillers- A fortunate thing for rural development. *Cooling India*. https://www.coolingindia.in/bio-chillers-a-fortunate-thing-for-rural-development/. Accessed on 8 November 2020.

Shecco. (2020). World guide to transcritical CO_2 refrigeration. Part II. https://issuu.com/shecco/docs/r744-guide-part2.

SimGas. (2020). A biogas-powered milk chiller to increase income of small-scale dairy farmers and to eliminate milk spoilage. https://simgas.org/projects/biogas-milk-chilling/. Accessed on 8 November 2020.

Singh, P.L., Jena, P.C., Giri, S.K., Gholap, B.S. and Kushwah, O.S. (2016). Solar PV powered cold storage system for improving storage quality and reducing wastage of horticultural produce. *Akshay Urja*, *9*, 37–39.

Sinha, V. and Tripathi, A. (2014). Integrating renewable energy to cold chain: Prospering rural India. *2nd International Conference on Sustainable Environment and Agriculture, IPCBEE*. Vol. 76, IACSIT Press, Singapore. doi:10.7763/IPCBEE.2014.V76.20.

Teverson, R., Peters, T., Freer, M., Radcliffe, J., Koh, L., Benton, T. and Braithwaite, P. (2015). Doing cold smarter. Project Report. The University of Birmingham. https://www.birmingham.ac.uk/Documents/college-eps/energy/policy/Doing-Cold-Smarter-Report.pdf. Accessed on 8 November 2020.

University of Birmingham (2019). Newsflash. UK and China scientists develop world-first cold-storage road/rail container. *Agriculture for Development*, *36*, 10.

Valeo. (2021). CryoFridge. Liquid nitrogen unit for rigid trucks and trailers. https://www.valeo-transportrefrigeration.com/global_en/Products/Cryogenic-Systems/CryoFridge. Accessed on 30 January 2021.

Vourvoulias, A. (2020). How Efficient Are Solar Panels? https://www.greenmatch.co.uk/blog/2014/11/how-efficient-are-solar-panels. Accessed on 8 November 2020.

Wang, F., Maidment, G., Missenden, J. and Tozer, R. (2007). The novel use of phase change materials in refrigeration plant. Part 1: Experimental investigation. *Applied Thermal Engineering*, *27*(17–18), 2893–2901.

Zalba, B, Marin, J.M., Cabeza, L.F. and Mehling, H. (2003). Review on thermal energy storage with phase change: Materials, heat transfer analysis and applications. *Applied Thermal Engineering*, *23*, 251–283.

8 Refrigerated Transportation in Marine Containers and Cold Chain Transport Logistics

Eduardo Kerbel
Carrier Transicold

CONTENTS

8.1 INITIAL QUALITY

The highest possible quality potential of fruits and vegetables is determined at harvest and the initial quality must be sustained throughout the postharvest handling system. Therefore, harvest criteria and maturity indices are key elements for realizing quality potential, and initial quality at harvest and at subsequent loading of the cargo into a marine reefer container will determine maximum transit life potential. Even though we cannot improve the inherent quality of a fruit or vegetable after we harvest it, we can maintain the quality of the batch of fruit that we select to ship to destination markets. To do that, we need to follow the best postharvest handling practices and comply with an effective cold chain process that provides proper temperature management.

The potential for the best possible initial quality at harvest is affected by biological and physical factors including plant nutrition, age, maturity, size, tree or plant position, microbiological pressure in the field, and climatic conditions during the latter part of the growing period and at harvest, environmental stresses (flooding, drought, and heavy wind), handling and protection from direct sun in the field and at the packinghouse, seasonality, and time from harvest to cooling.

DOI: 10.1201/9781003056607-10

This chapter will focus on proper temperature management as the most important variable for maximum transit life potential of any fresh commodity, and CA as the best supplement to the proper temperature management for maximum transit life extension. However, adequate management of these two variables would not be meaningful if the product being shipped was not at its optimum quality potential when shipped.

8.2 PACKINGHOUSE OPERATIONS

After harvest, fruits and vegetables need to be prepared for market. That preparation involves several operations regardless of whether the packing house process is simple or sophisticated. The key parameters which need attention in the packing house operations include sanitation, removal of the decayed product, proper handling, cooling the product as soon as possible after harvest to remove field heat, maintaining the product at the recommended minimum safe temperature, and expediting the process of packing and loading to the refrigerated container for transport.

Packinghouse sanitation, including its immediate surroundings, is imperative to avoid any potential contamination of the product with harmful microbes viz. protozoans, molds, and bacteria. This includes the use of chlorinated water, and thorough sanitation of equipment, personnel, packing materials, cold rooms, and refrigerated containers. Proper sanitation and handling of the product with care reduces the potential for decay and spoilage. Handling of fruits or vegetables has to be kept to a minimum at the packinghouse during the process from harvest to loading into the container, as double or triple handling can cause physical injury, discolorations, and can enhance decay and premature ripening that will affect transit and shelf life potential. For successful transit life potential, fresh produce needs to be at its best possible quality and condition when dispatched from the packing house to the shipping lines.

8.3 TEMPORARY STORAGE – CARRY OVER

To obtain the maximum transit life potential, harvest and all postharvest handling practices including processing, precooling and loading into the container should occur on the same day, and the cold chain maintained until the product reaches its destination. In some instances, however, the product needs to be carried over to the next day because containers cannot be filled on the same day of harvest. Reasons for this include low volumes during the day, the products to be filled in the container need to be received from various farms or fields and it takes a longer time, or untimely container arrival when the product has already been processed. If products need to be stored for one or more nights at the packing house, the products would have to be held in cold rooms or containers that are maintained at the recommended low safe temperature and relative humidity. Palletized or breakbulk product needs to be stowed in such a way to maximize the efficacy of cool air circulation around and through the product for best keeping quality.

It is imperative that on the day the container is loaded, the product is at the same temperature whether precooled, stored overnight, or freshly harvested and processed. Mixing cold and warm products into a container must be avoided. Fruits that are not traditionally precooled, such as bananas, citrus, and some tropical fruits need similar treatments as other precooled fruits in terms of loading the product at the same temperature and avoidance of mixing warm and cold products. Mixing cold and warm products can affect the condition of previously cooled products resulting in loss of firmness and reduced transit life potential.

8.4 PRECOOLING

Removal of field heat is vital for maximum transit and shelf life potential. The most effective way to remove field heat from the product is by cooling immediately after harvest as quickly as possible to their minimum safe temperature. The minimum safe temperature is the coldest temperature a product can sustain without risking the development of chilling injury symptoms. Minimum safe temperatures also known as optimum handling temperatures (cooling, storage, transit, staging) are available for each fruit and vegetable.

The preferred method for precooling is forced air cooling although hydrocooling and vacuum cooling can also be used for certain products. Forced air cooling is more universal, simpler to implement, and less expensive than hydrocooling or vacuum cooling. Room cooling is passive and slow and should only be used to maintain a product's temperature once it has been precooled properly with one of the precooling methods mentioned. Forced air cooling involves building a tunnel that is assembled with pallet stacks of boxes of product in an array that creates and maintains pressure inside the tunnel forcing cool air through boxes and products (Figure 8.1).

If a stand-alone forced-air precooling room cannot be built, then one alternative is to build a forced air tunnel assembly inside a standard cold room that can maintain the desired temperature. An example of a practical portable forced air cooling tunnel that can be assembled inside a regular cold room by stacking boxes or pallets is shown in Figure 8.2. This tunnel would create a positive pressure by means of a tight assembly and an adequate size high-speed fan on one end to pull or push air through the stacks of boxes.

It is crucial to build precooling tunnels and operate them properly in order to create and maintain the highest possible pressure inside to force the cooling air through the product preventing the cooling air from bypassing the product. Carton, paper, or foam should be used to block the open area of the wooden or plastic pallets facing the walls of the precooling room in order to force the cooling air above the surface of the pallet and through the product rather than allowing air to flow under the pallet surface. Tarps or canvas should cover the entire top surface of the stacks of pallets extending from side to side over the entire tunnel. This protection should be extended to the floor from the two pallets closest to the room doors in a tight manner against the pallets. This would prevent gaps between tarp and boxes that would short cycle the cooling air around the product rather than forcing it through boxes and product. Precooling tunnel assemblies need to be tested to ensure a consistent temperature of the air and flesh of the product in any position along the tunnel. As such temperatures would be consistent at the top, middle, or bottom of pallets and on pallets closest to and away from the fans.

FIGURE 8.1   Precooling forced air room.

FIGURE 8.2 Improvised precooling assembly inside a regular cold room. The black arrows illustrate cooling air being pooled through the load. The blue arrow illustrates warm air being pulled out of the load.

When building or implementing forced air tunnels, it is important to use the recommended refrigeration equipment and fan speed to provide forced air uniformly through the product to reduce the temperature efficiently. Excessive and fluctuating relative humidity would cause condensation on the evaporator coils that would trickle on cardboard boxes resulting in a weakening of their stacking strength and consequent collapsing.

Pallets on which products are stacked may have longer and shorter length dimensions. When placing pallets inside the precooling room, it is important to orient them in the most efficient direction either on their shorter or longer dimension being parallel to the flow of cooling air through the product. Stacking arrangement on the pallets, box dimensions and orientation must enable the fastest airflow through boxes and product without blocking air on one side of the pallet which would result in uneven and longer precooling times. Precooling needs to cool down the product to setting temperatures in a matter of a few hours and should be completed thoroughly before the product is loaded into a container.

Sometimes when a product is harvested on the same day the vessel needs to sail to its destination, the grower is under pressure to complete the precooling process and dispatch the container to the port. The product may not be precooled thoroughly or not pre-cooled at all because of lack of time. Under this scenario, products will have less transit life potential and may ripen and/or decay in transit before arriving at destination markets.

The product that needs to be treated with hot water or vapor steam to comply with quarantine requirements should, ideally, be hydro cooled (cold showers is the preferred method) after the hot treatment to remove heat from pulp and peel as quickly as possible. Water is more efficient in removing heat from products than air because it has a higher specific heat capacity as well as a higher thermal conductivity capacity. After allowing treated and hydrocooled products to rest for a few hours, the product should be packed in its commercial box and be precooled with forced air to bring down the temperature to optimal transit setting values.

Once products are precooled, they need to be staged or temporarily stored under refrigeration at the same temperature they were precooled and should never be exposed to warmer temperatures until reaching the destination market. Exposing products to warmer than precooled temperatures after this process will only reduce their transit and shelf life potential.

Due to packing house logistics, it is sometimes not possible to precool a product as soon as possible after harvest. If delays in processing product from harvest to loading are unavoidable, it is important to keep in mind that it is less detrimental for the product to wait longer to precool than waiting a long time after precooling before loading into a reefer container or temporary cold room.

8.5 PACKAGING

In addition to protecting the product and allowing unitization of commercial volumes of product, packaging plays an important role in precooling effectiveness and efficiency as well as in cooling the product in the container. The composition of the various layers of paper that go into making boxes need to allow for a box that is strong enough to withstand stacking on pallets to various heights. The box also needs to tolerate high-humidity environments, as normally, during ocean transport, the humidity inside containers is high, typically in the range of 85%–95%.

Overpacking a box with a product can result in physical damage to the product especially in the bottom tiers of palletized cargo. The lid of the box provides some physical protection to the product, but it will not always prevent damage that may result from overpacking. Boxes need to allow effective circulation of cooling air through the product, and for that, they need to have the stipulated amount, size, and positioning of vent holes to allow free circulation of air through the product without blocking the pathway of the air. The design of the box needs to consider the footprint of the pallet on which it is stacked to ensure bottom vent holes and openings allow cooling air circulation upward through the product. Additionally, boxes should exactly fit the footprint of the pallet as shorter box dimensions will create air chimneys and larger box dimensions creating overhanging, which will affect cooling air circulation adversely.

Packaging materials such as plastic bags or liners, paper, plastic sheets, and cardboard are often used inside boxes to better protect the product during handling and transportation. It is important for growers and shippers to realize that all these packaging materials can interfere with adequate cooling, and therefore, they need to be designed to protect but not to impair proper cooling. Thick plastic bags and shrouds that are used for modified atmosphere packaging (MAP) where the bags regulate the gas exchange and levels of O_2 and CO_2 inside to extend the transit life and shelf life of the product can impair proper cooling and therefore product may need to be pre-cooled properly before packing in these thick bags or shrouds. If a product is properly precooled to optimum setting temperatures before loading in a container and the cold chain is complied with, modified atmosphere (MA) bags or shrouds should not represent an issue in cooling in the container in transit, as the entire cargo just needs to be wrapped in cooling air to prevent external heat from reaching it by convection and is not necessary to move cooling air through boxes and product.

If the product is shipped in a container using a controlled atmosphere system (Container CA), and the product must be packed in plastic bags, liners, or clamshells, these need to be perforated. Sealed bags or MAP cannot be used as the product will very likely asphyxiate under double MAP/CA.

8.6 LOADING THE CONTAINER

As mentioned above, most fruits and vegetables will be precooled before loading into containers. If the product is precooled and is transferred to a container inside a refrigerated cross-dock, as long as the temperature of the product and the temperature of the dock are the same or very close, then the refrigeration unit of the container can be operating while loading, provided that the connection of the container to the dock is ambient air temperature tight and maintains the cooler temperature between dock and container. However, if the product is loaded in open ambient conditions without a properly refrigerated cross-dock, the container should not be precooled beforehand nor should the refrigeration unit be operating while loading, because the evaporator can pull warmer air and ethylene from outside the container which can have, could negatively impact on the transit life potential of the product by enhancing premature ripening, loss of texture and firmness, and decay. The warmer air being introduced can cause excessive condensation on the inner surfaces of the container as well as on the evaporator coils and can form ice causing airflow restrictions and reducing cooling effectiveness. Besides, this high condensation can cause water to drip down to the cargo risking collapsing of the cardboard boxes.

Refrigerated docks for transferring the product to the containers will use electrical pallet jacks or forklifts. Forklifts need to be electrical, as those operated by compressed natural gas, gasoline, diesel, or liquid petroleum gas (LPG) will generate ethylene gas. Other potential sources of ethylene around packing houses include decaying plant material, combustion engines, and smoke from burning organic material.

Containers should be loaded with same type of products as much as possible. If mixing of different products into the same container is unavoidable because the customers at destination markets have ordered a mixed container, it is important to only load fruits or vegetables that are compatible in having the same or very similar optimal temperature, are not sensitive to potential ethylene produced by any of the commodities loaded, have similar humidity requirements, and quality in terms of flavors and aromas are not affected by other products loaded.

Products can be loaded into a container either breakbulk or palletized. If the product needs to be loaded breakbulk, boxes need to be stowed as tight to each other as possible and in a format that reduces to a minimum the number of chimneys or gaps that are left inside the container. The most common loading and stowage format, however, will be palletized cargo. Pallets come in a few different dimensions such as 101.6×122 cm, 100×120 cm, 102×120 cm, and 80×120 cm. Pallets need to be stowed close to each other following an orientation format to avoid leaving vertical chimneys or gaps inside the container (Figure 8.3).

Poor air circulation of cold air throughout the cargo can occur due to bad stowage. Chimneys and gaps will favor cooling air to short cycle through these, bypassing the product and reducing cooling effectiveness. If breakbulk or pallet stowage tightness and format still leaves chimneys or gaps at some point inside the container, these need to be blocked with thick plastic, heavy paper, or cardboard over the T-floor (T-shaped decking which allows chilled air into the container) at the floor level under the boxes (breakbulk) or pallets (palletized cargo) and also between the top two tiers of boxes in the perimeter of the chimney/gap. This is to force the cooling air that escapes through box vents into these chimneys or gaps, back into the boxes through those box vent holes. The paper, carton, or plastic used for this top blocking needs to be sized to avoid covering box top or bottom vents or openings.

The first rows of boxes when loading breakbulk, and the first two pallets inside the container when loading palletized cargo, need to be positioned tight against the bulkhead and sitting on top of the baffles at the bottom of the bulkhead to prevent cooling air from short cycling between the first two pallets and the bulkhead. It is important to make sure that pallets are placed on top of those baffles smoothly and carefully without striking the baffles into the T-bar floor blocking cool air circulation along the T-bar floor and resulting in very poor cooling of the cargo. Pallet stowage formats will usually result in the last two pallets by the door not being flush with each other but being dephased where one is closer to the door than the other. This is true except for one type of pallet stowage format typically referred to as "pinwheel" where pallets are stowed in alternate directions in groups of four creating a chimney in the middle of them, which is not a desirable format to follow.

When loading a container, if the last two pallets come flush and the "pinwheel" stowage format was not followed, this is a sign that pallets were not stowed tightly against each other and that at least one chimney or gap was left inside between pallets at some point. If this is the case, the load should be retrieved and reloaded to ensure pallets are stowed tight and no undesirable and avoidable chimneys or gaps are left inside. The cargo stow should cover the entire floor of the container and should not project beyond the T-bar floor to facilitate an effective flow of return air by the doors.

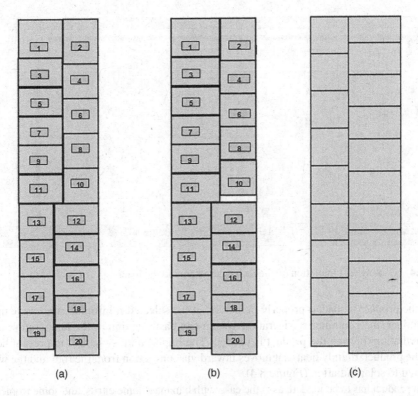

FIGURE 8.3 Adequate pallet stowage patterns for optimum air circulation through cargo. A.10×10 101.6×122 cm US pallets, and 10×10–100×120 cm EU pallets. B. 10×10–102×120 cm EU pallets. C. 9×11– 101.6×122 cm US pallets.

The height of the stow should be uniform along with the container. The height of the loaded cargo needs to comply with the maximum height of the red mark that is shown on the walls closer to the doors. The cargo must not be stowed above that height line so that enough pressure is built up in the space above the cargo to pull the air that is moving upward cooling the product, back to the evaporator.

Containers should always be fully loaded with cargo. Partially loaded containers can have inferior temperature control, larger void volumes that can favor cooling air bypassing cargo and can alter reduction/accumulation of O_2/CO_2 from product respiration which is not conducive to optimizing atmosphere management. Airbags, cartons, paper, foam, and other structures that are used to fill open spaces on partial loads to prevent the loaded cargo from moving during transit but can prevent proper air circulation When trans-shipments between containers are necessary and product from one container needs to be consolidated with product in a second container, two important recommendations need to be followed: (i) Never expose either one of the products to warmer than minimum safe optimal temperatures and (ii) Do not mix cold and warm products.

8.7   CONTAINER COOLING

Reefer containers are not generally designed to cool down the product and to bring down pulp temperature effectively because they do not have the refrigeration, nor the air circulation capacity required. Reefer containers are designed to maintain the temperature of the already cooled product. Therefore, if the product is loaded warm, it will take several days to cool it down to setting values. The transit life potential of the product is significantly reduced when it takes many days to finally bring down pulp temperature to recommended setting values. In order to maximize transit life potential, the product needs to be loaded into a container already precooled to the recommended setting temperature.

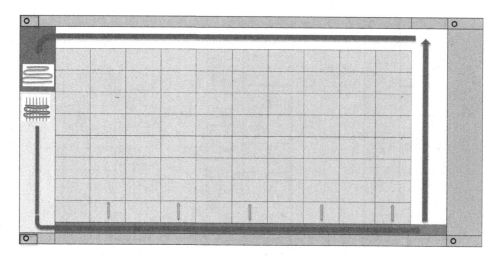

FIGURE 8.4   Cooling air circulation on precooled cargo around the load.

When the product is loaded properly precooled at the selected minimum safe low temperature into a container, the container refrigeration system can easily maintain the temperature and very low air circulation through the product is needed. The important task here is to prevent heat from reaching the product; mainly heat that moves inward via convection from the roof and the walls that are subjected to solar radiation (Figure 8.4).

If warm product has to be loaded, as is the case with bananas, some citrus, and some tropical fruits, or because of export infrastructure or scheduling, adequate cooling air circulation through the product is crucial. Air circulation in a reefer container takes place from the bottom up. Bottom air delivery is very effective for cooling because it provides better air circulation and forces upward movement of cooling air through the cargo. When supply air is delivered under the bulkhead toward the doors, under the T floor and pallets, the pathway of least resistance for this air is to move fast to the door, travel in the space between the last two pallets and the doors, and move along the space between the top of the cargo and the ceiling and back to the evaporator. Less cooling air moving this way goes upward through boxes and product, making the cooling of warm (not precooled) product slow. In order to speed up and improve the cooling of the product, it becomes important to place a blockage by the doors at floor level or the top of the last two pallets to make cooling air bounce back via the space between pallets and doors and force it through the cargo to return to the evaporator (Figure 8.5).

Top blocking is more effective than floor blocking because floor blocking still allows some cooling air to escape boxes through vent holes into space between the last two pallets and the doors losing cooling effectiveness, while top blockage bounces all cooling air moving up between the last pallets and the doors and forces all this air back through boxes and product. This blockage can be effectively achieved using thick plastic or paper, cardboard, or foam.

When planning to load a container with the product that is not precooled thoroughly, when stowing pallets, and when assembling blockage by the doors to improve air circulation through the product, it is important to consider the following:

1. The warmest part of the cargo can be above the required temperature for several days due to the limited heat transfer available with small temperature differences and low airflow rates.
2. Any warming in transit makes it very difficult to recover to the set temperature.

Different refrigeration units used on reefer containers can vary in their performance. An effective cooling system is considered to be one where the temperature range of the supply air is close to ±1.0°C.

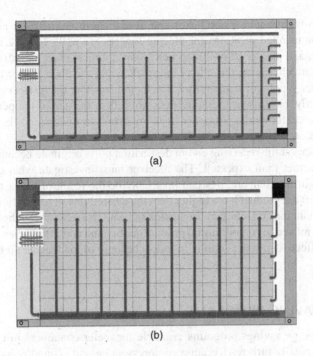

FIGURE 8.5 Blocking at the doors to improve cooling air circulation through cargo. (a) Floor blockage. (b) Top blockage.

8.8 TEMPERATURE RECOMMENDATIONS

The key factor in proper postharvest handling and transport of fresh produce is temperature maintenance. Keeping fresh products at the optimum temperature to avoid exposure to temperature fluctuations can ensure optimum quality at destination markets. In general, although the shipping lines may offer advice as to the carrying temperatures of the cargo, the shipper decides on the recommended shipping temperatures as they know the nature of their products based on type, variety, maturity, origin, and growing season conditions.

Reefer containers can be set to vary the temperature during transit as required by the cooling requirements for specific commodities. Variable temperature protocols are sometimes used, for example, to allow the commodity to arrive slightly more mature or ripe, or perhaps to slow down starch conversion into sugar which tends to contribute to pulp browning upon cooking of some commodities.

Each type of commodity and variety may have a specific minimum safe temperature under which chilling injury may develop depending on the condition of the particular commodity and on the duration that commodity is exposed to chilling temperatures. One of the potential benefits of the use of CA as a supplement to proper temperature management is that CA can mitigate chilling injury occurrence and symptoms and, therefore, allow the use of transit temperatures that are colder than traditional minimum safe temperatures used under regular air, on certain fruits.

Packaging materials used to pack a product inside a box may offer some protection against chilling injury with some limitations, especially if for a short time and this time is commodity specific.

8.9 MONITORING TEMPERATURE DURING TRANSIT

Cargo temperature can be recorded with the USDA probes that are fitted in the container box during transit. Alternatively, temperature data loggers can be used to monitor product temperature during transit. The warmest spot inside a loaded container is going to be the center or middle of the container. In order to monitor product temperature, the temperature data loggers need to be placed very close to the product inside boxes and at least in three different positions within the container. The recommended positions for placing loggers are pallets 1 or 2 bottom tier, pallets 10 or 11 middle tier, and pallets 19 or 20 top tier. Many shippers place one data logger on one of the last two pallets, attaching it to the side of the stack facing the door. In this location, the logger will certainly monitor supply air temperature which confirms that the refrigeration equipment has been working properly, but this temperature does not reflect the actual temperature of the cargo especially not in the center of the container. It is important to understand the usefulness of this data logger by the door.

Many quality claims at the receiving end in destination ports originate because arriving product temperatures are warmer than expected. The receiver must investigate what caused the warmer temperatures. The warmer temperatures can invariably be the result of either (i) the refrigeration failing in transit or (ii) the product has a higher than the desired rate of respiration and consequently ripening occurring with the concomitant production of heat. The reason why the refrigeration could have failed could be related to (i) poor initial product quality and condition when loaded at the packing house or (ii) ineffective product cooling in transit because of inadequate air circulation through boxes and product.

8.10   ENERGY SAVING PROGRAMS

There are several energy savings programs available for reefer containers in the market. Energy savings programs typically turn refrigeration compressors on and off and reduce the speed of the fans by 50% to save energy. As the compressor is turned off and the fan speed is reduced, the temperature can slightly increase and to compensate for this and bring down the temperature to setting values rapidly, most programs introduce supply air that is at a lower than the minimum safe temperature of the product, and this action could, in some cases, result in chilling injury on susceptible products. A good energy savings program will manage the temperature at ±1°C when turning the compressor on and off.

8.11   RELATIVE HUMIDITY

Most commodities require a relative humidity (RH) of around 85%–95%, depending on the type and variety, to prevent significant water loss and dehydration during transit. Water loss results in weight loss which can result in quality problems and can reduce the monetary value of a product. It can result in quality problems like wilting, denting, loss of integrity, discoloration, and uneven ripening. The RH inside a container with fresh produce is dependent on factors such as respiration and transpiration of the product which provides humidity to the air, ventilation of the container with ambient humid air during ocean transit which raises humidity, and the cooling process itself which removes humidity of the air inside the container through condensation on the evaporator. Because of these factors controlling the humidity inside the container, no further control of humidity is really required by the reefer container. Packing products in plastic liners, bags, and clamshells can maintain humidity around the product and reduce water loss. However, if the product is exposed to suboptimal temperatures at destination markets, the high humidity inside liners, bags, and clamshells can condense water on the product and accelerate decay and spoilage.

Marine reefer containers are capable of adding humidity to the atmosphere if fitted with humidity misters, but this is not a common practice because at sea the humidity provided from the outside is sufficient to keep humidity above 88%–93%. Reefer containers can dehumidify the air inside though to prevent germination and decay, for example, when shipping onions, garlic, and ginger which are susceptible to high humidity. During ocean transport adjusting the vent of the container to the recommended air exchange of 10–60 m³/h (CMH – cubic meter per hour) depending on commodity requirements, will typically maintain a relative humidity between 85–95%.

8.12 VENTILATION

The rate of fresh ambient air ventilation for fresh produce should be specified. Container vents need to be checked and maintained in good operating conditions. When shipping fresh produce without CA, the vent should always be adjusted between the recommended $10–60\,m^3/h$ depending on the commodity. Different commodities have different requirements for ventilation, and it is important to optimize venting for each commodity. For vent settings, cubic feet per minute (ft^3/min) can also be used. The vents of the various refrigeration units available for marine reefer containers are not standardized and so measuring ventilation in percentage (%) or air changes per hour (ACH) can lead to the inaccurate calibration of vents and should not be acceptable. Overventilation can cause dehydration of product from the excessively hot or cold air that comes in through the vent. Additionally, ventilation with hot air can strain the refrigeration unit (mainly the compressor) as it may have to work harder to maintain the recommended low temperatures.

The vent setting is typically checked and adjusted as a pretrip procedure before the container is sent to the farm for loading. The roads to many farms are not paved, so the container is driven over rough roads which loosen the bolts and nuts and cause closure of the vents. Therefore, vents should be rechecked at arrival to ports from the farms to make sure they are opened, locked, and maintained at the required venting setting. Vents that are inadvertently closed completely can result in deprivation of oxygen to the product inside the container causing symptoms from off-flavors and aroma and brown or black discolorations to loss of integrity and severe decay.

When ambient air outside the container is extremely hot or when the container is staged in an area where strong winds can introduce dust or other materials through the vents, it is good practice to close the vents of regular containers (containers that are not under CA). It is important to limit the time these containers have their vents completely closed to less than 72 hours to avoid any risk of accidental asphyxiation of the cargo if the container happens to be airtight.

8.13 QUARANTINE COLD TREATMENT

For a number of fresh commodities, the USDA and equivalent entities in other importing countries demand and have approved cold treatment protocols during transit to eliminate the potential presence of insects in the cargo and inside of the container before arriving at destination ports. It is important that the entire cold treatment protocol is completed before arriving at the destination port. It is important to keep in mind that for many products for which the cold treatment is required, the temperature that needs to be maintained to comply, is below the commodity's minimum safe temperature and can likely result in chilling symptoms.

Most reefer containers can sustain the required low temperature to meet the requirements of the cold treatment, and if the treatment is managed in automatic mode, the reefer unit will automatically raise the set point to the recommended transport temperature of the specific commodity once the cold treatment protocol is fulfilled.

8.14 COLD CHAIN

Proper temperature management is the single most important factor in maintaining the quality of fresh produce. If breaks occur in the cold chain, product condition and integrity will be compromised, and the commodities will be more susceptible to premature ripening, senescence, and decay. Compliance with the cold chain is crucial for the realization of the maximum transit life and shelf life potentials for fresh produce. Visibility throughout the chain can minimize disruptions and allow for action to be taken if there is a breakdown in processes or procedures.

The cold chain starts at the packing plant. Products need to be cooled as soon and as quickly as possible after harvest to remove the heat load they have been accumulating in the field or orchard for many weeks during growth. Compliance with the cold chain is the responsibility of all those involved in this process: farmers, growers, shippers, forwarders, transporters, port personnel, receivers, and handlers at destination markets. Many problems can be overcome if people involved understand their place in the cold chain and understand the needs of the product. Once the container is dispatched from the packing house to the port, subsequent handling operations such as (i) using generator-sets to transport the precooled product to the port, (ii) connecting the container to a power outlet as soon as possible after the container arrives at the port, (iii) consolidating containers with more than one type of fruit when needed as quickly as possible and under a refrigerated environment, and (iv) disconnecting from the port and loading the container onboard the vessel and reconnecting it to a power outlet onboard as expeditiously as possible. These are all mandatory to comply with the cold chain before sailing to destination markets.

The cold chain process demands that once the product is pre-cooled it cannot be exposed to higher than optimal recommended temperatures. The cold chain process needs to continue at destination markets, except for a few destination markets such as North America and a few cities in Europe and Asia where the containers are dispatched to end-users such as distribution centers and other central locations. In the majority of other destination markets, the product is unloaded from the container at the port of arrival into local trucks or trailers, which then take the product to distribution centers, ripening facilities, and stores.

Trucks and trailers may maintain the product at the same optimum transit temperature at which it was shipped in the reefer containers from the country of origin. However, distribution centers and retail and wholesale stores do not have storage rooms for all the temperatures needed for all commodities. Although a few commodities can be maintained at the recommended temperature, most are held at suboptimal temperatures and this can be detrimental to the shelf-life potential of fresh produce.

8.15   CONTROLLED ATMOSPHERE

A controlled atmosphere (CA) can help extend transit life and shelf life potentials. CA is the best supplement to proper temperature management and can offer many benefits to a large number of fruits and vegetables. The mode of action of CA is by slowing down respiration and metabolic activity, thereby delaying undesirable changes in the commodity. CA can offer many benefits to fresh produce beyond those of temperature management alone, but the degree of how beneficial CA will depend on the type of commodity, O_2/CO_2 concentrations, whether or not adequate postharvest handling practices were followed, and that the commodity shipped under CA has the optimum quality and is in optimum condition.

Avocado, banana, blueberry, cherry, lime, mango, mandarin, asparagus, strawberry, apple, kiwifruit, guava, lychee, peach, nectarine, plum, grape, and others have been shipped with several different CA systems and have resulted in arrivals to destination markets with product in optimum quality and condition. In CA, the levels of O_2 are reduced from regular air levels of 20.9% to between 2% and 10%, and the levels of CO_2 are increased from the regular air levels of 0.03% to between 2% and 20%. For most commodities, the recommended levels of O_2 and CO_2 are given in ranges for each of the two gases rather than a single specific O_2 and CO_2 value because the optimum level for each can vary and will depend on variety, maturity, season, temperature, trip length, and country of origin. Holding or transporting a commodity under its recommended minimum-O_2 levels or beyond its maximum-CO_2 levels can lead to injury. The potential extension of transit life by using CA, in addition to proper temperature management, can be in the range of 1X–3X in optimum quality and condition (Figure 8.6).

CA can offer benefits to maintain quality and optimum condition including slowing down ripening, retard decay, maintain freshness, maintain firmness and texture, reduce dehydration and weight loss, protect against chilling injury (within limits) when product is shipped at lower than minimum safe temperatures, retard compositional changes, retard color changes, retard internal browning of pulp, protect against the negative effects of ethylene, improve ripening uniformity at destination, and delay senescence.

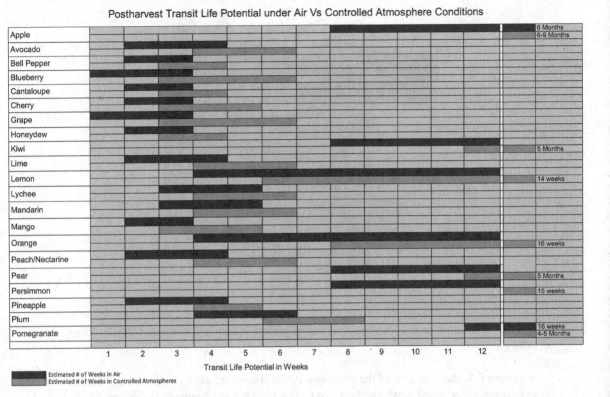

FIGURE 8.6 A potential extension of transit life when produce is transported under CA. Blue bars under regular air. Green bars under CA. (UC Davis Postharvest Facts Sheets, USDA Handbook #66.)

CA can offer **commercial** benefits beyond what proper temperature management can do including reaching more distant destinations, the opportunity for opening new markets allowing a greater selection of fresh produce to consumers around the globe, the opportunity for successful quality arrivals to destination markets for several sensitive organic commodities that cannot be treated with fungicides or other chemicals to extend their transit life, protection to products when routes are extended because of additional port stops, protection to products when sailing is slower to save on fuel resulting in longer trips, help to maintain more uniform and consistent quality at arrival to destination markets, the opportunity for switching from air freight to more cost-effective ocean transport.

CA differs from the MA in the degree of control. CA offers faster O_2 pulldown and better control in maintaining the recommended levels of O_2 and CO_2. During the transport of fresh produce, CA can be offered in reefer marine containers essentially in three formats based on how CA is established: passive CA, active CA, and fresh air management.

In passive CA, O_2 is pulled down through commodity respiration and is increased by fresh air intake from the outside. CO_2 buildup is also by commodity respiration, but excess CO_2 can be pulled down by scrubbing it or forcing it out through some type of sieve or filter. The time to reach recommended setting levels of O_2 and CO_2 will depend on the respiration rate of the commodity, the concentrations of O_2 and CO_2 that need to be established, the temperature, and the leak rate of the container. Scrubbing excess CO_2 actively can be a more effective approach, and if an activated carbon scrubber is used for this purpose, it can also remove ethylene very effectively as it removes excess CO_2. An example of such a CA system is shown in Figure 8.7. Depending on the commodity, other systems that do not remove hydrocarbons may need ethylene filters if the cargo is very sensitive to them.

FIGURE 8.7   An effective CO_2/ethylene scrubber as part of a passive CA system. (Carrier Transicold Corporation.)

In passive CA, the leak rate of the container (O_2-rich ambient air leaking in) can significantly affect how fast O_2 is pulled down and how fast CO_2 is built up, to recommended setting levels, during transit. An effective way to confirm that the container is airtight enough for passive CA is by measuring the drop in pressure with an inclined manometer or Magnehelic gauge in a given time.

Based on their rate of respiration, commodities can be classified into three categories: high, moderate, and low respiring (Figure 8.8).

In passive CA:

High-respiring commodities – can elevate CO_2 fast enough to target levels (2–3 days). They take a little longer to bring down O_2 to target levels (3–5 days). Moderate respiring commodities – depending on what concentration of CO_2 is optimal within the recommended range of CO_2 levels, and depending on shipping temperature, these commodities may or may not be able to reach O_2 and CO_2 target levels fast enough (<4–5 days). Some may need initial purging with N_2 or injection of CO_2. Low respiring commodities – will need initial purging with N_2 and/or injection of CO_2 to reach recommended setting levels of O_2 and CO_2 fast enough during transit.

Active CA provides true control of O_2 and better control of CO_2 levels. In this system, regular ambient air is introduced to the container through a separating system like hollow fibers where N_2, O_2, and CO_2 are separated by their molecular size, and N_2 of very high purity between 95% and 98% is introduced into the container displacing regular ambient air. The stream of N_2 contains very low O_2 in the range of 2%–5%. The stream of high N_2 with very low O_2 will displace and pull down O_2 inside the container with cargo faster than passive CA and will maintain more effectively recommended O_2/CO_2 concentrations during transit. Figure 8.9 illustrates an example of an effective active CA system.

The active CA systems do not remove ethylene from the atmosphere inside the container, but if the system maintains O2 and CO2 levels very precisely like the system shown on Figure 8.9, it is seldom necessary to add external ethylene filters nevertheless, growers and shippers may or may not add external ethylene filters or scrubbers as extra insurance. It depends much less on commodity respiration to bring down O_2 to setting levels, but the increase of CO_2 to setting levels to a large extent still depends

Low	Apple, Bell Pepper, Blueberry, Cantaloupe, Cherry, Grape, Kiwi, Lemon, Lettuce, Lime, Mandarine, Mangosteen, Orange, peach, Pear, Persimmon, Pineapple, Plum, Pomegranate
Moderate	Apricot, Beans (Green/Snap), Blackberry, Dates, Dragon Fruit, Honeydew, Lime, Lychee, Mango, Papaya, Raspberry, Strawberry, Rambutan, Tomato
High	Avocado, Banana, Broccoli, Durian Passion Fruit
Very High	Asparagus, Flowers

FIGURE 8.8 Commodity classification based on the rate of respiration. (Modified from UC Davis Postharvest Technology of Horticultural Crops, Publication 3311.)

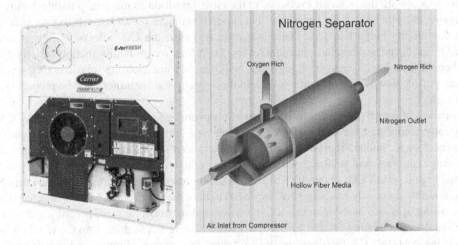

FIGURE 8.9 An example of an effective Active CA system (EverFRESH). (Carrier Transicold Corporation.)

on product respiration. Active CA creates a positive pressure inside the container as the stream of high N_2 with very low O_2 is injected, and therefore, the leak rate of the container becomes a minimal factor affecting, reaching, and maintaining the required O_2 and CO_2 levels. Control of excess CO_2 is regulated by displacement with the high N_2/very low O_2 mix being introduced as needed as the system does not remove CO_2 actively.

Inactive CA O_2 pull down to target levels is controlled by the system injecting high purity N_2 with very low O_2, as required, for all commodities from low to high respiration rates:

High-respiring commodities: CO_2 can be elevated by respiration fast enough to target levels (typically between 2 and 3 days).

Moderate respiring commodities: Depending on CO_2 target values, some may elevate CO_2 to target values fast enough (2–3 days). Other commodities may take longer to reach CO_2 target values so the initial injection of CO_2 might be necessary.

Low respiring commodities: Depending on CO_2 target values, many commodities may not elevate CO_2 fast enough (less than 2–3 days) and so will need an initial injection of CO_2.

Active CA does not require an initial injection of pure (100%) N_2 to displace O_2 for low respiring products or commodities with a moderate rate of respiration, as passive CA does need it, to pull down O_2 to optimal setting levels.

Fresh Air Management is offered in two formats: Manual and Automated.

Manual fresh air management – is difficult to control, can be unreliable, and is limited. Reaching reefer units to manually adjust the fresh air vent during transit onboard the vessel can be very difficult and unsafe for workers, and this forces the fresh air exchange to be selected and set only once at the time the container is loaded onboard the vessel. This system can lead to mistakes in the adjustments of the vent ports. Excessive air exchange settings especially in hot and humid tropical environments will force extra refrigeration capacity to cool and remove humidity from the incoming air, which can result in elevated carrying temperatures and fluctuating humidity levels inside the container. Manual fresh air exchange does not precisely control O_2/CO_2 levels within the container. When the fresh air exchange vent is closed, the respiratory activity of the commodity and the production of ethylene can potentially result in the development of injurious levels of O_2, CO_2, and ethylene in the cargo space.

Automatic fresh air management –The vent opens and closes primarily based on the levels of CO_2, but it can also be based on O_2. There are several different systems of this type offered by industry.

In all systems, a purge valve is used to try to manage the atmosphere in the container. All systems try to maintain the fresh air exchange in the closed position as much as possible during transit, opening the vent only when necessary. This system manages the CO_2 levels to tolerance limits of the commodity being handled. Therefore, in most cases, the CO_2 tolerance limit is reached before O_2 is reduced and reaches optimum levels. Therefore, a commodity that potentially derives more benefit from low O_2 than from high CO_2 will benefit from avoiding injurious CO_2 levels but not from pulling down O_2 to optimum levels. In this system, the container must be properly sealed and airtight.

Regardless of which CA system is used, the atmosphere inside a reefer container under CA can vary greatly according to several variables including type of product, variety, commodity maturity and ripeness stage, amount of product, load mix, the temperature of the product, packaging, loading and stowage patterns, void volume, a leak rate of the container, and time in transit. Some commodities need high levels of CO_2 above 10% to control mold growth during transit and since this group of commodities are mostly low respiring, they need to have the required CO_2 levels injected initially before sailing. It is also possible to inject CO_2 on demand in transit for more precise control of CO_2 levels like the option being offered by the EverFRESH active CA system from Carrier Transicold. When using container CA for shipping fresh produce, the grower, shipper, and the shipping line need to specify the levels of O_2 and CO_2 that need to be used.

Several fruits and other fresh commodities have been shipped successfully from different producing countries to various destination markets utilizing container CA, such as avocado, banana, mango, blueberry, grape, lime, mandarin, asparagus, kiwifruit, flowers, guava, and others.

Operators and personnel involved in handling containers using CA need to be instructed that CA produces an atmosphere that can be deadly to humans and that breathing an oxygen-depleted atmosphere, even for a few seconds, induces rapid unconsciousness and may result in death. Adequate warning and safety systems must be in place. The safety requirements should extend to those unloading cargoes. Proper ventilation before entering containers and training of workers are both necessary.

MAP can be used as an alternative to container CA and can offer some benefits over container CA such as:

1. maintains the atmosphere through handling at the destination.
2. maintains high humidity through handling at the destination.
3. has a lower cost.

On the other hand, MAP also has some disadvantages and limitations including:

1. levels of O_2 and CO_2 that can be established are not optimal.
2. temperature fluctuations during handling favor water condensation and decay.
3. compromising of adequate precooling and container cooling.

4. removal of ethylene from around the product is not very effective.
5. when not properly sealed or punctured, the atmosphere is lost and the enclosed environment favors decay and premature ripening.
6. increase of the cost of labor at the destination as it needs to be removed for proper ripening.

Ozone: The use of ozone gas can be considered an active atmosphere system as it is introduced actively during transit into the container. However, ozone cannot be considered genuinely a CA or MAP system as control of O_2 and CO_2 to establish optimum atmospheres is impaired by the fact that the vent of the container needs to be opened continuously during transit to a certain minimum % opening to allow ambient air in for conversion into ozone. Ozone can significantly reduce decay in transit as it kills airborne mold and bacteria. Ozone can also act upon mold and bacteria on the surface of a commodity but has the limitation that it cannot penetrate beyond the peel, so any wounds, cuts, or abrasions on a commodity will still be exposed to decay even in the presence of ozone. Another action that ozone can have is to decompose ethylene in the atmosphere. However, ozone is probably not as effective in removing and controlling ethylene levels inside a container as is the use of an effective active CA system or ethylene scrubbers or filters. It is important to consider that ozone is a strong oxidant, and over time, it can have a corrosive action on the aluminum material of T-bar floors of containers

8.16   ETHYLENE MANAGEMENT

Ethylene is the simplest hormone that can be detrimental or beneficial to many commodities depending on when the commodity is exposed to it. Some commodities are ethylene producers and when triggered by higher than optimal temperatures, decay, or physical damage can start producing significant amounts of ethylene. Other commodities are very sensitive to ethylene in very low concentrations, and they need to be protected from being exposed to this gas during transit.

On regular air, non-CA containers, the vents must be kept open during transit and are continuously exchanging fresh air from the ambient in and air atmosphere from inside of the container out. There are several ways to control and remove ethylene gas inside the container while in transit including using an effective active CA system or ethylene filters, most of which use potassium permanganate as the active ingredient, but there are other materials available that can be used, ethylene sachets and pads, and ethylene scrubbers such as those that use activated carbon. The ethylene competitor, 1-methylcyclopropene, can also be used to treat commodities before shipping them or by adding elements to the boxes of products for slow release during transit.

Ethylene gas is emitted by combustion engines and is present in smoke resulting from burning fires of organic matter such as grass, bushes, or any organic material. If any of these events take place close to the packing house and the loading docks are opened to ambient air, ethylene gas could be introduced to containers when loading cargo if the refrigeration unit is running while loading.

When placing containers onboard a vessel, it is important to consider the positioning of these containers based on whether the commodities loaded are ethylene producers or ethylene sensitive. If the commodities loaded are ethylene sensitive, these containers need to be loaded preferably toward the front of the vessel. Containers with ethylene-producing commodities need to be loaded preferably toward the rear of the vessel. All containers should be loaded with their refrigeration unit toward the rear. This is because the air coming out of the vents in regular non-CA containers will flow from front to rear as the vessel sails. This arrangement is important because in the event commodities that are ethylene producers start ripening, the ethylene produced exiting through the vents is not drawn through the vents of containers with ethylene sensitive commodities. This is the best way to isolate potential premature ripening fruit from the rest of the containers.

8.17   CA - READY REEFER CONTAINER

Reefer containers that are used for ocean transport under CA regimes need to meet certain requirements and undergo a pre-trip inspection to make sure they are ready to sustain and maintain adequate CA conditions.

The following is a checklist to confirm that a reefer container is ready for transport with CAS:

1. Adequate condition (structure and integrity).
2. Container is clean and the floor free of debris.
3. Baffle plates of the bulkhead are correctly positioned and won't block the cooling air supply.
4. The refrigeration unit and its elements work properly.
5. The evaporator drain hose water trap is correctly filled with water.
6. Water level of thawing drains is correct.
7. Fresh air vents must be completely and securely closed.
8. Unit access panels are fully secured.
9. Seals and gaskets are undamaged.
10. Unit/container joints are in good condition.
11. Floor drains are securely plugged.
12. Existing relief valves are closed and sealed.
13. An intact plastic curtain can be installed at the door and is sized to seal properly.
14. Tracks for the plastic curtain are undamaged.
15. The container is pressure checked to make sure it meets the positive pressure requirements.

8.18   MONITORING AND TRACKING VARIABLES IN TRANSIT

It is very important to monitor variables such as temperature, RH, levels of O_2 and CO_2, and absence of harmful ethylene, in transit, to confirm that products are shipped and maintained in transit under optimum recommended conditions, established prior to shipment. The refrigeration unit of a reefer container has the capacity and ability to monitor and control the temperature. While the refrigeration unit of a CA-ready container has the capacity and ability to monitor and control levels of O_2 and CO_2 as per the recommendations. The precision to control of O_2 and CO_2 will depend on the type of CA system being used in the container.

RH can also be monitored in transit. However, unless fitted with a humidifier, which is not common, a reefer container does not normally add humidity, but it may have the ability to reduce humidity, i.e., dehumidify, as needed. Monitoring harmful levels of ethylene in transit can be much more difficult because the level of detection would need to be very sensitive as only parts per billion ethylene can trigger a negative response by most fresh commodities. Refrigeration units record data during transit on the variables mentioned via the microprocessor and then downloaded it upon arrival at the destination. However, it is desirable to be able to monitor these variables live and remotely during transit for the peace of mind of shippers and receivers to confirm proper refrigeration and CA equipment performance and to help know in advance in what potential quality the product is expected to arrive at destination so that the receivers can make better decisions on how to handle, display, and sell the product to their customers. The refrigeration unit, reefer container, and marine shipping industries can provide available remote monitoring options.

Disclaimer: The views and opinions expressed in this chapter 'Refrigerated Transportation in Marine Containers and Cold Chain Transport Logistics' are those of the author exclusively and do not necessarily reflect the official policy or position of the Carrier Transicold. Any content or opinion provided by the author is not intended to malign any religion, ethnic group, club, organization, company, individual or anyone or anything.

Section III

Cold Chain Development, Capacity Building and Case Studies

9 Capacity Building for Cold Chain Development

Amanda Brondy, Lowel Randell, and Madison Jaco
Global Cold Chain Alliance

CONTENTS

9.1   INTRODUCTION

The COVID-19 pandemic revealed the critical nature of the global food supply chain and its need to be resilient and adaptable. While frontline workers in the food system have worked tirelessly to ameliorate disruptions, images of bare shelves combined with reports of product rotting due to a lack of cold storage underscored the critical importance of temperature-controlled logistics. Even before the pandemic and its associated challenges, the problems associated with postharvest food loss and waste (FLW) and its negative impacts on nutrition, climate change, and economic development were widely acknowledged. This resulted in increased attention from governments, international donors, business leaders, and academics on how best to combat FLW.

The good news is that a solution to these challenges exists through the cold chain. The cold chain refers to the temperature management of perishable products as a means of maintaining quality and safety from the point of slaughter, harvest, or production through distribution networks to the final consumer. The impacts of temperature control on reducing food loss throughout the supply chain while enhancing food safety and quality are well documented, and the need to develop cold chains in emerging economies is frequently cited (Rezaei and Liu, 2017; Kitinoja and Thompson, 2010).

While cold chain should be recognized as a solution, it must also be acknowledged that the actual development is complicated. As a chain, it begins once the product has been harvested with postharvest handling practices and that critical but often overlooked step of removing field heat. It includes the transportation required to get the product off the farm, perhaps to further processing, and onward

DOI: 10.1201/9781003056607-12

119

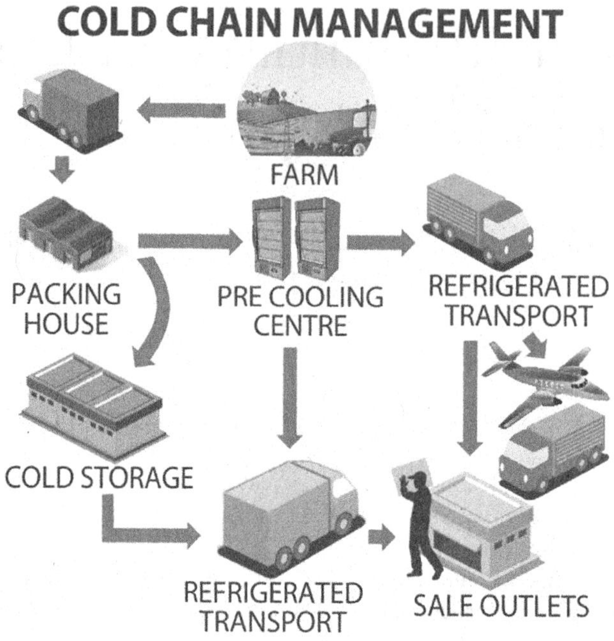

COLD CHAIN MANAGEMENT

FARM

PACKING HOUSE

PRE COOLING CENTRE

REFRIGERATED TRANSPORT

COLD STORAGE

REFRIGERATED TRANSPORT

SALE OUTLETS

FIGURE 9.1 Overview of the cold chain.

into storage, and then secondary distribution and finally into retail where it can be sold for consumption (Brecht et al., 2019). Too often, a cold chain is equated only with cold storage. This is important because taking a product that has been temperature abused and placing it into cold storage will mitigate further deterioration, but it will not fix the product. Also, if field heat is not removed through precooling before putting the product directly into a refrigerated truck or cold storage, it taxes the equipment which is designed to maintain temperatures, not to remove the field heat. Because it is a chain, what occurs at each link is critical for the next step, as demonstrated in Figure 9.1.

The cold chain is also costly. Making the necessary investments, whether it is building a refrigerated warehouse, purchasing trucks, or providing the necessary infrastructure, such as roads, water, and energy, requires capital. As it costs money to provide temperature control for a product, these costs are eventually passed on to consumers where, in many countries, only the wealthy citizens can afford to pay for products that have passed through the cold chain.

Finally, as a chain, it requires specific knowledge about commodity storage, handling, packing and packaging, refrigeration maintenance, warehouse operational practices, and appropriate business and financial models. Considering the different links of the chain, successful capacity building involves working with farmers, processors, transporters, warehouse workers, and even retail to ensure proper handling, storage, and food safety practices are implemented. It can require educating representatives from government, financial institutions, and academia on cold chain practices to facilitate a positive enabling environment, especially in areas where the cold chain is in a nascent development phase.

In recent years, there has been a push to search for new technologies and innovations to solve the challenges associated with FLW. In addition to widespread interest, this has generated some really interesting ideas and concepts, several of which have been demonstrated and tested on pilot projects. These efforts may increase with the challenges highlighted by the COVID-19 pandemic. However, the critical importance of accompanying technologies with foundational best practices cannot be overlooked. Introducing a technology may be as much as 50% of the solution, but it will not be enough without appropriate capacity building for all of the actors involved with the cold chain.

This chapter examines the critical role of capacity building for cold chain development, based on the experience of the Global Cold Chain Alliance (GCCA). It is written from the perspective of the GCCA based on our experience in growing the cold chain globally through international development work, education programs like the WFLO Institute, and our advocacy work with governmental policies and regulations. This chapter will begin with a description of cold chain drivers, followed by a deeper examination of the experience of GCCA's foundation, the World Food Logistics Organization (WFLO), with cold chain capacity building via member-based education and specific international development projects. Finally, it will close with a deep dive into lessons learned and subsequent recommendations.

9.2 COLD CHAIN DRIVERS

Indicators such as per capita gross domestic product (GDP) growth, trade flows, poverty, and education impact the development of the cold chain. On average, adding temperature control to a product increases the price by approximately one-third, so strong economic growth is a good measure for success. In addition to economic data, auxiliary issues critical to driving cold chain development include consumer demand, government regulations, and access to finance.

9.2.1 CONSUMER DEMAND

As stated above, cold chain infrastructure is expensive. In addition to basic infrastructure, including access to clean water, a reliable power source, and good roads, constructing facilities and purchasing transport solutions require large capital expenditures sustained by extensive operating costs. Some of the costs may be recouped by improvements to product quantity and quality as food loss is reduced. However, at least a portion of the cost must be passed on to the consumer. Therefore, a question of fundamental importance is whether consumers, foreign or domestic, are willing to pay more for products that have benefitted from the cold chain.

Typically, the export markets lead the demand for cold chain in countries with access to these markets. On the domestic side, the GCCA experience has reflected that cultural food preferences are important for consumer demand. In one Central Asian country, meat was sold in open-air fresh markets. Temperature control was only applied when the meat was beginning to rot, leading to a general distrust of any product that had been in cold storage. In India, many consumers prefer to shop in markets daily, choosing products that have arrived straight from the farm. Many American consumers are expressing similar preferences as can be seen by the increased popularity of farmers markets in urban settings.

It is important to acknowledge the implications that the COVID-19 pandemic may have on open-air, wet markets. We may start to see a greater push from consumers and even governments to shift to more organized retail as a means to better control and curtail the spread of diseases and enhance food safety.

9.2.2 ACCESS TO FINANCE

Shortage of capital is a constraint to small- and medium-sized enterprises (SMEs) looking to invest in the cold chain. In Nigeria, business owners struggle with high interest rates on loans needed for capital-intensive investments, often leading them to seek assistance from development projects. While projects can and often do provide valuable sources of funding, there are at times certain strings attached via deliverables and reporting that may detract, or at least distract, from the fundamental core business of managing cold chain logistics.

The persistence of financing gaps is common for SMEs. Even if alternative financing options exist via family and friends, microfinance from banks, or statutory bodies, the cost may remain out of range, making profit elusive. This can incentivize shortcuts in best practices and deviations from standard operating procedures that lead to lower quality products. Where cold chain does not yet exist as a regular practice, loan officers understandably have little prior knowledge of the business models and are reluctant to loan large amounts of capital for what is perceived to be a risky and little-understood service.

9.2.3 GOVERNMENT REGULATIONS

Regulatory issues involving the cold chain cover a variety of areas including food safety, worker safety, refrigerant policies, sanitary transportation, subsidies for infrastructure development, and favorable tax policies. Not all of these exist in every country, but they often begin with food safety.

The lack of clear food safety regulations and cold chain standards is often cited as a concern in developing countries. GCCA experience finds that most countries do have regulations for certain food safety protocols, but existing regulations are often not enforced or are enforced inconsistently. In many cases, the lack of enforcement is not malicious; it is the product of a struggling government working with limited resources. Enforcing regulations will increase the cost of food, which will likely be passed on to the consumer. In countries where the population is food insecure and vulnerable, investments in food safety enforcement may not be seen as the highest priority.

In some cases, GCCA has witnessed enforcement practices that break the integrity of the cold chain. For example, while the sampling of food products is an important step to monitor food safety, opening refrigerated containers or trailers and leaving them open for extended periods breaks the cold chain, leading to temperature fluctuations that damage the product. Besides, while enforcement of regulations is important, it is equally important to consider how the policies are enforced, whether the government plays the role of policer with fines and damages for any infringement or whether they take the position of the promoter, facilitating the private sector to understand the regulations and actions that need to be taken to mitigate situations before levying a hefty fine.

In addition to inconsistent enforcement, existing regulations are sometimes copied and pasted from another government or taken from an entirely different industry, leading to regulations that are unreasonable and unrealistic. In one South American country, when entering cold storage, a person must step into a sanitizing food bath. While this is common for food processing facilities, the practice is counterproductive for cold storage as it can create frost. Food safety practices written into regulations should be considered from the lens of industry best practices and enforcement should follow accordingly. Engagement with the private sector or industry associations is a valuable way to learn about industry best practices.

9.3 THE GLOBAL COLD CHAIN ALLIANCE AND WORLD FOOD LOGISTICS ORGANIZATION

The GCCA is an alliance of international business associations comprising companies within the perishable foods supply chain and logistics industry. The alliance is made up of four core partners which include the International Association of Refrigerated Warehouses (IARW), the International Refrigerated Transportation Association (IRTA), the Controlled Environment Builders Association (CEBA) and the WFLO. Total GCCA membership exceeds 1,300 companies from over 80 countries and includes refrigerated warehouses, distribution, transportation and logistics companies as well as academic, civic, and business leaders.

In addition to its headquarters in Arlington, Virginia, GCCA has offices in Brazil, Europe, India, Latin America, and South Africa. In each of these locations, GCCA responds to the needs of members and provides services to facilitate the operations of their businesses. As a member-driven organization, the association's activities and initiatives are focused on how the association can best serve its members. While GCCA has members in over 80 countries, the majority of those members are in North America and Europe where the cold chain is the most developed. Recognizing that members can contribute to cold chain growth and development, GCCA also supports international cold chain development projects through our nonprofit foundation, the WFLO.

The WFLO is dedicated to the proper handling and storage of perishable products and the development of systems and best practices for the safe, efficient, and reliable movement of food to the people of the world. It serves as the technical assistance arm of the GCCA providing training, education, and research services to companies and organizations concerned with producing, processing, shipping, transporting, and storing goods requiring temperature control. It is through the WFLO that GCCA members participate in international development projects. While the specifics vary per project from training and education to technical consultations, feasibility studies, and market assessments, much of the work revolves around building the capacity of local actors.

9.3.1 CAPACITY BUILDING THROUGH THE ASSOCIATION

To promote cold chain education and training, WFLO founded the WFLO Institute in 1965, with the participation of 76 industry members at the University of Maryland. Over the years, the location moved to different cities, but in the United States, it has always been located in partnership with a university. By its 13th year, Institute had enrolled over 1,000 total students since its inception. The program continued to grow so much that GCCA expanded into Latin America, added a second Institute in the United States, and 2019 launched the first Institute in Australia. WFLO Institutes have now reached over 8,000 industry members globally.

The program covers 4 days and has been designed for professionals engaged in temperature-controlled logistics, offering more than 40 classes taught by leading experts in the industry. The curriculum takes students through cold chain management, customer service, employee safety, food safety, warehouse operations, transportation operations, and professional development. Students from every professional area come to the Institute to enhance their speciality while developing a 360-degree understanding of their company and the refrigerated warehousing industry. The program is tailored for current or potential refrigeration engineers, supervisors and managers in Operations, Administration, Engineering, Information Technology, Construction, Sales, Marketing, and Human Resources. In 2018, GCCA added a refrigerated transportation track (Figure 9.2).

FIGURE 9.2 GCCA trainer teaches at WFLO Institute 2020.

Institute courses are led by a faculty member who is either an active industry expert or an academic scientist with a passion for sharing knowledge and experience. The program is self-sustaining as participants pay a fee that covers room, board, and registration. This fee is covered by the companies themselves as they recognize the value and necessity of training their workforce. It is a considerable investment to make in employees. For that reason, GCCA does not need government subsidies or development funding.

What is most interesting about WFLO Institute is that it is driven entirely by the industry. It is taught by practitioners, GCCA members working in or managing warehouses. This means that students from one company may be taught by their competitors. Members recognize not only the value of training their staff but also the importance for the industry to work together to reduce common challenges. In the case of the Institute, our members know that dissatisfaction or failure with an industry competitor could result in a frustrated customer deciding to build their facilities as well as tarnish the reputation of the entire industry.

The North American Institute continues to be one of the most highly ranked association programs among industry members, and there is potential for this model to be successful in additional locations around the world. As mentioned, WFLO Institute programs are offered in Australia and Latin America (Mexico). The Latin American experience is further detailed below as it provides valuable lessons for future expansion.

9.3.2   LATIN AMERICAN INSTITUTE

Efforts in Latin America began with a series of short courses and programming focused on cold chain executives. As a result of these programs, it was determined that sufficient demand existed to implement a full Institute program for the region. In the beginning, most of the students were company executives, who participated to gain a better understanding of the programming. This resulted in increased support for the Institute across the Latin American membership and helped grow the Institute. Executives went back to their companies and were better able to identify team members who would benefit from the program.

In 2014, GCCA opened its first Institute for Latin America. For the first 3 years, it was held in Panama, with the idea that this location would be central and easy to get to for participants from Mexico, Central America and South America. However, despite high interest, low attendance numbers caused GCCA to reconsider its location.

In 2017, the location was moved to Mexico City as Mexico is the location where GCCA has the highest number of members in Latin America. Immediately after this move, registration increased by 57%. As of 2020, 223 students have attended; 30 have completed the third year and graduated, earning an official certificate of completion from the WFLO Institute. Students have come from ten countries including Mexico, Guatemala, El Salvador, Costa Rica, Colombia, Chile, Uruguay, the Dominican Republic, Venezuela, and Panama.

Instructors for the Latin American Institute come from a mix of countries. Some are US-based, but others are from GCCA members in Chile, Colombia, the Dominican Republic, Mexico, Panama, and Peru (Figure 9.3).

Latin American Institute offers a different experience to that of the United States. While the need exists to train workers on standards, it has taken some time to build participation at the level of the worker that most needs it. Original attendees were the owners and executives from member companies, not the mid-level workers who typically attend the US-based Institute. As the Latin American Institute has become more established, the participation of mid-level workers is increasing.

Also, the approach has evolved in a manner that is slightly different from the United States. While the US Institute is heavily geared toward education and training, in Latin America, GCCA's approach has been focused on the formation of a cold chain talent program. This program has provided young professionals with opportunities to specialize and improve their technical knowledge through a cold chain immersion experience with regional exposure. This modified approach has enabled Latin

FIGURE 9.3   Participants from the Dominican Republic at WFLO's Latin American Institute in 2019.

America to attract a broader subset of food industry workers, not only the warehousing and logistics industry. Students attend the Latin America Institute to gain knowledge and connections, and it has the potential to become the cold chain program for career exploration, training, and jobs.

Retention has proven more difficult, as evidenced by the number of students who will attend 1 or 2 years, but who do not complete all 3 years. This is partly due to higher turnover rates, with students changing companies before completion of the program. However, it is also important to stress to companies and students that the Institute is a 3-year program that allows students to apply what they have learned to their job in between annual sessions. This helps them to stay engaged with the knowledge that it is an ongoing development. The students also develop relationships over the 3 years that enable them to call on their peers and instructors when facing a difficult problem in their facility.

Finally, in the United States, the Institute is always housed within a University setting, but in Latin America, the location was better received when it was offered in a hotel. Part of this related to the differing attendee levels. Location can impact price and should be considered very carefully as different levels of participants may have different expectations about the experience. Senior executives may not be as comfortable in a student or academic-focused setting.

The different experiences between WFLO Institute in the United States and Latin America have provided very interesting lessons that have facilitated a broader understanding within GCCA and which helps to guide our approach to exploring the establishment of Institute programs in other regions moving forward.

9.4   CAPACITY BUILDING THROUGH INTERNATIONAL DEVELOPMENT PROJECTS

International and national donors can provide much-needed funding to support capacity building through training and education. Much of the work that GCCA executes through WFLO projects involve training. Some of this is via one-to-one consultations or informal training at a facility.

Where GCCA does conduct formal training, the information is usually extracted from the WFLO Institute coursework, or in some cases where it involves postharvest education, through our affiliation with Dr. Lisa Kitinoja of The Postharvest Education Foundation.

9.4.1 FORMAL TRAINING

In response to requests for formal training, WFLO has created our Short Course program. In-person, this program takes 3–4 days, depending on on-site visits, and covers ten foundational modules taken from Institute and adapted to the local context. GCCA completed a Short Course program in India and South Africa with assistance from the United States Department of Agriculture and is working on providing them in Central Asia for 2020–2021, although the design for Central Asia is a remote-based learning program due to safety concerns and travel limitations from the COVID-19 virus. The Short Course programs were also the foundation for introducing the WFLO Institute into Latin America.

Participants take a pre-and post-test to demonstrate learning and receive certificates of completion. WFLO developed some additional courses that are not or were not offered at Institute, such as introduction to transportation and postharvest handling. A transportation track was added to the Institute in 2018. There is great potential with remote-based learning to offer additional courses to all GCCA members as electives or one-off training opportunities (Figure 9.4).

9.4.2 STUDY TOURS

A second option for semiformal training is a study tour. These can be to North America or Europe, depending on what the learning objectives are, but there is great value to visiting a country within the region. For example, in 2018, GCCA organized a study tour in collaboration with our affiliate partner, the Cold Chain Association of the Philippines for a group of stakeholders from Indonesia and some individuals from the more remote islands of the Philippines to visit temperature control logistics service providers in Manila and Cebu. In addition to allowing participants to see the application of new techniques and innovations, the study tour included multiple networking opportunities and built up a small resource group that is still active today. The study tours have also been popular in the Dominican Republic where they were organized to Panama, Mexico, and the United States to cover postharvest, processing, cold storage and transportation. From the association perspective,

FIGURE 9.4   WFLO Chairman leads a cold chain training workshop in India in 2019.

FIGURE 9.5   A cold chain study tour for Indonesians took place in the Philippines in 2018.

these are great opportunities to involve GCCA members and to demonstrate the value of working together despite natural competition (Figure 9.5).

9.4.3   CONSULTATIONS

While formal training and study tours offer a specific learning experience, particular value can be gained from business to business consultations. Ideally, these would occur at a facility and would involve enabling specialized responses to questions. These may be on design/build or construction, operational practices, or fleet management. The benefit to such meetings is that participants may feel more comfortable asking targeted questions about the running of their business that they do not want to ask in front of their peers. It also enables targeted information that is directly applicable to those participating.

The difficulty with consultations, of course, lies in inaccessibility to a wide number of workers. In GCCA experience, these work well to demonstrate to the owners and CEOs some of the issues that could be ameliorated with broad training of the workforce and may encourage buy-in for more formalized training (Figure 9.6).

In almost all projects that the GCCA has supported, the goal is to enable cold chain training and education that is self-sustaining. GCCA applauds this, but in our experience, it is a long-term goal that requires years of careful investment. The challenges that we have encountered while endeavoring to build long-term sustainable education problems are further discussed in the Challenges section.

9.4.4   REMOTE AND ON-LINE LEARNING

The COVID-19 pandemic necessitated a shift to remote capacity building. While the GCCA had developed online learning and provided educational opportunities via webinars, these were supplemental

FIGURE 9.6   GCCA member consulting with solar-powered cold storage company in Nigeria in 2017.

to the in-person Institute. GCCA's approach to remote and online learning is still being refined, but this will offer some flexibility in reaching an audience that otherwise might be inaccessible.

9.5   CHALLENGES FOR SELF-SUSTAINING CAPACITY BUILDING PROGRAMS

Ideally, GCCA would be able to launch a self-sustaining WFLO Institute program in specific countries, such as India, or following the regional model of Latin America. This is certainly a goal to be worked toward, but several challenges prevent it from taking place in a short time frame. While the conditions in no two countries are same, we also faced a variety of challenges while offering sustainable training to build the capacity of cold chain stakeholders.

First, there is often little interest or commitment to continued training. In many countries, owners and operators are constantly working to manage products or source customers. Training is not always considered a high priority. This is especially the case when only a few employees are hired, and day laborers are responsible for many of the primary functions of the business. With limited resources, why invest in workers? Corresponding to this is the notion that training workers provide them with skills that they then leverage into their next job opportunity. This is an ongoing concern for employers in all countries. The case can certainly be made that while training an employee who leaves is expensive, it can be more experience to suffer devastating business impacts from untrained employees.

On the donor and project side, holding a one-off training can be a beneficial activity, but over time, many donors lose interest as the long-term impact is more difficult to monitor and follow. The sheer number of people trained appears to be output in and of itself and one that is none too exciting.

Rather than focus on training, many donors and development actors have taken up the call for innovations and new technologies. There has been a myriad of challenges designed to support innovations to reduce postharvest losses, many of these involving some aspect of temperature control.

We have also witnessed projects that have introduced technology into an area, such as on-farm solar-powered cold storage. While innovations and technology are critical and offer the potential for developing economies to potentially leapfrog over some of the practices that were followed in countries like the United States, the introduction of these technologies must be paired with a fundamental understanding of best practices or there is a great risk that the effort will fail. This type of capacity building is more than training on how to care for and maintain the technology. When it comes to cold chain, it is important to know how to maintain and repair refrigeration units, but it is equally important to know the best practices in commodity storage and operations. The information helps maximize the use of technology, thereby preventing the need for future costly maintenance.

In cases where training on best practices was not included, we have seen the blame shift to technology. In one case, a solar-powered storage unit was piloted in East Africa. It was used to store cabbage that was not pre-cooled and was packed into the unit, allowing for little to no airflow. The cabbage wilted and technology was blamed. This is dangerous for the cold chain as it provides farmers, transporters, and other stakeholders whose actions along the chain are so critical, with the idea that the cold chain does not work. No technology and innovation will act as a panacea to solve food loss without appropriate training and capacity building.

Beyond the lack of interest in training, there is often little willingness to pay for it by those attending, reducing the potential for sustainability. This is one of the tensions that a program like WFLO Institute will face when exploring its feasibility in emerging economies. Where development projects often offer training, it is not only for free but, in some cases, participants may earn money through per diem or transport to attend. Besides, the value and expense of that training are not always fully explained or understood. Participants may receive the training for free, but the training in itself is not free. There are often substantial costs for trainers, travelers, and facilities. In the case of the WFLO Institute, a registrant fee covers these costs. On projects, they are typically borne by the donor. In countries with an appetite for training, where owners have decided to send their employees to attend or attend themselves when asked if they would pay a fee to attend, the response is typically negative, or it is very small.

A self-sustained approach to training will face competition from donor-funded projects. This may be unavoidable. Each project has its objectives. While those seeking to support private sector development but focus on how to instil a sustained approach to business where profits are made without the assistance of development, other projects are targeted to working with smallholder farmers to improve nutrition. The idea that a farmer or business should pay for training may not factor in the training approach in the same way.

Finally, when considering how to build capacity for the cold chain, it is not as simple as training one group of people. For the cold chain to work, everyone working within it must understand the best practices and why these are required to prevent breakdowns. The vast number of individuals along the cold chain who need capacity building can be overwhelming and working up and downstream may be beyond the scope of one project. However, to establish a fully integrated cold chain, it is not enough to focus on one group. Learning is ongoing and adaptive, and the process never truly ends. With such a daunting task, business as usual is sometimes a more attractive option.

9.6 LESSONS LEARNED AND RECOMMENDATIONS

The various challenges with cold chain capacity building encountered by the GCCA over the years have generated several lessons learned that can be translated into recommendations.

First, when working with logistics companies, it is important to target the owners and CEOs. This is important for a few reasons. Although the training itself may consist of more basic information that is most relevant to warehouse workers, the owners may not yet have received all of the fundamental information. In addition to ensuring they have the necessary education, this will also enable them to decide whether the training is a worthwhile investment for their workers. Even if

FIGURE 9.7   In 2019, GCCA targeted CEOs through a cold chain investment seminar in Kazakhstan.

the training is provided by a donor, the time it takes for staff to be away for training is a cost to the business. It is much easier for the owners to make this decision if they fully understand the value of the information that is taught (Figure 9.7).

The structure of the training that is devoted to CEOs and owners can include a focus on how investments in workers – not only in equipment or construction – can enhance the profitability of a business. GCCA has observed in some countries that companies may make significant investments in new technology but neglect training and education. In this case, companies do not realize the maximum benefit of the technology, and the lack of trained operators has resulted in a loss of investment in the new equipment. The buy-in of the top-tier of the cold chain to commit to investing in their workforce is critical.

Second, when working with farmers, effective strategies include bringing in buyers and providing technology-oriented demonstrations. Retail, especially organized retail, can explain the critical need for high-quality and safe foods, and their willingness to pay. For those farmers who are growing or who wish to grow, for larger retailers or more formalized buyers, bringing these two groups together can be effective for both sides to determine how best to achieve the quality and quantities demanded. At the same time, demonstrating the quality enhancements and reduced food loss that can be achieved by incorporating simple low-cost/no-cost investments are important.

Third, to avoid perpetuating the idea that there is no value to training, that when offered without a registration fee, it is "free," include the value of the training on invitations and in other locations to ensure that the value is tied to the content. This can be done by a simple statement that "this training is valued at $100 but due to the generous support of our donor/sponsor, the registration fee is waived." If projects can charge for training, this could be experimented with although this may get complicated. Some projects have paid for participants to attend the WFLO Institute, but the participants must cover the cost of transport for the training. This is another way to demonstrate that training and education are not always free.

Fourth, the value of training can be conveyed easily where the training results in a certification. Certification generally refers to an earned credential that demonstrates the holder's specialized knowledge, skills, and experience. There is a necessary process to establish a certification program with a specifically developed curriculum, in which deviation is not allowed. WFLO Institute is a certificate training. Upon completion of the third year, participants receive a certificate of completion. WFLO is in the process of transitioning the Institute from a certificate to certification, but this process will take some years to complete.

Fifth, GCCA advocates for a mix of international and local trainers. There is an appetite for knowledge of global best practices, trends, and innovations and technologies. International speakers can provide this and can provide some authenticity or credibility to areas where training is offered frequently. However, without local experts, it is easy for participants to believe that the principles taught do not apply to their country. While a European may advocate for refrigerated transportation, the reality may be far from best practices, and it may seem so difficult that it just could not apply. When the information is coming from a neighboring country, the actions to achieve this level of the cold chain may seem more realistic.

In areas where the cold chain is just beginning, it may be difficult to find local experts. In these cases, suppliers – those who manufacture the equipment needed to build and operate a successful cold chain – can be helpful, although they must understand the difference between providing foundational education and making a sales call. If identified local experts are from these supplier companies, it is important to work with them on the presentation to ensure it does not devolve into a sales pitch. This can be done as there is value for the suppliers to meet potential customers, establish relationships, and understand the local context. Many of GCCA's supplier member companies are valuable education providers as they understand the nuances and needs for technologies and can convey this.

Sixth, training throughout the cold chain must occur at each link along the chain. It is understandable for fatigue to set in when there are so many actors and topics that need to be relayed. It is a complex undertaking that should be coordinated among a variety of donors over a long period.

Finally, there are no one-size-fits-all whether working in a specific country or a region. While the basic premise of educational service delivery is critical everywhere, the model through which it is provided may differ or need to be tweaked in each location. While some new models are under consideration by the GCCA, such as sponsorships and scholarships by cold chain suppliers eager to expand into a new market, the one lesson that stands above all others in GCCA experience is that each new day brings the opportunity to learn, adapt, and apply something new to our approach.

9.7 CONCLUSIONS

Increased consumer demand, access to finance, and discrepancies in government regulations are all auxiliary issues linked to driving cold chain development. The GCCA and the WFLO seek to increase cold chain access and efficiency in countries all over the world. The Association is actively participating in capacity building through such training courses as the Latin America Institute. Additionally, international development projects including formal training, study tours, consultations, and remote and online learning options promote cold chain advancement. Building the capacity of local actors to understand and successfully operate cold chain systems is critical to food safety and the reduction of food loss. The lessons learned by the GCCA over more than 50 years will hopefully provide some guidance to those seeking to advance knowledge in these areas.

REFERENCES

Brecht, P. E., Brecht, J. K., and Saenz, J. E. (2019). Temperature-controlled transport for air, land, and sea. In Yahia, E. M. (Ed.) *Postharvest Technology of Perishable Horticulture Technologies*. Woodhead Publishing, Cambridge, UK, pp. 591–636.

Kitinoja, L., and Thompson, J. F. (2010). Pre-cooling systems for small-scale producers. *Stewart Postharvest Review*, 2(2), 1–14.

Rezaei, M., and Liu, B. (2017). Food loss and waste in the food supply chain. *FAO Nut Fruit, 71*, 26–27.

10 Historical Perspectives on the Cold-chain in India

Pawanexh Kohli

National Centre for Cold-chain Development (NCCD)

CONTENTS

10.1   INTRODUCTION

The use of natural cold to safeguard food is an age-old practice across the world. In cold regions, fishing and hunting folk would bury their stockpile in the snow, akin to storing unused food in your home refrigerator for later consumption. However, such storing of food does not make a cold-chain, which is about enabling a supply chain to support commerce.

Ice itself was probably the first commodity to be trafficked in the cold-chain. In India, ice was shipped south from the Himalayas and sold, for hundreds of years before the advent of refrigeration. Ice cut from glaciers and frozen ponds were also traded in other parts of the world. The natural ice trade is said to have become global when Frederic Tudor of New England, USA, exported ice cut from frozen ponds in Maine to the Caribbean in the early 1800s. There were many challenges and costs associated with engineering ships and maintaining ice holding depots, but this cross-geographical trade was far more profitable than selling ice locally.

The lure of higher profits led to the expansion of markets, and the ice trade from the United States to India was initiated in 1833. The first shipment of ice (about 180 tonnes) left Charlestown, the USA, on 12 May 1833 and, after 16,000 miles at sea, arrived in September at Calcutta, where the British rulers had permitted its duty-free import. Despite one-third of the ice melting in transit, the trade was still profitable. With the increasing demand and markets for ice, the supply to the port cities of Bombay and Madras commenced. Monopolies were created to limit new competitors, and the earlier indigenous ice traders were slowly pushed out. Even 200 years ago, the cold-chain functioned to open new markets, which were eagerly captured.

Tudor's most profitable market was India, but the trade was disrupted by the Indian Mutiny of 1857 and subsequently the Civil War in America. However, with a market having been developed, steam-powered artificial ice-making plants were able to replace the trade in natural ice. In 1874, the International Ice Company was founded in Madras, followed by the Bengal Ice Company in Calcutta, in 1878. By 1925, British India had 66 ice-making plants across its territory. Locally produced and readily available artificial ice eventually contributed to the demise of the natural ice trade and its associated cold-chain.

However, the ice trade allowed other experiments, by hitching small loads of dairy and apples to the supply of ice, which gave birth to the cold-chain of fresh food. Initially, over short distances, this extended to transport by barges, rail wagons and small barrel loads on ice-carrying ships. With more innovation, bulk volumes of meat were shipped across continents in the ice-cooled holds of ships. This market expansion ensured that the cold-chain never looked back since.

Most of these initial developments were confined to meats, dairy and certain temperate fruits, which are amenable to storage with ice, at temperatures close to 0°C. However, in the case of tropical fruits and vegetables, the cold-chain made sense only some decades ago, as knowledge on their unique postharvest precooling and handling practices was refined.

India's population of 1.35 billion is culturally vegetarian and even if meat is consumed, it occupies the place of a side dish. Unlike other parts of the world, most Indians consider themselves non-vegetarian even when they consume small amounts of meat occasionally. Despite this peculiarity, it is estimated that nearly 35% of Indians are pure vegetarian, which amounts to nearly 470 million people and more than all the vegetarians in the world combined. Further, the larger share of the plate of non-vegetarians is still occupied by vegetables. Consequently, the country has the largest consumer base of fruits and vegetables, globally. Associated with this predominantly vegetarian diet is the world's largest production and consumption of milk, as it provides nutrients that would otherwise come from animal foods. Thereby, India's dietary preferences determined the cold-chain systems it requires for its food supply.

The basic categories of cold-chains are classified by the temperature range and include cryogenic or ultra-cooled, frozen, chilled and mild-chilled. The categories differentiate the services and utilities to handle different products such as liquefied gases, biological samples, meats including fish, milk and dairy products, vaccines, partly or pre-cooked items, fresh produce and flowers (Figure 10.1). The cold-chain categories overlap over the different product segments.

In any country, the cold-chain development is aligned to its food habits and changes in demand demographics, which can include those from foreign geographies. India somehow strayed from a holistic approach in its cold-chain for fresh produce but took major strides in most other user segments. A historical review of the strategies that shaped such requirements will help to understand the trajectory of India's cold-chain development.

10.2   BRITISH INDIA'S FOCUS ON PERISHABLES

Until 1947, when the Union of India was born, the region had a British led government. On 23 April 1926, about 20 years prior to India's independence, George V, King of the United Kingdom and Emperor of India, commissioned his cousin Victor Alexander John, Marquess of Linlithgow and nine others to examine and report on the status of British India's agricultural and rural economy and to make recommendations to improve agriculture and the welfare and prosperity of the rural population.

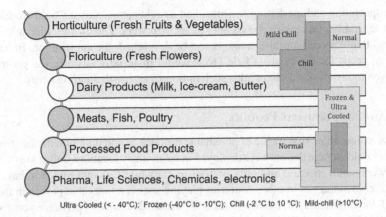

Ultra Cooled (< - 40°C); Frozen (-40°C to -10°C); Chill (-2 °C to 10 °C); Mild-chill (>10°C)

FIGURE 10.1   User segments of temperature-controlled supply chains. (Author.)

The so-called Royal Commission on Agriculture in India submitted its final report on 14 April 1928. Nearly a century ago, this Commission made observations that remain pertinent even today with regard to the cold-chain. It emphasised the importance of cooled transport for moving perishable produce to markets, the packaging aspects to safely move the produce and cooling systems on steamer ships and at railway depots. The primary aim indicated, was to connect fresh produce from farms with markets and to expand markets.

10.2.1   COMMUNICATION OF PERISHABLES

In their report, the Commissioners stated that the presence of 'good markets are of little help to the cultivator', and their presence is meaningful only when the producers can effectively connect and supply to these markets. The Commissioners also said that time assumes special importance where price disparities tend to induce supply variations and stressed that speedy transport of produce to markets was essential and integral to the marketing of perishable produce. Their Report stated that 'Good communications, in combination with efficient marketing arrangements, enable produce to be moved cheaply and quickly to places where the demand for it is active...' and 'difficulty of communications may leave the cultivator entirely at the mercy of the local dealer who alone has at his command enough pack or cart bullocks to undertake the transport of produce to the nearest market.'

The Commission also explained that transportation meant that producers could begin to supply alternative markets to their advantage, and the provision of good connections could make the cultivation of several crops profitable. The Report mentioning that 'ice-cooled vans for the carriage of the more delicate kinds of fruits have already been brought into use'. They recommended that robust steps be taken for the 'extensive employment of refrigerator or cold storage vans' and that this form of traffic would depend on establishing 'cold storage depots at suitable centres'. The phrase 'cold storage' was used as an omnibus term to qualify the technology and means at the facilities; and the phrase 'communication' refers to the physical conveyance of goods.

10.2.2   FRESH PRODUCE AND PROSPERITY

The Commission opined that until the extraction of produce from the area of production was secured, it had no commercial basis. The Commission wrote 'transport is an obstacle to the extension of the growing of vegetables for marketing in a fresh state.' At that time the main production areas of fruits and vegetables were 'in the immediate vicinity of cities and large towns' and even though the conditions necessary for producing certain fruits were favourable in other areas, production on an economic scale was precluded for want of suitable transport facilities. It also noted that adequate transportation and marketing facilities had a higher impact as it would stimulate demand for fruits and vegetables, increasing the number of such growers. The Commission added 'All experience suggests that where favourable markets can be developed, the substitution, in part or whole, of horticultural crops for the existing field crops would materially advance the prosperity of the cultivator.' Even today, the cold-chain adds such options for growth and prosperity.

10.2.3   TRAFFIC OF PERISHABLE PRODUCE

The Commission remarked that traffic of perishable produce would undoubtedly be stimulated provided that more cold storage depots were deployed in up-country locations and at ports. At that time, such depots were small in scale and only present in erstwhile Calcutta and Bombay. The Report conveyed that 'cold storage is in other countries playing such a remarkable part in in the marketing of goods, both for export and internal consumption', to the advantage of private enterprise as well as farmers and that they 'do not doubt that sooner or later there will be similar development in India'.

The Commission felt that successful export of fresh fruit would require 'cold storage on steamers'. However, it said the potential demand of the home market was far greater than any development of export trade in the near future and that it would be wise if the main efforts of government officers were towards opening up the home market.

10.2.4   PACKAGING OF FRUITS AND VEGETABLES

Considering the climate and long distances, the produce would have to be transported the need for standard methods of packing, and 'suitable containers for fruit and vegetables' was emphasised. Standardised containers (packaging), it said, would add to business efficiency and reduce marketing costs. This obviously relates to operational efficiencies and food loss as is commonly understood today.

10.2.5   SPECIALISED KNOWLEDGE

Noting various developments worldwide, this Commission expressed that 'We trust that those concerned will keep abreast with the research on the subject of cold storage which is being carried out in other countries, and will, when the time comes, prosecute in India any investigations required to adapt the modern practice to local conditions.' Here again, the phrase cold storage does not mean cold store, but refers to the technology and knowledge of what today is referred to as cold-chain.

The Royal Commission also spoke of the need to understand the state of logistics between producing and consuming centres and demand assessment. Most importantly, it expressed a strong opinion that such research in itself will not be sufficient unless the whole question of marketing is dealt with by expert officers. The importance of having experts with a combination of sound knowledge and practical experience is regularly mentioned in the context of perishables marketing in this 1928 Report. It made obvious then, as it is now, that specialized knowledge in postharvest management of agricultural produce, especially for perishables is a prerequisite for success.

10.2.6   SUBSEQUENT DEVELOPMENT

Shortly after the Royal Commission adjourned, seven new cold rooms were built at a research centre in Poona in 1932, primarily to research the optimal storage and handling conditions for fruits and vegetables produced in the country.

Over the next 15 years, the cold-chain traffic across British India was limited to small loads of sensitive fruits in ice-cooled vans, chiefly on the North-Western Railways. A few ice-making plants added cold stores to their facilities, primarily to stock more hardy commodities like potatoes and dry fruits for delayed sales.

It is pertinent to note again that the phrase 'cold storage', as used in this Report, was a qualifier to denote the technology and special practices. 'Cold storage' did not mean a cold store, nor used in its plural even once to refer to multiple storing facilities. Instead it described the structural units when explaining the type of depots, vans, wagons or ships, and implied 'cold-chain' (the terminology was not yet in use).

The Report of the Royal Commission on Agriculture noted the importance of specialised logistics practices for perishable produce and was the first such wide recognition at the highest level by the government in India. It was made clear in their recommendations that farm produce with short postharvest usable life must be transported safely and timely to markets, and 'cold storage' technology must be deployed to aid such farm-to-market connectivity.

10.3 INDEPENDENT INDIA'S APPROACH TO COLD-CHAIN

On 15 August 1947, India became an independent country within the British Commonwealth and underwent a turbulent period. India's food production had suffered somewhat, from the commercial thrust to supply non-food crops like cotton for British mills, and this combined with other disruptions, brought about a phase of acute food insecurity. This was eventually eased in certain parts of the country with the advent of intensive farming, known as the 'Green Revolution'. This period witnessed a heavy focus on irrigation, hybrid seeds and the use of chemicals to increase the production of food grains.

Among vegetables, the potato was viewed favourably, being amenable for storage in cold warehouses for longer durations and being less demanding in its handling. There was much done to increase milk production also, and fortuitously, equal attention was given to building capacity across its supply chain. Therefore, cold storage systems continued to focus on holding potatoes, with some partial uses in the fisheries sector and as a platform for marketing milk.

In 1950, India declared itself a Republic and initiated its integrated economic programs by way of a series of centralised 5-Year Plans. The planning process entailed the regular stock-taking of various developments, which provides records on how the cold-chain progressed in the country.

10.3.1 EARLY DEVELOPMENT

In 1955, towards the end of the First Plan period (1951–1956), the country is reported to have created 83 cold storage depots or warehouses capable of storing 43,000 tonnes of perishable goods. While there is no detail on the number of chambers or temperature zones, they were mostly used for storing potatoes, for the short-term holding of some fruits and dairy products, and to make ice. Eight of them, totalling 300 tonnes in size, were in service of fish and one for frozen fish. A few remain in operation today.

In the Second Plan period (1956–1961), the marine fisheries sector received greater attention, and by the end of this period, the transportation of fresh fish in refrigerated wagons to places like Delhi was reported. This was fresh chilled fish (with ice), and hence, coequal impact came from the speed of rail transport. However, inadequate use of refrigerated transport, especially of the rail wagons on return trips, was a handicap. The frozen fish trade (below −18°C) was not prevalent.

During the Third Plan period (1961–1966), the emphasis was on refrigerated road transport of fish. In 1965, India also set up its National Dairy Development Board that became the driving force of the milk revolution or 'Operation Flood' as it was called. This led to the organisation in the milk supply chain and a continually increasing demand for cold-chain interventions in the dairy sector. The government also began to control the creation and operation of cold stores, regulate the technology used and safeguard the interests of farmers and depositors. In the fresh produce segment, the easy ability to store potatoes attracted private entrepreneurs and many cold stores were created. In 1964, the government began to subject the cold storage of food commodities to specific regulatory controls to prevent monopolistic practices. Even so, the number of cold stores doubled between 1965 and 1970 with an average size of about 1,500 tonnes. By 1970, India had created about 1.6 million tonnes in cold storage space.

10.3.2 NATIONAL COMMISSION ON AGRICULTURE, 1970

In 1970, the country had greater food security and the Indian government appointed the National Commission on Agriculture to review and recommend future strategies for the agriculture sector. The report was completed in 1976 and, like the Royal Commission of five decades earlier, also discussed aspects related to perishable produce.

This Commission noted that the production of potatoes was not attempted by small farmers in villages far from the cities and towns where demand was concentrated. Hence, it stated that 'only when cold storage and marketing facilities become widespread that the area under the crop could be extended to interior villages'. It also recommended studying the impact of cold storage on the prices of fruits, as it 'may give some lead to growers on the types of fruits that should be grown'. This demonstrates that the integration between developing cold-chain and production was understood.

The Commission stated that the efforts to increase the production of foods, such as milk, fish, meat and eggs, should be intensified in a phased manner and correlated to the availability of 'adequate infrastructure for cold storage, quick means of transport and marketing facilities'. It stated that without these means for efficient delivery to consumers, the production will have a set-back. This fact is relevant even today, and any increase in production of perishable items, including fruits and vegetables, without commensurate development of cold-chain, only adds to the distress of the producers. The Commission said that adequate measures needed to be taken to provide cold storage facilities in production areas to facilitate the storage and transport of perishables like fruits and vegetables. It also noted that a few cold stores had come into use for apples and hence observed that 'the keeping quality of apples, therefore, assumes great importance'. This fact remains pertinent even today, and more so for apple varieties that are intended for low oxygen storage.

Acknowledging that time is of the essence, the Commission recommended increasing the number of refrigerated and insulated wagons attached to faster-moving express or mail trains, to facilitate the speedy transport of fruits and vegetables. It also emphasised reducing transhipment delays on railways. Regarding handling and marketing of milk the Commission bemoaned that in several instances, chilling plants were set up indiscriminately without considering the viability of its economics. For efficiencies, the Commission recommended that the availability of milk (local production) and the operational economics (cost of distributing to demand) should determine the size of milk chilling plants to be installed. In that period of India's cold-chain history, the milk supply chain was already being organised by establishing village collection centres, to pool the fresh liquid produce. Such first-mile aggregation is most important from the perspective of any perishables trade, as it helps evolve the subsequent capacity to efficiently move the produce up the market chain to consumers or processors.

With regard to the supply of fish within the country, this Commission adopted suggestions that to save investments in actively refrigerated transports, this could be undertaken on 'insulated' transport systems, supported by 'cold and freezing storage plants'. Even today, the bulk of the domestic fresh fish trade is trafficked in non-refrigerated insulated trucks. This Commission also said that refrigerators be provided to ease access to bovine semen, encourage 'cold storage equipment for sale of fresh pork and pork products', and provide 'cold storage facilities for preserving raw hides and skins'. The multi-volumed Report of this Commission provides references to perishable produce and cold-chain that remain useful even today.

10.4 FOCUS ON COLD STORAGE

The two Commissions mentioned above, given the language and science of their times, used terms like refrigeration and cold storage to generically refer to the technology. More germane was their appreciation of the purpose, to rapidly and efficiently transport perishable foods (whether produce, meats, fish or milk), to consumers, to expand markets, and to wisely balance such supply. The two

Reports[1] encouraged the services and the outcomes of what today is known as the cold-chain. The purpose of preserving was mentioned only once, in reference to animal hides. The phrase 'cold storage' was used to describe cold storage steamers, cold storage vans, cold storage depots, cold storage equipment, and cold and freezing storage plants.

Even India's Cold Storage Order defines cold storage to mean 'any chamber or chambers insulated and mechanically cooled by refrigeration machinery for storing foodstuffs'. This means any warehouse structure or depot could have space to serve ordinary storage as well as a few chambers with the means for cold storage. The English language also differentiates between storage and a store, the first as the act or method of storing and the latter as the structural facility for storing. However, these finer distinctions were lost in translation, and the concept of a multi-product, multi-utility, dry and cold logistics hub was forgotten. The repeated use of the phrase 'cold storage' was readily noted, the logic was ignored.

India, in the following decades, focused on cold storage with a vengeance, which came to mean structural entities or storehouses licensed exclusively and in entirety for storing refrigerated goods. These were technically designed to suit a specific commodity with limited financial models and purpose, and were encouraged by government agencies. The misconception spread that having more 'cold storages', the plural also came into use, would fix any and all the maladies that plagued the country's perishable trade.

This mindset pushed India's perishables logistics services sector into a somewhat dark period, setting back the development of proper cold-chains in the country. It was only in the dairy sector, where the independent thinking of Dr. Verghese Kurien, the then chairman of the National Dairy Development Board, prevailed and an efficient and holistic cold supply chain was progressed. The meats and fish sectors also managed to develop, in accordance with real requirements of private sector, to suit the domestic and export markets. Today the total meat and fish production in India is less than 20 million tonnes. But the fresh horticultural production, which amounts to more than 313 million tonnes, does not have an appropriate cold-chain, only bulk storage of potatoes, dried chillies and apples.

10.4.1   COLD STORES PERCEIVED AS A BOON

By 1975, India had 1615 cold storage units for the food chain with a holding capacity of 1.99 million tonnes, an increase of 1,532 in number and 1.95 million tonnes in holding capacity from that in 1955. In the next 20 years, the country added another 1,552 cold stores of larger sizes which contributed another 6.6 million tonnes in size. In the subsequent 20-year period, by 2015, India had created an additional 24.3 million tonnes in cold storage space. As of the end of the financial year 2018–2019, India had accumulated a total holding capacity of 36.8 million tonnes in 8,038 cold stores (Figure 10.2).

Even the development of cold storage capacity was not without hitches. Almost all such capacity was created by the private sector who had to comply with Cold Storage Orders of the government, one in 1964 and the other in 1980. Notified under the Essential Commodities Act of 1955 meant to control of the supply and prices of food items, these orders would also define the technical design, operation and even the tariff charged by cold stores. The construction of a cold store or the expansion of an existing unit was prohibited without prior permission. Regulations notwithstanding, the creation of cold stores proceeded undeterred, a sign of the profitability of the sector for those connected with licensing officers. The regulations also served as an entry barrier restricting competitive dynamics.

This situation continued until 1997 when in a complete reversal, the provisions of compulsory licencing, rent control and other regulations were repealed to encourage demand-driven growth in this sector. In 1999, the government also commenced a capital investment subsidy scheme to fund

---

[1] Report of the Royal Commission on Agriculture in India, 1928 and the Report of the National Commission on Agriculture, 1976.

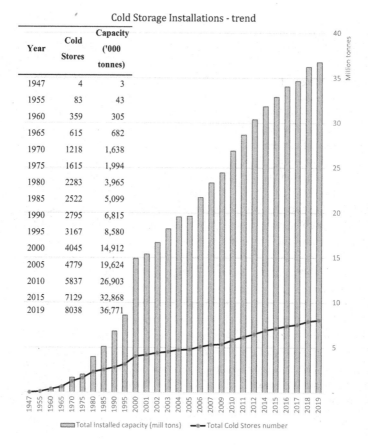

Year	Cold Stores	Capacity ('000 tonnes)
1947	4	3
1955	83	43
1960	359	305
1965	615	682
1970	1218	1,638
1975	1615	1,994
1980	2283	3,965
1985	2522	5,099
1990	2795	6,815
1995	3167	8,580
2000	4045	14,912
2005	4779	19,624
2010	5837	26,903
2015	7129	32,868
2019	8038	36,771

FIGURE 10.2   Cold stores created from 1947 to 2019 in India. (Compiled from Economic Surveys & reports of Govt. of India.)

the creation of more cold stores, including an income tax rebate. In the following decade, about 13 million tonnes of new cold stores were added, growing almost three times faster than in the previous decade. This aggressive creation resulted in India being host to the world's largest capacity in refrigerated warehousing.

Nevertheless, much-needed capacities in precooling and packhouses at the first mile and associated refrigerated transportation were ignored and remained abysmal. Without these two components, the modern concept of the cold-chain, especially for the country's large horticultural production, remained incomplete. Without these two components, the cold warehouse is as good as a large-sized refrigerator and is effective only for a small basket of produce.

Meanwhile, universities, government agencies and equipment providers kept hailing cold 'storages' as the 'boon' for farmers, the postharvest experts remained invisible and horticultural farmers saw no change in their situation. Those producing crops like potatoes found storage, and the others were left in distress. This remains a critical flaw in its cold-chain development and the critical weak link in India's ambitions to take full advantage of its agricultural prowess.

10.4.2   OTHER COLD-CHAIN DEVELOPMENT

Except for the fresh produce of horticulture, the cold-chain progressed substantively in other segments. For fresh milk, India's distribution network is extensive and exemplary. The country has produced nearly 188 million tonnes of milk (182 billion litres) in 2018–2019. Since the greatest demand is for liquid milk (fresh or pasteurised), fresh milk is collected and supplied daily and, at times, twice a day. Typically, nearly half the production (46%) is consumed within the producing areas. The remainder is supplied to non-producing areas (urban) with about 77% is sold as milk and clarified butter (ghee) and the rest delivered as other dairy items including ice cream, butter, whey, yoghurt, cheese, flavoured milk and ultra-high temperature milk. Depending on market circumstances, it is only the surplus after meeting ready demand which is processed into milk powder, whole or skimmed, for long-term storage.

An estimated 45,000 insulated milk tankers are deployed, and 200,000 milk collection centres are in use at the first mile of this supply chain. The food loss in milk is estimated at less than 2%. During the early 1950s, India had to import milk powder to meet urban demand. Today, it generates surpluses and growth is no longer restricted because of the cold-chain but because of demand and productivity concerns. The current cold-chain needs of the dairy supply chain are more related to modernisation and cost optimisation, with new development required only in regions where milk production is scaled up.

In the case of meats and fish, the domestic consumer favours fresh (unprocessed) meats and fish, which defines the cold-chain needs. Meat is harvested on-demand to suit local demand patterns and has near-daily selling cycles. This practice mainly requires the chilled range in the cold-chain. Similarly, consumers prefer whole fresh fish, and hence the bulk of such supply is still undertaken using flake ice. The deep-frozen cold-chain, with blast freezers, and −18°C transport and retail cabinets, exists, but the volumes are insignificant in comparison and its growth will continue to abide with consumers' food preferences and buying habits.

On the other hand, to suit export markets, the cold-chain was aligned with consumers' demand and is well established to serve the frozen category. India has been the world's largest exporter of beef and carabeef (buffalo meat) and also exports smaller quantities of goat, sheep and pork. In the case of fisheries, it also is among the top exporters and has a matching cold-chain. Here again, the development and growth of the cold-chain are market-led.

Poultry is the favoured meat in India and this coupled with the demand for eggs has resulted in a better-organised industry with almost 77% of production coming from poultry farms and the rest from backyard breeders. During 2018–2019, the broiler production was approximately 4.7 million tonnes (carcass) and 109 billion eggs. Yet, about 90% of the poultry is sold at fresh markets; consumers prefer to buy live poultry at small establishments that then dress it in their presence. This buying practice also allowed for the regional spread of poultry farms since live chicken cannot tolerate lengthy transportation. About 10% of poultry is sold from processors who supply frozen broilers and use the frozen cold-chain for distribution.

India also has 30,000 small cold stores for vaccines to support its immunisation program. Through this network, it undertakes the routine vaccination of nearly 60 million children and expectant mothers, conducted via 1.2 million vaccination sessions. This cold-chain plays an important role to help the country safeguard the health of its most vulnerable. Cold-chain is also utilised to supply gel capsules to pharmaceutical factories, for electronics shipments, to maintain animal and human sperm cells and other biologicals.

In general, the above segments have progressed well and their cold-chain is aligned with market requirements. Yet, the cold-chain for the handling of fruits and vegetables, despite a high market demand, did not develop accordingly. It has attracted a larger share of government support and financial rebates which distorted the demand dynamics. Other segments also received government support, but in the case of the fresh produce segment, the scale was overwhelming and almost exclusively for building cold stores, and somehow nobody thought beyond storage.

10.4.3 Cold-chain for Horticulture (Fresh Produce)

The cold-chain for horticultural crops is the most intricate, and unlike other products, fresh produce needs much more care than simply temperature and insulation. It requires specialised attention to sustain a living, breathing produce in the right environment until consumption.

Worldwide by the 1990s, there was a better comprehension of commercial postharvest cold-chain care of fresh tropical fruits and vegetables. India did not stay abreast of such developments with its fixation on refrigerated storehouses. Lacking sagacity of this unique domain left India repeating past mistakes in the first decade of this millennium. There were a few remedial opportunities, but India remained unlucky and disconnected from the advances. As a consequence, the cold-chain to cater to fresh produce remained incomplete.

10.5 MISTAKES PERPETUATED (2000–2012)

At the start of the 21st century, the Indian government realised it must reconsider its agricultural sector and provide it with a forward-thinking strategy. It constituted another Commission to provide comprehensive medium-term strategies to advance the country's food and nutritional security, to provide methods to enhance productivity, profitability and sustainability of major farming systems by harnessing frontier technologies, to bring synergy between technology and public policy, and to recommend comprehensive policy reforms to trigger agricultural growth led economic progress, among others.

This Commission completed its report by 2006, but the frontier technology of cold-chain was given short shrift. The empowering aspects of the cold-chain to enable market access and support market expansion for economic growth, gainful productivity, profitability and sustainability skipped the attention of this Commission. Its report recommends creating markets next door to farms, rather than empowering farms to connect with better markets regardless of location. It perceived agri-logistics (transport and warehousing) as a cost, rather than as a lever for economic growth. These are finer distinctions that merit a deeper understanding, especially when dealing with perishables.

The Commission members, unfamiliar with the cold-chain systems for perishables, put out umbrella statements such as cold storage facilities are required to store perishable produce like a tomato; create cold storages at villages and processing and collection in larger towns and cities; at markets vegetables (cold) storage facility for unsold vegetables is lacking; controlled atmosphere ultra-oxygen [sic] technology may be used instead of cold storage; the cold storage capacity is grossly inadequate as well refrigerated trucks; cold storages be constructed to ensure reasonable prices to farmers; etc. The statements were not elucidated but used as oft-quoted clichés by many others. Unfortunately, not once did the reports by this Commission mention packhouses or precooling to enable aggregation and for organising the market access of perishable crops.

The reports submitted by this Commission indicate their strategy was from the limited perspective of input costs, reducing power tariffs, increasing subsidies, provisioning of funds, but without considering enterprise aspects of market-led gainful productivity and demand expansion for growth. This Commission merely bolstered the prevailing trend to treat cold-chain systems as subsidy supported infrastructure development. Leaving other more important and associated infrastructure components unnoticed, it only recommended that cold stores be spread evenly across the landscape. Without a holistic supply chain orientation, it offered no change-direction in the development and other misconceptions were allowed to remain in this regard.

The strategic aspects of access to markets and increasing demand, though mentioned in the Reports from previous Commissions, were not expanded in the context of new enabling technologies. In 1928, the urban population of India was a little over 11% and now it was almost 30%. This changed circumstance requires enlarging the farming systems' ability to connect with larger urban markets and to build contextual tools, infrastructure and services.

10.5.1 A PARODY OF ERRORS

In March 2007, some cold-chain stakeholders voiced their concern that government agencies were slow in executing policy and did not have a holistic approach when implementing cold-chain support schemes, which discouraged greater private sector participation in cold-chain development. To attract a better response from the private sector and accelerate cold-chain development, they suggested creating a nodal agency so that the government's policy would be better implemented. Taking the cue, the government promptly set up a Task Force in May 2007. The primary task assigned was to 'develop the terms of reference for nodal agencies to be constituted for policy execution and to ensure participation of crucial stakeholders in cold-chain development'.

However, instead of formulating the guiding terms for the suggested agencies to administer and monitor policy execution, this task force took upon itself to direct the policy itself. In all eagerness, it even chose to extend its opinions on various technical aspects of the cold-chain.

This Task Force on cold-chain development compiled its report based on secondary reviews and inputs from the industry, who mostly were equipment manufacturers and not cold chain practitioners. The bureaucrats and officers involved had no actual cold-chain expertise. As a result, technical catchphrases were picked and stitched together as best possible, but with the lack of input from experienced cold-chain professionals, the output only confounded the purpose.

In its report, completed in 2008, the task force incorrectly mentions that most of India's cold stores were based on the principle of evaporative cooling. It disregarded ammonia refrigeration because it is 'old technology'. It advised that fresh produce be precooled on fields, and thereafter the load be transported in reefer trucks to distant packhouses for subsequent sorting, grading and packaging – literally interpreting that 'field-heat removal' is meant to happen on the field. The Report recommended having ripening chambers at first-mile, presumably so that ripened fruit can be directly trucked to faraway markets. It also suggested packhouse locations at city markets; perceiving that packaging would add value at the last mile – whereas, the primary purpose of packaging is for efficient logistics and to minimise handling losses.

Indiscriminate terms like, 'end-to-end logistics for cold chain', 'cold storage network', 'cold chain infrastructure and logistics', 'cold chain network', intermingle in this report. For this task force, 'logistics' referred to transport, and 'infrastructure' meant cold stores. It suggested establishing a special transport grid to direct freight of perishables on pre-planned corridors. Lacking clarity, it also said reefer containers should be attached to express passenger trains and proposed half a billion dollars for procuring solar-powered reefer containers which would run independently of fuel.

The report of this task force contained many other mixed messages, including that terminal city markets should have cold stores with controlled atmosphere (CA) systems, little realising the implications and confusing CA with modified atmosphere packaging (MAP) technology. Consequently, it resulted in the careless and disproportionate allocation of funds to support the creation of CA storage systems, in all parts of the country. It emphatically recommended subsidy support for mobile precooler trucks, disregarding the operational backdrops of Indian farms.

Further, this task force felt that subsidies for cold stores be increased, stating that 'low subsidy has not encouraged the use of energy-efficient but capital-intensive modern technologies'. It not only recommended higher subsidies and duty exemptions for such energy-efficient equipment but simultaneously also proposed subsidising the electricity to cold stores. It rated infrastructure at exorbitant costs and the revenue stream from infrastructure was primarily assessed from the perspective of space rentals and, to support such a model, asked the government to increase all forms of financial support. At that time, the government was already providing up to 33% of the capital cost, with other favourable financing options. The task force recommended subsidies be enhanced across the board up to 55% of capital cost, along with interest subvention on credit, waiver of import duties, and that cold-chain be given the status of infrastructure so as to avail themselves of added income tax benefits, besides endorsing improper cold-chain practices and protocols.

The task force's report does correctly note that past policy and efforts had focussed on the creation of cold storage, as opposed to the development of cold-chain. It also expressed the need to have integrated cold-chains, yet such integration was given an infrastructure dimension only. The functional integration in supply chain activities and actors was not understood. In this confabulation, the integrated cold-chain became muddled with single ownership of all cold-chain assets.

The task force made many circumstantial statements without explanations, factual relevance or transparent context. The report does contain some interesting snippets copied from contributors which were of interest to new players, but for experienced professionals, it was inconsistent and the essence was lost in the compilation. The task force also disclaimed any responsibility for the usefulness, completeness and accuracy of its report.

The report of this task force ended with a list of expected outcomes, which are better crop management, higher farm-gate prices, stable supply and employment opportunities. These are generic secondary fallouts, while direct outcomes were not targeted, and probably not understood. With this, the bureaucrats who had merely read about the cold-chain, along with some non-practitioners and engineering companies, misshaped the next few years of India's cold-chain development. India futilely waited to see the results in practice, in hope of the accessory outcomes that were listed.

10.5.2 Repetitive Development

These last two reports, though not entirely acted on by the government, kept India's cold-chain development along its linear path, at a time when real change was needed. Unfamiliar with the context and the real gaps in the cold-chain, they had emphasised infrastructure instead of supply chain services, focused on costs instead of efficacy and professed modern technology for archaic business models. Those who felt higher capital support from the exchequer was crucial to success in business were vindicated.

The capital subsidy was increased but that too only continued to feed cold store creation, perpetuating cost inefficiencies. This also resulted in academic callisthenics by economists around facets such as the ratio of cold storage capacity to population and the geospatial density of cold stores. The core rationale to streamline the supply of perishables, across space (geographies) and time (periodicity) to cover more markets and capture optimal value, was mislaid.

For the layperson, a cold store was perceived as a panacea, tantamount to a cold-chain. 'Make more cold storages' became the catchphrase, where the uncountable 'storage', originally used to qualify the means or act of storing, transplanted the countable noun 'stores'. More cold stores, now commonly called cold storages, were installed, ignoring other developments and real requirements.

10.5.3 Cobra Effect

The 'Cobra Effect' is a scenario where a particular strategy or solution is not mindful of the consequences and, in turn, adds to the dilemma. This singular focus on building cold stores as a solution for the predicaments in marketing perishable produce is comparable. The growing presence of cold stores allowed a false sense of confidence, adding to misconceptions. Concurrently, horticultural production, especially fruits and vegetables, increased and exceeded the total output of all food grains in 2012. But the food loss from such produce was also seen to rise. Without the use of pre-cooling, most of the fresh produce that directly entered a cold store would still perish within a few days. This was equivalent to placing warm vegetables in a domestic fridge. Cold store owners asked for more financial relief from the government to continue their businesses and equipment providers joined hands asking for more exemptions.

By the close of 2010, India was home to 6,100 cold stores, totalling nearly 29 million tonnes of space. Yet, in December 2010, a paper by the National Spot Exchange stated that India required a total of 61.1 million tonnes in cold storage. The previous year the country had produced about 204 million tonnes of fruits and vegetables, and it assumed 30% of these commodities production be stored for out of seasonal consumption. This 'study' added to the clamour to double the refrigerated warehousing capacity. Without any thorough assessment by experts, this view was regularly regurgitated in 'knowledge papers' by banks, industry, researchers and even by policymakers, as fact.

However, the only solution advocated in these knowledge papers was the creation of more cold storage so that more of India's perishable goods could be 'preserved'. It was assumed that this would free the farmers of their distress and stop food loss. The fallacy in their assumptions was overshadowed by the overpoweringly attractive idea of investing in capital assets in the name of the farmers. Every year, the government could declare its achievement in terms of expenditure undertaken and numbers of cold stores created.

More cold stores did not transform the perishables trade, with little impact on the returns to the farmers. Without the enabling components and organisation at the first mile to prepare the produce to enter the cold-chain, and for it to serve as a conduit to markets, the tropical produce of Indian farmers was not benefiting from cold stores. However, the imported fruits and vegetables, that had entered the cold-chain at foreign packhouses, used the cold stores in India as platforms to access the Indian consumers. The subsidies did not change the state of Indian farmers but boosted imports which multiplied during that decade.

The perceived solution had become a problem and the Cobra Effect was rampant. The second or higher-order effects were not thought of, and worse, the necessary prerequisites for a cold-chain were not even considered. Distressed at the lack of results, the same one-dimensional thinkers moved on to identify another solution and the new catchphrase became value-added products, which is another cobra story in waiting.

With the cold store made synonymous with cold-chain, the cold-chain development in India was left hamstrung. The logic behind the cold-chain and the solution it was meant to the outcome was lost; a correction was awaited and commenced in 2012.

The next chapter shares the course corrections that India adopted in the period thereafter.

10.6 ANNOTATIONS

Over centuries, cooling systems and logistics abilities have merged and developed into what is now called the cold-chain. The principal benefit from deploying such a cold-chain has always been to provide perishable commodities with a safe conduit to markets, across time and/or distance. This facilitation allows the produce to reach more consumers, and to expand sales, which empowers farmers to intensify their production efforts. This facet was well understood and so deployed in many parts of the world. However, in India, the need to store food to offset frequent occurrences of famines had conditioned its strategies to place greater importance on developing cold stores. The circumstances also ensured that such development was seen from the perspective of infrastructure creation only and not in the associated development of suitable knowledge, practices, and business models.

Around the start of the 21st century, India's cold-chain development took a wrong turn. It became disconnected from cold-chain advancements happening around the world, lagging in understanding the more unique requirements of fresh produce, and lacking such expertise India's cold-chain development went off-track. The future of India's cold-chain development will depend on how well it comprehends its past failures and adjusts its trajectory. It will also require it to adopt a supply chain orientation in future development efforts and take full advantage of the cold-chain to advance its enormous agricultural base. Realising this, India took certain initiatives to bring about improved comprehension and correct its cold-chain development strategies, which are discussed in the subsequent chapter.

BIBLIOGRAPHY

DAC. (2008). Report of the task force on development of cold chain in India. Delhi: Ministry of Agriculture, Government of India.

DAHDF. (2019). *Basic animal husbandry statistics*. Annual Statistics, Delhi: Department of Animal Husbandry Dairying and Fisheries, Government of India.

Dickason, D. G. (1991). The nineteenth-century Indo-American ice trade: An hyperborean epic. *Modern Asian Studies*, 25(1), 53–89. doi:10.1017/s0026749x00015845.

Kohli, P. (2018). Marketing, agri-logistics and agri-value system. National conference – Agriculture 2022 – Doubling farmers' income, Ministry of Agriculture & Farmers Welfare, Government of India, Delhi (19–20 February 2018). doi:10.6084/m9.figshare.12318842.

National Commission on Agriculture. (1976). Report of national commission on agriculture. Delhi: Government of India.

National Commission on Farmers. (2004–2006). Reports of the national commission on farmers. Delhi: Government of India.

Royal Commission on Agriculture in India. (1928). Report of the Royal Commission on Agriculture in India. London: H.M. Stationery Office.

Weightman, G. (2003). *The Frozen-Water Trade: A True Story*. New York: Hyperion.

Wilson, D. (1980). *The Colder the Better*. New York: Atheneum.

11 Progress and Status of Cold-chain in India

Pawanexh Kohli
National Centre for Cold-chain Development (NCCD)

CONTENTS

11.1 INTRODUCTION

The previous chapter gives an oversight of various perspectives and circumstances that shaped India's understanding and its agenda on cold-chain development. In summary, despite wide recognition of the need for a cold-chain, the country's extensive efforts resulted in the creation of cold stores, resulting in the largest capacity worldwide. However, cold stores alone do not make a cold-chain, and in India, their use has been limited to a handful of produce categories.

Yet, India has a large and diverse basket of fresh produce, for which mere cooling during storage was not sufficient. The large variety of tropical fresh produce farmed in India required preconditioning facilities in the form of packhouses with precoolers, but this had not captured the relevant attention of development planners.

By the close of the first decade of this millennium, the fresh fruit and vegetable production was equal to the production of traditional staple foodgrains, and from 2012, it has exceeded foodgrains. Lacking the relevant cold-chain systems to effectively connect with the enormous home market, the growth in production did not translate into the expected economic growth for the farmers. Instead, an inverse relationship between production and income had surfaced with higher production resulting in less income to farmers. The situation was aggravated by increasing food loss, especially in fresh produce. Surprisingly and simultaneously, the Indian market absorbed a spurt in imports of fruits in the previous 5 years and a 10- to 15-fold growth in imports in the recent decade

DOI: 10.1201/9781003056607-14

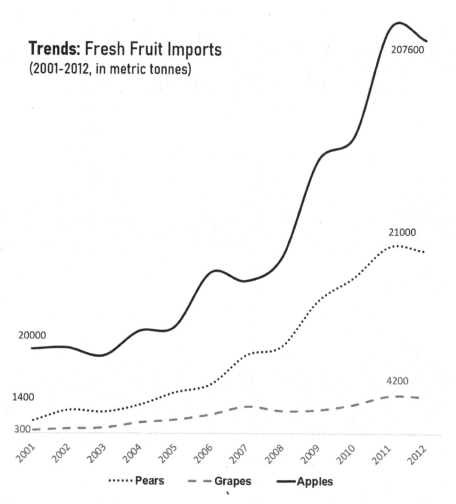

FIGURE 11.1   Import of Fresh Produce into India (2001–2012, in metric tonnes). (Kohli, 2013.)

(Figure 11.1). This situation was reviewed and then discussed throughout 2012 and at the National Horticulture Conference in 2013.

Imported fruit travelled to India in the cold-chain where the country's large base of cold stores served as a cross-dock to access local consumers. However, the same fruits produced in India were not able to access the cold-chain due to a lack of packhouses at the first mile limiting the benefit from their subsequent entry into cold stores. It was obvious that India's large base of cold stores was helping imports but not domestic farmers. India's cold-chain development is incomplete and was not planned in a complementary manner.

Nevertheless, until 2012, the common perception remained that India should create more cold stores. It was assumed that this strategy was correct, and it was translated into increasing the subsidies for building cold stores.

11.2   RESURGENCE IN INDIA'S COLD-CHAIN DEVELOPMENT

The government's task force on cold-chain development of 2007, mentioned in the previous chapter, had recommended that the government create a new organisation called the National Centre for Cold-chain Development (NCCD). It had proposed that NCCD aid in implementing the government's ongoing subsidy scheme for cold-chain and develop a mechanism for the fast track clearance of projects; directly handle interstate projects; invite projects and approve the grant of subsidies, etc.; and, over time, also come to address cross-cutting issues related to the sector's growth, including policy, technology, protocols and more. Essentially, the task force had suggested that this dedicated agency should shoulder the burden of executing the government's scheme.

NCCD was accordingly registered as an entity in January 2011. However, in subsequent discussions with certain domain experts, the Ministry of Agriculture reviewed whether pursuing this tangent would be the most appropriate. It was opined that the government officers are expected to be accountable for managing public funds, but the proposed devolving of the subsidy approvals to an external body would abdicate them of this responsibility. It was suggested that the subsidy implementation should remain with government-run agencies, whereas the policy and technical direction be formulated by domain practitioners. The discussions also noted that the task force of 2007 had approached subsidies as a funding mechanism and not as an incentivising tool. It had primarily focused on increasing the existing subsidies, which did not bring any real change in the development strategy. The Ministry realized that the more urgent need was to review and overhaul the existing strategies and schemes which, as was pointed out, lacked systems thinking and domain relevance.

Despite regular mentions in the past reports emphasising the importance of practitioners and domain experts, cold-chain experts were not among the policymakers. So deprived, some misconceptions about cold-chain had perpetuated and this along with institutional shortcomings, despite copious expenditure by the national exchequer, resulted in critical inadequacies in the cold-chain. The government schemes to support cold-chain development were not strategic in their interventions nor did they encourage modern supply chain-oriented business models. The biggest gap in India's cold-chain development was relevant expertise in policy formulation. Hence, it was felt that it would be more appropriate to task NCCD as an expert body, first to restructure the ongoing cold-chain development support programs, and next to elaborate a long term national policy for cold-chain. Concurrently, to pioneer excellence by undertaking wide-scale knowledge dissemination and capacity building.

On 9 February 2012, the NCCD, after suitable modifications in its structure, was approved by the country's Cabinet. Subsequently, NCCD was installed as an expert advisory body to the government, structured as a public–private partnership (PPP), and given autonomy in all its affairs by requiring it be distanced from any direct government involvement on a day-to-day basis.

At the core, this partnership between government and non-governmental stakeholders was not about bringing together public and private funds, but for pooling together of genuine sectoral knowledge and to serve as sherpa for the task. By the end of 2012, a new roadmap for the NCCD was devised and consequently the body was strengthened in 2014. Kept divested from regular government officers, NCCD was entrusted the liberty to correct government strategy, making it unique in the annals of India's cold-chain. A paradigm shift in how India approached cold-chain development was enabled and a period of correction unleashed.

11.2.1 Breaking the Stockholm Syndrome

Concurrently, in May 2012, a committee of the Planning Commission of India reported there were considerable wastage and spoilage of fresh produce despite indications of unmet demand in the domestic market (Planning Commission of India, 2012). This committee assessed agricultural supply chains and distribution of farm produce to note that "the push to build up storage capacity through cold chains has not been successful in vegetables and is limited for fruits". It observed that the benefits of demand expansion were not passing on to the farmers, nor were the consumers benefiting from the increased production. The Committee stated that instead of a "straight-jacketed approach" to cold-chain development, a matrix approach be adopted to suit the varying requirements of fresh produce. It noted a lack of understanding of the manner and extent of cold-chain requirements covering a wide variety of produce and that "no such study has been done for cold chains as distinct from cold storage". The Committee strongly recommended that the Ministry of Agriculture strengthen NCCD and issue the body the relevant operational guidelines for it to provide an enabling environment for future cold-chain development.

Some other such well-intentioned analyses and reports had attempted to set the proper context for the future development of the cold-chain in India. But, context can be lost in clerical interpretations and the result is counterintuitive outcomes. In cold-chain development, this was readily evident, seen in policy actions that belied domain logic and in the distortion of the basic premises.

In India, almost all the cold-chain assets are owned and managed by thousands of private operators. However, it was mainly the engineering industries and technology providers who interacted with policy makers. The policies therefore focused on provisioning technology and not on facilitating effective services. The bureaucrats could excuse themselves by claiming they had duly taken forward industry demands, and the industry consultants could say they had merely represented their views to the authorities. The produce owners, users, postharvest experts and cold-chain service providers were left stymied.

NCCD began to highlight how engineers and inexperienced government officers, having assumed the mantle of cold-chain experts, had hijacked the agenda of cold-chain development. Held captive, the real cold-chain sector had no say, but could only rely on its kidnappers for sense and succour, and like in the Stockholm syndrome feel grateful for any pittance.

NCCD pointed out that the insistent calls to keep increasing capital subsidies were not signs of healthy development and only perpetuated a vicious cycle, and that development funds were poorly deployed, merely to offset the cost of inefficiencies. The need was for intelligent subsidies, along with technical knowledge support and other strategic changes, to support solutions and not just to maintain a status quo.

11.2.2 CLOSING THE KNOWLEDGE GAP

Cold-chain development for fresh produce, unlike in the case of the milk sector, had lacked a systems approach. With minds fixated in comfortable patterns, a re-education of sorts had to be attempted. Policy reviews require meticulous detailing with easy-to-understand explanations. NCCD took on this onerous job to edify, elucidate, educate about the inaccuracies and omissions in the government schemes. Domain specifics and associated logic were repeatedly discussed at the apex level with senior government officers who were willing to consider other opinions and who had the vision to affect positive change.

NCCD had to describe how the cold-chain market had been misrepresented, misreading the service market while clubbing the revenue of equipment suppliers, and that various researchers, economists and bureaucrats had not understood the supply chain orientation, and hence were satisfied in merely setting targets for the creation of cold storage capacity. However cold stores are only a middle link in the larger concept of the cold supply chain and do not suffice for the cold-chain needs of all possible products and especially for fresh produce. Crucial missing links needed to be bridged and the market ecosystem required a change in structure.

NCCD also explained how inordinately high subsidies linked to excessive normative costing had attracted opportunists and encouraged capacity and cost overruns. The mechanism failed to incentivise good practices nor strategically shape future direction. The government scheme for cold-chain could be revamped to incentivise private sector to invest in critical areas, interlink the investment with other activities in the chain, encourage standardisation in handling of perishables, give direction for inter-modal movement, guide innovation, adopt technologies, and, most importantly, maximise the evacuation of perishables from farm-gate to market-gate. With visible pitfalls and reasoning explained, the Ministry then entrusted NCCD with the required freedom to suggest the full gamut of the most "necessary" changes.

In April 2013, NCCD entered a historic memorandum of understanding with Cemafroid of France, to collaborate on knowledge-sharing programs. This was done under the aegis of the Indo-France Joint Agriculture Working Group. By end of 2014, a regular series of training modules to suit Indian requirements was initiated. The course curriculum provided holistic training on key aspects of the cold-chain to heterogeneous groups from the private sector and state government officers to share knowledge to improve government to industry interactions. More than 15 such courses were held over 3 years. Importantly, both governments of France and India recognised the importance of this subject as an example in inter-governmental cooperation and shared the costs. In this same period (2014–2017), other knowledge-sharing works were commenced in the country. These included a short course on advanced technologies and energy efficiency, more than 200 roving workshops across the country for entrepreneurs, and various pilots to demonstrate the business advantages of the cold-chain.

Knowledge outreach and exchange of ideas were taken up with a wide range of participants, including included farmer groups, investors, logistics service providers, traders, bankers, government stakeholders, researchers and academia. The primacy given to cold stores was corrected and surprisingly, there was ready acceptance for a new approach from the veteran cold store federations, while the most intractable were the policymakers and desktop researchers. Today, these concepts are echoed freely and have become common parlance.

11.2.3 COURSE CORRECTIONS

By mid-2013, NCCD was reviewing ongoing schemes and proposing necessary changes. It observed that the weakest development was in the fresh produce segment, and so the NCCD prioritised its efforts in the chill/mild-chill category. At the same time, other institutional mechanisms were reviewed and all horticulture development agencies were included under an integrated agenda bringing convergence to the previously jointed efforts. This also required standardizing the language so that policies and understanding could be harmonized.

The basic types of cold stores were redefined from standpoint of usage and location in the supply chain. The generic multi-commodity storage was revised to mean segregated temperature zones in multi-chambered designs and volumetrically sized for the purpose. The misapplied class of modified atmosphere cold store was done away with.

Cold stores with controlled atmosphere (CA) systems – where the internal atmosphere is purged actively to bring about rapid and extreme change in air composition – were henceforth to be supported only at producing areas and for selected pome fruits. Before this, such cold stores were subsidized at up to ten times the capital cost of a regular cold store, and expected to store all types of fruits and vegetables, including at city markets located thousands of kilometres from production zones.

To bring a shift in attitudes that were conditioned to see the subsidy as upfront project capital, other rules and mechanisms were enforced. For this, only fully financed projects were considered eligible to apply for subsidy, and the subsidy amount could no longer be included as part of initial capital outlay. The government subsidy was hence not guaranteed in advance, so banks would now evaluate projects as commercially viable for credit, independent of subsidy. The subsidy amounts would only be released after completion of the project. This was designed to offset the interest burden over the full term of the bank loan, and to incentivise the early completion of projects. In 2016, NCCD recommended exempting cold-chain activities from service tax to foster a service orientation and incentivise higher throughputs in the cold-chain and so approved.

The prevailing capital subsidy for cold stores was adjusted for the economy of scale with larger cold stores fetching lower subsidies. Various other admissible cost norms were revised and the subsidy rate was also reduced by 5% across the board. Cost norms, a normative assessment to plan and allocate government resources, were no longer benchmarked to declared equipment invoices. Often, these exaggerated the real market price of equipment. The cost norms for subsidies were now independently checked and set with the strategic purpose to encourage investment in less developed and more critical areas of the cold-chain, and to foster better and sustainable business operations. The government subsidy was no longer intended as an assured source of capital to offset stated expenditure, but as an incentive to drive specific investments and efficient operations.

In rationalising the unjustifiably high subsidies and rates, much wasteful expenditure by the exchequer was also curtailed. It also allowed for the addition of 13 new components to the support matrix including modern packhouses with precoolers, dock shelters with levellers, inter-modal reefer containers, packaging lines and automation systems, retail merchandising systems, alternate energy options and critical infrastructure to aid safe handling of produce, higher capacity utilisation and greater efficiency of the cold-chain. The subsidy applications were simplified, and the process made more open-ended, reducing subjective, selective practices.

In 2014, the government approved the launch of a new program for horticulture under the Ministry of Agriculture, which included holistic support for the cold-chain. Notably, it said that NCCD would guide policy as well as revise technical standards and adherence protocols when improved technologies and efficiencies are introduced. By placing such onus on an expert body, it reduced the scope for inexpert notional changes in policy.

The Government of India took notice of the logic behind developing appropriate infrastructure and modern postharvest management (PHM) practices for fresh produce. In the 5 years until 2012, cold-chain development was allocated only 8.2% of INR 4942 crore (about USD 900 million then), the horticulture development budget. But during the next 5 years, not only did horticulture see a three-fold jump to receive a budgetary outlay of INR 16,840 crore (about USD 3 billion), but the share for PHM and cold-chain development was increased to 25% of this total. To match production growth, cold-chain development had to be ramped up and its development budget was allocated a ten-fold increase.

A gamut of opportunities opened for cold-chain operators to expand and improve their business models and operating practices. In this period, instead of continuing erstwhile controlling of equipment specifications and narrowing of system-wide choices, NCCD authored the country's minimum system standards and guidelines. Components of the cold-chain were elaborated on result-based principles, without imposing limiting specifications or material types. The idea was that the outcome should be the benchmark to target. For example, if the desired outcome is to insulate a space from heat ingress for 4 hours without power, there was no need to regulate which building material or dimensions to use. This reduced the scope for undue harassment and allowed for individual innovation.

Entrepreneurs were advised to look beyond a one-time capital subsidy or stop-gap electricity tariff waivers. Under the revised support program, they could now install the right equipment for the right job, modernise their operation, build to suit their actual needs, and achieve self-sustaining and resource-efficient capacities. Even solar panels with net metering to offset energy costs were supported. Cold store owners could now use government support to acquire reefer trucks, or build and operate packhouses in production zones, and expand their business models. Cold-chain concepts and supply chain linked asset creation were explained so that ripening chambers were built at the last mile and not at farm-gate.

A slow but steady shift was seen, focusing on increasing inventory turns or throughputs, to profit from cold-chain services, rather than from infrastructure creation and rentals. The lure of quick capital was no longer the prime driving force for cold-chain development. Instead of infrastructure developers, genuine entrepreneurs and supply chain managers were incentivised to enter this domain. Some long-time associations of cold storage owners decided to demonstrate this change by renaming themselves as associations of cold-chain owners. The Federation of Cold Storage Associations of India also acknowledged the change by initiating the process to rename itself as the Federation of Cold-chain Associations.

Cold-chain capacities are correctly recorded by volume and not by mass, and changes were initiated to harmonise this data. The country did not have a true understanding of the operational capacities of its cold chain, as the data was limited to the aggregate number and size of facilities. Desegregated data was missing about the number of chambers and temperature zones, age and type of refrigeration technology, kind of material handling equipment, energy consumption and the skillsets available. NCCD initiated such data capture by designing a baseline survey, first of cold stores in the country, with the intention to replicate the process later for packhouses, reefer transportation and ripening units. The country's first baseline survey of cold stores, which included geo-mapping each with extensive operational and infrastructural information, was completed in December 2014. The comprehensive data collated in this survey was used to analyse the current cold-chain in detail at multiple levels and will help when formulating a holistic national policy on the cold-chain.

As asked, only the most 'necessary' and urgent changes were attempted initially by NCCD. It was considered prudent that further corrections in government programs be taken up after revising the performance metrics used to assess cold-chain development. Normally, government agencies demonstrate their performance and achievement by way of listing numbers for activities undertaken or expenditures incurred. NCCD proposed that in future, at least for those in charge of developing cold-chain, performance and achievement would be measured by way of measurable results, such as the volumes handled, food loss mitigated, market expansion and farmers' income share in such outcomes.

The long run objective was to encourage a bigger shift, away from primary capital support for infrastructure towards support that fostered cold-chain as a service, to empower cross geographical market expansion for farmers and for an ecosystem where value captured is equitably shared.

11.2.4 BREAKING THE MYTH ON COLD STORAGE CAPACITY

In 2014, India elected a new government, and wishing to give quick impetus to cold-chain development, it set up another Task Force to review the status of cold-chain projects. By the middle of 2014, India already had 32 million tonnes of space in refrigerated stores, but with merely one-tenth of this capacity in reefer transport, and negligible packhouses with precoolers at the first mile. This Task Force, taking a cue from a 2010 study – which had said that India required more than 60 million tonnes of cold storage capacity – was planning investments to target the creation of another 20 million tonnes in cold stores within 4 years.

At the first meeting of this Task Force, NCCD pointed out that building more cold stores was "barking up the wrong tree" and that the 2010 projections were cursory at best and grossly incorrect. Yet, attracted by involved expenditures it was regurgitated by bankers and researchers who lacked the capacity for original research. The Task Force was also informed that a comprehensive study on the status and gaps in India's cold-chain was under process at NCCD. This task force was positioned under the Ministry of Food Processing Industries and the logic around cold stores and their position in the cold-chain was explained, as had been done in the previous year at the Ministry of Agriculture.

Meanwhile, instead of planning for 20 million tons more of cold stores, NCCD recommended that around 7.5 million tonnes in cold-chain capacity be targeted. Even with inefficient monthly turns, this would generate a handling capacity of 90 million tonnes annually. Soon, this task force came to a consensus that a decision on further infrastructure creation be delayed until NCCD completed its assessment of India's cold-chain assets. Its report listed some of the weaknesses highlighted by NCCD in the earlier report that claimed a large shortfall of cold stores in India.

The "Report of the Task Force on Cold-chain Projects" was discussed at the apex office in the Government of India. The resulting, clear directives stated that all the government agencies involved, "should treat cold-chain management as a part of the second green revolution and that they should address it 'end to end' connecting farm gate to the consumer in a seamless manner." This conveys that the efforts must shift from blind infrastructure creation to *cold-chain management*, to result in *end-to-end seamless connectivity*. Henceforth, development efforts included larger organisational aspects for delivering a safe and effective supply chain.

In early 2015, NCCD completed the country's first-ever assessment of all the main infrastructure assets. The conclusions were bound to ruffle many vested interests. For a few months, the All India Cold-chain Infrastructure Capacity (AICIC) Status and Gap study was reviewed and explained across ministries until it was officially released in September 2015. The findings in the study upended the blind focus on cold storage capacity and brought to fore that the country had a

TABLE 11.1

Cold-Chain Assessment, Status, and Gap in India (2015)

Type of Infrastructure	Requirement	Created	All India Gap	% Shortfall
Integrated Packhouse (no.)	70,080	249	69,831	99.6
Reefer Transport (no.)	61,826	<10,000	52,826	85
Cold Storage (Bulk) (MT)	34,164,411 t	31,823,700 t	3,276,962 t	10
Cold Storage (Hub) (MT)	936,251 t			
Ripening units (no.)	9,131	812	8,319	91

Infrastructure in number of units for predefined unit size or in MT (metric tonnes).
Source: NCCD (2015).

very high share of cold stores versus the capacities it should have in reefer transportation and modern packhouses (Table 11.1).

Closing these gaps would empower India's cold-chain with an annual handling capacity of nearly 180 million tonnes across various product types. The total production of fruits and vegetables in India is about 280 million tonnes (of the 313 million tonnes of horticulture), and meats, including poultry and fish, are another 20 million tonnes. However, all of this annual production does not use a cold-chain at the same time. Meats are normally harvested according to demand, and fruits and vegetables have a certain periodicity. Furthermore, the produce that can be utilised well within their normal saleable life, usually produce that supplies the local demand adjacent to production zones does not usually enter the cold-chain. Lastly, the capacities assessed have inherent elasticity in assumptions to err on the safer side and account for a possible high level of inefficiency.

After this study, the country no longer insisted that there was a shortfall in cold stores but began to focus on factors preventing the provision of cold-chain connectivity for fresh produce. This change had required the NCCD to expose many preconceived ideas and refocus them. conversations on the role of the cold-chain by moving perishables from source to destination over space and time.

11.2.5 Packhouses Are Important in Cold-chain

Every conduit has a point of origin, and in the cold-chain, this is where the load is first aggregated. It can be the ice cream factory, the milk pooling point, or in the case of fresh produce the packhouse. India's abysmal status in packhouses limits the safe and viable entry of fruits and vegetables into the cold-chain by limiting packaging and precooling. Hence, most of the cold stores could only come into use for produce types that do not require precooling before entering a cold space.

It is this author's assessment that more than 94% of India's cold stores are designed for single commodity storage, for crops such as potatoes. The presence of cold stores gives producers a false sense of security and they feel safe to produce more. In 2019, India's potato production was more than 53 million tonnes, far more than its total domestic consumption. With supply outstripping demand, the value of the crop was reduced. Blind creation of cold stores can lead to blind increases in production and the result can be a wide-eyed disaster. Likewise, the production of perishable produce should be intensified in tandem with the availability of cold-chain facilities, and cold-chain systems should be developed concerning market availability. Such integration across the supply chain should become commonplace among policy planners.

Meanwhile, without precoolers, the cold stores are unsuited for high-value fruits and vegetables. Without access to precooling and other cold-chain pieces to reach distant markets, most of the fruits and vegetables must be sold at markets close to productions areas and lower prices. From here produce they finally reach cities, barely after a haphazard route and the result is high handling losses and spoilage.

Simply put, the lack of integrated packhouses means that the radius of sales of perishables is limited to a 100 or so kilometres, or about a 48-hour radius from harvest location. Surplus, unsold produce will perish and spoil. The lack of packhouses also means that whatever is sold is in a loose format, unpacked or poorly packaged. Thereafter, if the supply is forwarded to secondary or tertiary wholesale markets, the produce suffers added handling injuries, and hence the fresh produce supply chain incurs a large share in food loss.

The abysmal lack of packhouses with precoolers, which are loading points for onwards transit to wholesale markets, also means that refrigerated transportation is mainly deployed for meats and ice cream in India. These products have abattoirs and ice cream factories as the loading points for reefer trucks. Now that the country is saturated with ice cream factories and the growth in reefer transport will be limited, unless the development of packhouses is scaled up.

The lack of packhouses and precooling facilities means that climacteric fruits (like banana, papaya and mango) travel warm to markets. Therefore, modern ripening systems, to evenly warm up and ripen the cold fruit that arrives in a cold-chain, do not get used.

Most importantly, when farms are small and fragmented, as is the case in India, the packhouse allows for the consolidation of small lots from nearby farms into viable large loads and travel more economically to markets. Packhouses bring organization to farm-gate, convince farmers to collaborate in production, consolidate their production and arrange to safely dispatch the aggregated value to distant destination markets.

The above situations clearly predict that the future of India's cold-chain hinges on its success in building thousands of packhouses. Without these first-mile facilities at the village level, there is not much relevant cause to invest more in reefer trucks or intermodal reefer containers, on more modern ripening units, merchandising units and even on more cold stores. The packhouse is the primary condition, the first-mile protocol for fresh produce and the mainstay of its cold-chain.

The capacity assessment study by NCCD found that about 70,000 packhouses were needed to safely and scientifically service the fresh produce consumption in India. The government has since announced that it will directly support entrepreneurs and organisations in creating 22,000 aggregation and preconditioning hubs in villages across India. These village-level units, along with other cold-chain components including reefer containers, rail terminals and port-based infrastructure are also featured as part of India's Rs.104 lakh crore (USD 1400 billion) national infrastructure pipeline.

11.3 REVIEW OF THE COLD-CHAIN CONCEPT

The cold-chain is colloquially seen as the conjoining of two separate nouns, Cold+Chain, and is interpreted from the singular perspective of cooling, the *Cold* taking precedence in a lay person's perception. Domain practitioners, however, know that it designates an enmeshed set of activities and procedures to safely transport perishable goods from source to final destination.

Just as the concept of a "greenhouse" (which regulates temperature, humidity, pests and the composition of air and light) is not confused with a green coloured house, the cold-chain also needs to be understood to involve more aspects of postharvest care and not merely cooling. It is with this purpose that the author insists on hyphenating the term "cold-chain," to emphasise the compound nature of this complex concept. This is not just minor semantics, as the finer distinction is of paramount importance and will be well recognised by the postharvest specialists, especially those who manage fresh fruits and vegetables.

Many intricacies in the cold-chain had to be elucidated before the government ceded that its support programs needed to be reformulated, and a few are shared in subsequent sections.

11.3.1 COLD-CHAIN COMMUNICATES VALUE

The food cold-chain is not merely about keeping food cold to preserve it, but about keeping food safe and consumable while shipping it to markets. Cooling may or may not intercede at each stage of this journey. The cold-chain for potatoes and dry onions only needs refrigeration when they are warehoused, and not before or thereafter. For many other produce, the cold-chain carry out preparatory activities, even before cooling is applied. Cooling may also not be necessary at the last mile if selling cycles are very rapid, as is the case in India. However, whatever the technology and effort, the purpose is to protect the value and then communicate the value to markets. Cooling, or the provision of a specific temperature profile, wherever used, costs effort, and should only be undertaken for a gainful outcome.

The first-order gain is that it helps keep perishables marketable for a longer period. Besides extending its marketable life, it also safeguards the quality and hence protects the value of the commodities. However, this is only temporal and all produce under care will eventually perish. A partial extension of saleable life is not a sufficient condition or the sole purpose.

The second-order reward comes from deploying the cold-chain to harness this extra time that cooling provides, to reach a shelf. This can be a retail shelf or the consumer's kitchen shelf. This means that the produce connects with its gainful end-use. This is important from the perspective of perishable produce, which is otherwise inherently constrained for time.

An even higher order of reward follows, when the cold-chain also uses that extra time, for demand expansion, by serving as a bridge over geographical space and time. So, besides extending the saleable life, the cold-chain uses the extra life for the produce to reach markets and connect with new markets.

From a supply chain perspective, the commodities have a certain marketable or saleable life (Figure 11.2). The time a commodity stays on a shelf is the shelf life and is a dynamic target in the cold-chain. Shelf life is only a factor of the total marketable life that was temporarily extended thanks to the process of cooling. The cold-chain practitioner will rarely make the mistake of referring to the time spent in a cold store or transit as shelf life of the produce.

The cold-chain for fresh produce includes preparatory activities such as sorting, cleaning, grading, packaging and precooling, collectively called preconditioning. These are functions of logistics, as the preconditioning activities do not alter the essential characteristics of the perishables, and the produce is merely readied for their onwards journey. It requires creating a specific handling capacity at the first mile. The cold-chain, therefore, brings more than cooling to agriculture, and perforce also brings organised postharvest management in its supply chain.

The benefits of time, quality, connectivity and market expansion each have their advantages, but collectively they make the cold-chain the most empowering intervention for fresh produce farmers. How the users will prioritise this scope depends on their capacity to fulfil the conditions, and the role of the government is to guide and bridge any gaps in their capacity.

11.3.2 Cold-chain Is Not Agro-Processing

Pre-conditioning is sometimes confused to mean processing. However, the latter involves transformative processes such as liquefaction, emulsification, cooking (boiling, broiling, frying, baking or grilling), mincing or macerating, dicing or slicing, pickling or preservation, canning, drying, refining, grinding, additives, etc.

Processing activities are akin to manufacturing, where the produce is the feedstock or raw material, and the output is a new product altogether and is no longer called produce of nature. Such

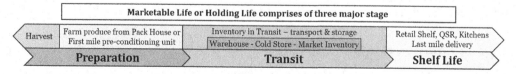

FIGURE 11.2   Life stages in supply chain of fresh perishable produce. (Author.)

activities may use cooling technologies, such as when creating a beverage, making ready-to-cook meals, manufacturing ice creams, producing medicines, extracting juices, etc. However, the application of cooling does not automatically make any production process a part of the cold-chain. Just like the production of milk, vegetables or meat in a temperature-controlled environment cannot be considered a part of the cold-chain.

Though food processing is not part of the cold-chain, it is commonly confused as such because of common parlance and blind correlation with the use of refrigeration technology. However, some manufactured or processed products may also need the cold-chain to service their transportation, distribution and retail, until consumption, like in the case of fresh produce. Cold-chain is about the safe and organised handling of value that was already produced, created, harvested or manufactured while communicating it to that point in time and space where that value can be fully realised.

Produce is natural harvest or capture, whereas products are manufactured by putting together various produce or ingredients. This clarity and distinction are most necessary when countries discuss public policies, create policies and allocate budgets. In India, almost all processed or manufactured products are taxable, whereas any agricultural produce is not. With the same logic and purpose, in the case of all fresh produce, the services in the cold-chain, i.e., pre-conditioning, loading/unloading, transporting, warehousing, are exempt from GST.

11.3.3 Cold-chain Is Not Just Cooling

It is emphasised to readers that the cold-chain does not rely only on pulling down or the lowering of temperature. When the outside is colder than tolerable limits, the cold-chain must warm up space, and when the external becomes warmer, the task is to cool down space. The latter method is more commonly associated with the term cold-chain, and for convenience, let us, henceforth, refer to all such process of temperature control as cooling.

The vital prerequisite for such control is to first segregate a designated space, from the external uncontrolled ambient. After creating an envelope to insulate the internal environs from outside thermal contact, steady control is effected by deploying some form of a temperature-modifying technology. The type of temperature modification (heating or cooling) technologies can range from the active, passive, evaporative or radiative thermal exchange and combinations thereof. The type of technology used depends on the characteristics of the goods, temperature desired and a cost–risk–reward matrix.

Nevertheless, the major efforts relate to isolating a space from inclement exposure and then extracting or adjust for any inherent or incremental temperature change or ingress. However, *cooling alone does not make for a cold-chain*. It is only *one* of the enabling tools within the cold-chain. In fact, cooling by itself brings along a host of problems.

Depending on the goods under care, especially for perishable produce, the side effects of cooling and storage also have to be considered. In time, the sealed insulated space will see changes in its air composition – the levels of oxygen, CO_2 and other gases will change from normal. In the fresh fruit and vegetable segment, this necessitates fresh air recycling in the enclosed cooled space. The exchange of cold air with outside air may seem an anathema to some, as energy was expended to cool down the air in the insulated space. Yet, if oxygen levels are not replenished, over time, the living breathing commodities will perish as food loss. For this reason, too, precooling of the produce is advantageous as it helps slow down such self-induced demands and minimise subsequent energy requirement while the produce is kept fresh and marketable.

Further, creating a temperature differential in any space will induce changes in humidity, either to dry the space or cause condensation. Dry air can cause shrinkage in most fruits and vegetables but is desired in the case of onions or dried commodities. Conversely, a cool, damp environment can be perfect breeding grounds for various pathogens and other unwanted flora and fauna, putting susceptible perishable produce at risk of disease. Infrastructure designs, whether for storage or carriage, need to counter these aspects.

The changes in an enclosed environment can be explained by relating to a car full of people. With windows closed and its air conditioner running, the living breathing passengers will slowly deplete the oxygen and increase the CO_2 levels inside. In time, they could suffocate in their exhalations. The process of cooling will also dry the air inside the car and drain the moisture exuded by passengers. Such modification of the atmosphere inside a truck container, room or cold store is common and self-induced, whether loaded with people, fruits or vegetables. Even a room full of iron ore will eventually induce similar changes. This needs to be understood and managed by cold-chain operators so that fresh produce is kept in a living breathing environment.

11.3.4 Cold-chain May Control the Atmosphere

Basically, in all cold stores, there manifests a slow, passive, self-induced modification (from normal respiration and physiological activity of the stored perishable goods) to the gaseous composition of an enclosed space. Exposure to low levels of oxygen or very high levels of CO_2 has a harmful and asphyxiating effect on living tissue.

However, in the case of certain fresh produce, low oxygen levels in tandem with low temperature are seen to induce an additional reduction of its physiological activities, compared to that which the process of cooling alone could provide. This, in some produce types, also translates into the greatly enhanced storage life of that produce. A low level of oxygen also kills insects, reduces microbial activity and inhibits pathogen reproduction, which aspects also aid in keeping quality while in storage.

Worldwide, a lot of research was conducted in quantifying the benefits of changing the atmospheric composition during the storage of fruits and vegetables. Such research continues, however, the key objective of greater life extension is not uniform across all fresh produce. In some cases, marketable life was extended by many months but, in most cases, only by a few days. Naturally, the use of any technology is defined by a risk-reward matrix; time gained and how it can be deployed in the prevailing market dynamics and it is not always justifiable for many produce types. However, depending on the produce, this technology offers specific commercial opportunities. Hence, the fresh produce cold-chain could also involve controlling the atmosphere, albeit for a few produce types, alongside the usual controlling of temperature and humidity.

Controlled atmosphere systems are essentially those where the air's normal gaseous composition, of 21% oxygen, 78% nitrogen and 415 ppm of carbon dioxide, is actively doctored and deviated from the normal. The other gases present are in too small a concentration to have a relevant effect on stored produce. The process of controlling the atmosphere involves forcibly replacing the mass of air in the enclosed and cooled chamber with an "inert" gas.

Natural nitrogen is commonly used as an inert gas since it is abundant in the earth's air and is not harmful to the produce. By using simple molecular sieves, the oxygen molecules are reduced, and the resulting air mixture is high in nitrogen. This is then pumped into the airtight cold chamber, purging the existing mass of air, until a predetermined safe and survivable oxygen level is achieved. The oxygen is not allowed to reduce to zero as it is essential for the produce to survive, for example, a range of 2%–5% oxygen is safe for apples and some pome fruits. However, as the produce still lives and breathes, the oxygen available in the space will continue to fall. If the purged space had 3% oxygen, it will start to drop and CO_2 will increase, and at this point, the air is "balanced" by piping in outside fresh air until the upper limit of oxygen, say 5%, is reached. Then the cycle repeats – the fruit breathes, oxygen falls below 3% and again a little fresh air is allowed inside. If oxygen drops too low, or CO_2 exceeds dangerous limits, then the produce perishes. The gases controlled are normally oxygen, carbon dioxide and ethylene and managed by purging or chemical scrubbing.

Globally, the use of this technology is commercially established for a few crops like apples, kiwi and pears for long-term storage gains, or in the transporting of fresh strawberries, cherries, bananas and lettuce.

There are complications to using this technology. Not all produce is suited for CA-enabled storage, and even among the apple cultivars variations are observed. Also, fruit from senile orchards or first bearing trees may not be able to survive the low oxygen conditions. Fruit picked too early will not store well in low oxygen conditions, nor will those that are past a particular stage of maturity. To maximise the benefit from low oxygen and cold storage, the produce must enter CA conditions in the early stage of its postharvest life. If delayed a few hours when the produce has aged, it cannot survive exposure to low oxygen conditions and, in fact, that will hasten its demise. Much like how an aged person cannot recoup from low oxygen exposure as well as infants can.

The produce must enter CA conditions quickly after harvest. Typically, if CA parameters (temperature and oxygen level) cannot be achieved within 36–96 hours of initiating the changes, the CA operation is annulled and normal cold storage parameters with fresh air replenishment is continued. Once in CA conditions, the chamber is opened only once, when it is to be evacuated. Once the produce has exited a CA environment, its metabolism quickly reverts to normal and reintroducing it back into CA environs will cause tissue demise. For this reason, globally, cold stores with CA systems are built at the start of the cold-chain, at farm-gate, and not at terminal destinations.

However, in India, attracted by an inordinately higher subsidy, they were being installed everywhere, including in desert regions and seaports, unquestioned and for non-viable uses. In many cases, the expenditure on CA systems stayed idle and the cold store was used as normal. Capital that flowed into non-viable technology diverted funds from where it was more urgently needed and brought the cold-chain a bad name. Technology should be encouraged and used where it fetches better returns. The government was advised that subsidy be approved only for commercially viable CA stores and only at production locations. This was enforced from 2013 itself and restricted initially to pome fruits only.

11.3.5 FRESH PRODUCE COLD-CHAIN IS KNOWLEDGE-INTENSIVE

In the case of fruits and vegetables, the finesse expected from the cold-chain specialist is of a higher order; these postharvest managers have to keep the produce fresh and living for longer (Kitinoja et al., 2019). If it gets colder than specific tolerances, then the living produce suffers a chill injury which can kill the living tissue. If it gets a little too warm, then the saleable life of the produce will fade faster. Therefore, the cold can kill and warmth can hasten demise, and a fine balance is expected. The fresh produce segment of the cold-chain, or the chill and mild-chill category, requires closer monitoring to balance humidity, air, temperature, and disease. Postharvest management and cold-chain of perishable horticultural produce is a science and fine art, similar to the medical care of living beings. If this is excelled at, then all other cold-chain categories are easily understood.

On the other hand, in the case of the frozen category, which caters to post-mortem products, the manager only needs to ensure the necessary cooling; the colder it gets, the safer they are. However, since −18°C is the acceptable norm for meats, ice cream, etc., this set point is maintained for energy optimisation. Some of the other above-mentioned aspects can also be of concern but are more manageable in extent. For food handlers, the frozen segment is easier to manage compared to the fresh produce cold-chain.

The cold-chain is the post-production supply chain of perishables and perishable products, and depending on the type of product handled, breaks in cooling are possible and sometimes desirable. A cold-chain does not always mean an unbroken chain of the cold. There are certain tolerances and within parameters, even ice cream, meats, etc., can tolerate a short breach in temperature. In chilled fresh produce also, some intermittent excursions are possible. The most vital aspect is the needed diligence at first mile, such that the fresh harvest does not remain hot for long, be it meats, fish, milk and most horticultural commodities. The initial rapid cooling has specific technical benefits that cannot be ignored. Subsequent variations are tolerable within limits, and cold-chain managers must have proper knowledge of tolerances and the impact of such excursions. Suitable knowledge dissemination must form a part of the cold-chain development agenda.

11.3.6 COLD-CHAIN INTEGRATES THROUGH PROCEDURES

No single technology or intervention makes a cold-chain. The cold-chain has to manage the safe handling and transfer of perishable goods from source to destination, such transfer includes buffering the supply and recovering any wastage or inefficiency in doing so. To do this seamlessly, the coordination among multiple actors in the supply chain is demanded. Such coordination means the transparent exchange of protocols for effective transfer of the goods in custody and other business transactions. Therefore, integration in the cold-chain implies shared operating procedures to manage the transactions between different entities, for streamlined and timely outcomes. The inherent challenges and important measures of the integrity of the cold-chain are about delivering in full, in quality, and on time.

Therefore, government policies need not blindly drive the objective of single ownership of all cold-chain infrastructure assets but be designed to encourage collaborations among the asset owners. This can happen by moving away from capital subsidies alone to support growth targets in terms of quantity-based throughputs and market expansion.

The quality of the goods, especially perishables, also depends on production, preharvest and harvest practices and others that lie outside the ambit of the cold-chain. The performance and quality maintained in the cold-chain can also depend on preharvest practices, i.e., if a crop is harvested too late, or if the field was irrigated shortly before harvest. Similarly, meat loaded at the right temperature from an abattoir, or a fine-looking load of ice cream, might have undergone a thaw and refreeze in the factory. The best of cold-chain care cannot change manufacturing faults, but the subsequent quality failure could be unknowingly blamed on the cold-chain. Hence, cold-chain integration also means it must have backward linkages with the producers, to closely collaborate with their activities, which are otherwise external to its services.

This external integration makes the cold-chain unique among all logistics services. While the ordinary logistics take responsibility for boxes or cartons "said to contain" a product, the cold-chain must get inside the box as it has to take charge of the specific product or produce. The cold-chain cannot add value to a product or produce but can totally destroy all value through poor management practices. An efficient cold-chain is not merely the proof of cooling technologies, but a successful proving ground of postharvest care and management processes.

11.4 COLD-CHAIN DEFINED

A cold-chain is defined as an environment-controlled logistics chain, consisting only of storage and distribution-related activities to ensure uninterrupted care from source-to-destination over time and space, in which goods under care are maintained within predetermined ambient parameters.

This series of activities involve multiple actors, even from different entities. Hence, the integrated cold-chain means the coordination of logistics activities among collaborating activity partners, to effect the seamless transfer of custody of perishable goods as they transit the thermally managed supply chain.

The cold-chain service integrates to collaborate activities with farms and/or factories as well as with the front end to ensure its objectives of delivering in full, in quality and on time. Detailing more aspects of the cold-chain, its economics and more will wait for another book, and here the time and space are to review cold-chain development in respect of India.

11.5 FUTURE OF INDIA'S COLD-CHAIN

The minimalism of storing produce in cold warehouses, holding to sell locally at a later date, had way-laid the more empowering aspects of the cold-chain. This was also fostered with the provision of government subsidies that had eased the creation of cold stores alone. Commencing 2012, and in greater strength after 2014, India's cold-chain development was delineated, rationalised and various activities and initiatives were commenced. However, the course corrections made so far were piece-meal, and they need to be continued to a logical conclusion. The strategy to enable cold-chains as a conduit to access and widen markets for perishables requires to be driven and any pause may push the development back a few steps.

Luckily, India already has an example in its enviable dairy network, especially for liquid milk. Even in the frozen category, India's cold-chain made it a global leader in the trade of beef and fish. In the chilled category, pharmaceutical and ready to cook or eat saw similar market-led develop-ment. For these segments, the future is steady and incremental.

In the case of fruits and vegetables, the fresh produce cold-chain needs to be actively pursued in India and developed holistically. Earlier, when local demand used to outstrip horticultural produc-tion, the storing of some produce was economically relevant. Today, with production being in abun-dance, the producer is always competing even for the local market and stays in distress because of perishability concerns. The producer must be empowered to connect with new demand destinations if his/her future growth is desired.

India ended 2019 with nearly 37 million tonnes holding capacity in refrigerated warehouses and the past trend of creating bulk cold storage is slowing down. In association, the throughput capacity of integrated packhouses is less than 2 million tonnes and about 9 million tonnes in reefer transport. The capacities in form of packhouses and transport have more than doubled since 2015. Yet, they still need to grow multi-fold in the coming years. Without them, India's cold-chain ambitions will have a diminutive future.

India must maintain a clear understanding that in the cold-chain, each component must comple-ment the carrying capacity of the other, more so than in another supply chain. Further, the cold-chain is not to be treated as an infrastructure business and that infrastructure is only a tool of the trade. Such a perspective has to lead its strategies to drive the trade, domestic and international, through the cold-chain. The Government in India shows all intent to do its best, but the key will be in how its visions are strategised and implemented. This last is in the purview of bureaucrats, but if a few among them fall back to the old ways, then the country will not see the transformative changes it yearns for. At the moment, India is yet to declare its national policy on the cold-chain. Such a policy should clearly define the long-term purpose, with medium-term objectives and measurable outcomes, to assure that future direction cannot be changed easily on unconsidered whims.

To remain in context, such a policy will also require focusing on sustainability aspects in the cold-chain. Cold-chain intrinsically provides sustainability benefits to various farming systems, by ensuring production is not spoiled before it reaches gainful end-use. This in itself can immensely reduce greenhouse gas emissions from food loss and waste, which was globally estimated at a CO_2 equivalent of 4.4 billion tonnes annually. Yet, there is the scope to ensure that cold-chain opera-tions also adopt innovations and enable a shift from fossil fuel dependence as well as make the power-hungry utilities more climate-friendly.

Most importantly, a cold-chain policy must ensure emphasis on both capacity building and train-ing. Suitable resources need to be allotted for heightened efforts in this direction. India has yet to develop and take up the appropriate level of extension education and training, to touch the wide variety of actors and activities in the cold-chain. All possible methods of outreach will have to be deployed and extensively in a country of India's scale and backdrop.

The Covid19 pandemic has brought to import the importance of supply chains and more essen-tial in case of the cold-chain, for food and medical supplies. Cold-chain is the sole delivery mecha-nism for high nutrition foods such as fruits, vegetables, all meats and dairy. Most countries are interdependent for their food supply, and this makes cold-chains even more relevant and indispens-able. The strategic strengths of a nation will, therefore, also depend on how well they manage and develop their cold-chains.

Meanwhile, the food cold-chain has the most extensive and reaches most residential regions across countries. This need not be seen merely for the traffic of ice cream, dairy or other foodstuffs. It can also be innovatively redeployed, in time of pandemics and other disasters, to quickly and efficiently reach vaccines, other medical supplies and succour to the needy. The cold-chain provides more opportunities than only for holding food.

Socio-economic progress requires the cold-chain to be developed to communicate the high value and sensitive produce from source to markets and to expand the markets for such produce. However, habits have a way to revert to comfort zones, whereas such development will require that we constantly challenge the horizon. Hence, the country will also need to keep abreast of global developments, ensure that the spirit of enquiry, scientific investigation and solution making is continually fostered and take the opportunity to lead cold-chain development.

11.6   ANNOTATIONS

The cold-chain is understood as the merging of various logistics practices that safeguard the perishable goods under care, in the post-production stages of its life cycle. The cold-chain for perishables is not just about the "cold" but refers to all logistical processes applied to maintain multiple parameters, involving handling, packaging, transport, storage and retail of sensitive perishable produce. Cold-chain enhances the saleable life of the perishable produce (fruits and vegetables) under care, which otherwise would perish at a faster rate in the open environment. In the case of other temperature-sensitive products, it maintains the goods and avoids spoilage.

The logic behind the cold-chain is related to time management – how to capture more time and how best to utilise the extended time. Depending on the motivation and the product, the cold-chain affects the transfer of value, which can be over a time period, and/or across geographical space. Extending of saleable life is the first-order effect and connecting with markets is the higher-order result from the cold-chain. Across all activities when handling and marketing perishable foods in the cold-chain, speediness is often the key to success.

The cold-chain does not alter or add value to the produce or product handled; however, it safeguards the value under its care and adds enormous value to the business involved. Cold-chain is a value-adding service to the perishables trade, without which its growth is limited. The success of cold-chain depends on how effectively it serves as a conduit for products that are sensitive to their holding environment (air quality and composition, temperature control, microbial load, etc.) from the place of origin to destination maintaining integrity in its care.

The post-production life cycle of fresh perishables includes a preparatory stage at first mile to precondition the harvested produce; transit stage in reefer transport and/or during temporary residence in cold stores and a delivery stage until the last mile for consumption or retail. The cold-chain typically involves a frequent transfer of custody of the goods and hence requires integration with separate entities. A system-wide procedural integration among all the supply chain activity segments becomes important to avoid a failure of the cold-chain system.

A few crop types, such as potatoes and dried chillies, do not require complex integration, and their mere residence in a refrigerated warehouse serves the purpose for subsequent trade. India is populated with such cold warehouses to hold produce like potatoes, dry chillies, tamarind, pulses and seeds. This segment is comparatively simple than the other which must manage the full cold-chain to market.

Lacking direction and not aligned with developments around the world, India's cold-chain had gone off-track. The initiative by India in 2012 to bring together policy implementers, domain practitioners, policymakers and hands-on industry expertise, under able mentorship, allowed for a change to happen in India's cold-chain trajectory. The future path of India's cold-chain development will depend on how well it applies knowledge and how strategically it deploys its resources to keep a holistic balance and a supply chain orientation.

Today, the cold-chain is enabling a commerce network that knits countries across the world and is already recognised to be essential for the continued progress of humankind. India has the opportunity to be a mainstay of this fabric, and it will have to continue to do the cold-chain right.

BIBLIOGRAPHY

DAC. (2008). Report of the Task Force on Development of Cold Chain in India. Delhi: Ministry of Agriculture, Govt. of India.

DAHDF. (2019). Basic Animal Husbandry Statistics. Annual Statistics, Delhi: Department of Animal Husbandry Dairying and Fisheries, Govt. of India.

Kitinoja L., Tokala, V. Y., and Mohammed, M. (2019). Clean cold chain development and the critical role of extension education. *Agriculture for Development 36*:19–24.

Kohli, P. (2013). Strengthening the Cold chain - 'Directly Linking Source to Consumer'. Delhi, 17 July. doi:10.13140/RG.2.2.13995.26407/1.

Kohli, P. (2014). Harmonising the Concept - Development/Component/Users. Task Force on Cold chain Projects. Presentation. Delhi, 15 9. Accessed 6 1, 2020. doi:10.6084/m9.figshare.12318851.

Kohli, P. (2015). Guidelines and minimum System Standards for Implementation in Cold-chain. 2. Delhi: NCCD. doi:10.5281/zenodo.3883873.

Kohli, P. (2018a). Demolishing the myths about India's cold storage shortage. The Hindu Business Line, Interview by Rajalakshmi Nirmal (4 11). Accessed 2020. https://www.thehindubusinessline.com/portfolio/commodity-analysis/demolishing-the-myths-about-indias-cold-storage-shortage/article25418538.ece.

Kohli, P. (2018b). Marketing, Agri-logistics and Agri-value System. Paper, Delhi. doi:10.6084/m9.figshare.12318842.

Kohli, P. (2019). Public-private knowledge partnership – redefining the cold chain. *Agriculture for Development 36*:11–15. doi:10.6084/m9.figshare.12318740.

Kohli, P. (2020). The Saga of India's Cold chain Development. Gurgaon. https://www.researchgate.net/publication/342396403_The_Saga_of_Cold-chain_Development_in_India.

MOFPI. (2014). Report of the Task Force on Cold Chain Projects. Delhi: Ministry of Food Processing Industries. https://nccd.gov.in/PDF/Final-ReportTFCP.pdf.

NCCD. (2015). Kohli P., Chopra S., Jindal K., and Premi B.R, 2015, All India Cold chain Infrastructure Capacity (Status & Gaps). Research and analysis, Delhi: NCCD.

Planning Commission of India. (2012). Report of the Committee on Encouraging Investments in Supply Chains Including Provision for Cold Storages for More Efficient Distribution of Farm Produce. Policy, Delhi.

12 Cold Chain Operations in the Caribbean
Opportunities and Challenges

Puran Bridgemohan
The University of Trinidad and Tobago

Majeed Mohammed
The Postharvest Education Foundation

CONTENTS

12.1 INTRODUCTION

With the growing demand for fresh fruits and vegetables throughout the Caribbean, the cold chain facilities and resources are being utilized in several segments of the supply chain. The rise in demand can be realized from the rapid growth and development of supermarkets; the hotel industry; the marketing boards acting as facilitators on behalf of registered producers; the oil rig platforms located in the sea but dependent on land caterers as supply agents; exporters targeting lucrative ethnic markets in metropolitan countries as well as the food caterers attached to various school feeding programmes. The types and sizes of cold requirements vary for different sectors. The supermarkets with chain stores (SWCS) are equipped with centralized low-temperature storage facilities, while supermarkets without chain stores (SNCS) have refrigerators large enough to handle their smaller volumes. The hotel industry requires larger volumes to maintain a more diverse range of perishables. The marketing boards need refrigerated facilities equipped with automated and semi-automated packing lines.

Postharvest losses across these outlets or distribution centres vary between 35% and 50%. The extent of losses depends upon the level of integration of the cold chain within the postharvest handling systems. The losses can be reduced by the continuous monitoring of temperature and relative humidity (RH) along the supply chain, sanitation protocols, produce compatibility considerations, packaging selection and stacking arrangements. The wide variation in cold chain monitoring and management together with a lack of facilities and cold chain equipment resulted in gaps or breaks

DOI: 10.1201/9781003056607-15

in the cold chain, the severity of which depend on the type of market, seasonality of production, maturity inadequacies at harvest, poor logistical factors associated with transportation linkages, produce type and cultivar selection (Mohammed and Kitinoja, 2016).

Any break in the cold chain conditions may affect the quality and/or shelf life of the fresh commodity or, in the worst case, make the fresh commodity unsafe to eat. Inconsistent cold chain conditions, whether caused by failure to follow established procedures or faulty equipment, can result in rapid deterioration of fresh produce, eventually causing these perishable commodities to be unmarketable and potentially unsafe for consumption. The gaps or breaks in the cold chain, such as a failure to comply with the maximum out of refrigeration time limit (MORTL), should be identified, corrective actions undertaken and records kept to show what happened and what was done.

This objective of a successful cold chain for fresh fruit and vegetables is achievable only when each link in the cold chain from the farm gate to the consumer is precisely recorded. It should include the initial time for which the commodity was not refrigerated; the environment in which the produce was held while out of refrigeration; a log of the temperature within that environment (e.g., the internal temperature of an air-conditioned receiving bay or ambient temperature); methods used to generate the temperature log; the time the commodity re-entered refrigeration and the total time elapsed out of refrigeration. These records of cold chain management are the quickest way to ensure that cold chain conditions for a perishable commodity remain intact. For example, enzymatic activity, which can develop off-flavours in some fresh-cut fruits and vegetables, only ceases at about $-18°C$ and might be spoiled even though it has been kept "frozen" (i.e., below $0°C$) throughout its journey. Only by keeping accurate records, the stakeholders in the cold chain will be able to tell whether or not such enzymatic activity might have taken place. Inaccurate record-keeping, on the other hand, does not assure that spoilage has not occurred and could cause the rejection of perishable produce consignments.

12.2   CAUSES OF POSTHARVEST LOSSES IN THE CARIBBEAN

The lack of reliable and adequate cold chain facilities in the Caribbean is one of the main causes of the losses of perishable products. These losses not only affect food security in all its dimensions but also result in loss of market opportunities, wastage of the natural resources (water, land and energy) devoted to producing the foodstuffs and can have a significant ecological footprint. Reliable and efficient cold chains contribute not only to reducing these losses but also to improving the technical and operational efficiency of the food supply chain. Cold chains thus facilitate compliance with quality and safety requirements and promote market growth, stimulating an increase in production (IIR, 2009). Development of the cold chain can, therefore, be considered a necessary step toward achieving food and nutrition security. Despite this need, however, cold chain development has not received the necessary attention from governments and development organizations and remains underused in comparison with the actual and potential needs of Caribbean producers and traders.

This chapter will focus on the gaps within the cold chain of various outlets in the Caribbean to highlight the causes of postharvest losses, methods to measure losses at critical loss points, recommend the required corrective actions and facilitate a fully integrated cold chain management system to reduce postharvest losses and wastage.

12.3   CASE STUDIES IN THE CARIBBEAN

12.3.1   Losses in Tomatoes at Supermarkets with and Without Chain Stores

Quantitative and qualitative analyses were undertaken to determine postharvest losses in tomatoes at supermarkets in the East-West Corridor of Trinidad (Mohammed and Craig, 2018). The investigation included 12 SWCS and 8 SNCS. All 20 supermarkets were equipped with cold chain facilities but with different levels of produce holding capacities and storage conditions. SWCS had a centralized cold storage facility with a range of both imported and locally acquired fruit and vegetables held at 4°C–7°C and RH of 70%–80% for 5–8 days. Tomatoes were stored in plastic crates and delivered to chain stores on the same day in non-refrigerated trucks (26°C–31°C, 55%–67% RH) which contained other fresh produce and processed food products. Postharvest losses of tomatoes in SWCS were 23.9%, and physical, physiological and pathological damages were recorded as 12.8%, 7.5% and 4.1%, respectively (Mohammed and Craig, 2018). The dominant physical damages were shoulder scars, bruising, abrasions, punctures and compression. Physiological losses which included pitting, surface discolouration, water-soaked areas and uneven ripening were attributed to chilling injury. Other physiological disorders, such as blossom end rot, desiccation and puffiness, were observed but in smaller quantities. Pathological losses were related to multiple infections such as bacterial soft rots, and fungal diseases, notably Anthracnose. The postharvest losses at SWCS resulted from a broken cold chain that occurred at four critical control points (CCPs) in the tomato handling system. First, it was the uploading and transfer of fruits from the centralized cold storage facility into poorly ventilated non-refrigerated trucks containing a combination of incompatible produce. The second was the unloading of produce including tomatoes at the SWCS under a designated shaded holding area (6–7 hours) under ambient conditions (28°C–30°C). The third circumstance was the placement of the consignment of chilling-sensitive and ethylene-sensitive produce together with frozen processed products in a single cold storage unit for an additional 1 or 2 days before transfer on refrigerated display shelves inside the supermarket. The time taken (1–2 hours) under ambient conditions at the point of consumer purchase to placement in the household refrigerator represented the fourth interruption of the cold chain.

SNCS, on the other hand, had postharvest losses amounting to 17.5%, that is, 6.4% less than SWCS. Here tomatoes were placed in cardboard cartons as well as in plastic crates and transported in the open air or tarpaulin-covered pick-up vans directly to the SNCS retail outlets by contracted producers or middlemen after overnight sorting, grading and packaging. Transportation although non-refrigerated was done in the coolest time of the day, that is, late evenings or early mornings or even nighttime. The tomatoes were placed in an air-conditioned room at the supermarket, and within 2–3 hours, they were displayed on refrigerated shelves. Produce managers at SNCS handled less produce which were more spaced out on the shelves with fewer overhead produce layers. Thus there was greater efficiency in the flow pattern of the cool air. Tomatoes were sold in 2–3 days. The nature of the losses was also dominated by physical damages, but the incidence was higher than that of the SWCS. Losses due to physiological damages like chilling injury were significantly less, and pathological losses originated as secondary infection on the physical damages were also lower than tomatoes at SWCS. SNCS had less overall losses due to fewer interruptions in the cold chain compared to SWCS.

12.3.2 MARINE SHIPMENT OF REFRIGERATED REEFER CONTAINER WITH PUMPKINS

A detailed case study was conducted on four marine shipments of pumpkins from the same exporter using a refrigerated 40-foot reefer container from Trinidad to Toronto, Canada (Mohammed et al., 2014). On each occasion, the container was loaded with other perishable commodities such as shadon benni (*Eryngium foetidum*), green mango (*Mangifera indica*), breadfruit (*Artocarpus altilis*), breadnut (*Artocarpus camansi*), cassava (*Manihot esculenta*), dasheen tubers (*Colocasia esculenta*), dasheen bush, hot pepper (*Capsicum frutescens*) and okra (*Abelmoschus esculentus*). The exporter maximized space by also including food products with longer shelf life, based on demand from buyers targeted on the ethnic markets in Canada. The exporter usually packed the container over 3 days before the eventual journey by the sea, which took another 12 days. Throughout these 15 days, the reefer cold chain temperature was maintained at 7°C–10°C and 75%–85% RH. Harvested pumpkins of different cultivars, shapes and sizes were packaged in poorly ventilated polystyrene bags and placed at the front end of the container. The pumpkins were not cured before shipment. Upon arrival at the retail market outlets in Toronto, mainly, Scarborough and Malton where large West

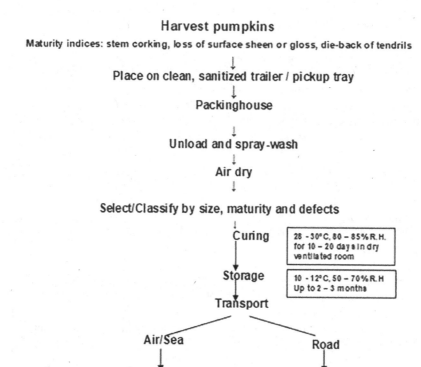

FIGURE 12.1   Postharvest handling system for pumpkins. (Mohammed et al., 2013.)

Indian immigrants exist, the pumpkins were placed in a storage room at 20°C–22°C and 60%–70% RH. Despite these cooling regimes, postharvest losses of pumpkins averaged 34% over the four shipments (Mohammed, 2013). Tracing and tracking studies were undertaken to determine the CCPs in the handling system. The first CCP was noted at the pre-harvest level where the farmer used excessive amounts of poultry manure. Poultry manure is high in nitrogen thereby boosting prolific vegetable growth with good vining which resulted in high yields but potentially short shelf life with increasing propensity to undergo softening. The second CCP was the failure to cure the pumpkins (typically, 10–12 days at 28°C–30°C and 80%–85% RH) before shipment (Figure 12.1). This investigation revealed that the produce quality could be negatively affected by pre-harvest and postharvest factors and highlighted the need for training and capacity building to offset such occurrences.

12.3.3   AIR TRANSPORT OF HOT PEPPERS FROM TRINIDAD TO CANADA

Studies were conducted to estimate the magnitude of postharvest losses of hot peppers from Trinidad to Toronto, Canada, using air transport (Mohammed et al., 2016). Harvested hot peppers at the green mature stage of maturity or with slight colour change with stem and with intact calyxes were packaged in cardboard cartons (15 kg) at exporters' facilities under shade at 26°C–28°C and 60%–70% RH. Packages of hot peppers were transported in non-refrigerated trucks covered with dark coloured tarpaulin to the airline collection point at the airport for loading into LD3 containers (airline container with 1.53×1.56×1.63 m). The initiation of the cold chain was delayed until produce arrived at the retail facility of the importer after the 6-hour flight plus another 6 hours after the customs clearance. Hot peppers were displayed on refrigerated display units in the same packages exported from Trinidad but, in some cases, were emptied on the display shelves in bulk heaps. At all

TABLE 12.1

Temperature and RH Profiles of Hot Peppers at Different Transfer Points in the Airport Cargo Handling Area (Mohammed et al., 2016)

Time-lapse (hours)	Transfer Points	Temperature (°C)	RH (%)
0.0	The arrival of hot peppers in cartons at the airport	31.0–33.1	60–70
2.0	Inspection at Plant Quarantine Office	33.3–33.9	60–70
4.5	Unloading, inspection at the airport cargo area	34.6–35.2	60–75
7.0	Loaded into LD3 containers	35.2–36.0	50–65
8.0	LD3 container loaded onto trailer trucks	36.4–38.2	50–70
12.0	LD3 container loaded onto aircraft	39.0–41.1	50–70

the retail outlets, there were multiple tropical and temperate fruit and vegetables, both climacteric and non-climacteric, displayed together in the store in an air-conditioned environment with temperatures of 19°C–22°C and 45%–55% RH. This represented the first and only time the hot peppers were subjected to a lower temperature. The surface temperature of the top layers of hot peppers reflected that of the store temperature, but underlying layers toward the bottom of cartons or those in the bulk display were almost 27°C. Postharvest losses averaged 53% (Bridgemohan et al., 2016). These losses were attributed to broken or decapped pedicels and calyces, shrivelling, water-soaked pericarp, bruising, abrasions, pitting, secondary infections and softening (Bridgemohan et al., 2017). Since the study was undertaken during the summer of July to August, the timing of the air shipment coincided with hot peppers harvested during the wet season in Trinidad.

The occurrence of losses in the quality of hot peppers occurred between two CCPs even before the produce left Trinidad. These were the exporters' packinghouse facilities and the airport cargo handling area where the cold chain was not initiated. Furthermore, the hot pepper consignment was required to be at the airport at least 5 hours before the proposed departure time. The stipulated 5 hours were extended sometimes to more, due to mechanical problems with aircraft, security checks and poor weather conditions. Such delays promoted quality deterioration in the absence of cold storage facilities. Evidence to support such occurrences is shown in Table 12.1 by the profile of temperature and RH between the point of arrival and the loading of cartons in LD3 containers onto aircraft, which took as long as 12 hours. Elevated fruit temperatures accompanied by low RH, favoured moisture loss. Also, fruits with latent infections and with delayed evidence of damage, undetected during sorting and grading operations at the packinghouse deteriorated rapidly. Other factors included lack of adequate ventilation in cartons and LD3 containers which were neither insulated nor refrigerated. The arrival of the already quality compromised produce at the importers' end could not be alleviated. The complete absence of cold chain facilities before air transport could not be compensated with the limited exposure to lower temperatures on display shelves among the ethnic market retailers in Canada.

12.3.4 Shipment of Fresh Produce in the Bahamas

This case study conducted by Mohammed and Craig (2018) highlighted how a well-coordinated cold chain of fresh produce sea shipped from Miami to the Bahamas could result in a reduction of postharvest losses, that is less than 3% when cold chain facilities, planning and logistics are managed and routinely monitored and implemented at all stages of the supply chain. Although this food service company supplied perishable food products to restaurants, hotels, supermarkets, resorts, caterers, convenience stores and government agencies, this case study was focused on the supply of fresh fruits and vegetables between the food service company and the hotel industry. Premium quality perishables were imported by reputable packinghouses in Miami where attention

FIGURE 12.2   Successful cold chain with high-quality produce.

FIGURE 12.3   Floor base of reefer container with debris from the previous shipment.

to maturity, precooling, sorting, grading, packaging and cold storage were done according to quality control protocols. The cold chain logistics were managed properly at all stages by qualified fresh produce managers to ensure no interruptions in the cold chain occurred. Figure 12.2 shows the cold temperature storage facilities at the foodservice company in the Bahamas with temperature and RH devices, quality sensors, packing arrangement and overall adherence to sanitation and food safety requirements.

Unlike the successful operations of this cold chain, there was another food service company in the Bahamas with fewer facilities that was managed by a centralized marketing agency where fresh produce was procured from local sources and stored in refrigerated reefer containers and subsequently distributed to small- and large-scale supermarkets. Closer examination of the container showed decayed produce from the previous load being lodged at the base of the container which would produce volatiles such as ethylene that would accelerate senescence of incoming produce (Figure 12.3). Produce from this cold chain were of poorer quality and incurred higher postharvest losses.

12.3.5   Oil Rigs Located in the Sea

A major food catering company located in North-East Trinidad was contracted to supply several offshore oil rig platforms in the Atlantic Ocean and was experiencing high postharvest losses of fresh fruits and vegetables. Investigations were conducted to trace and track the handling system from the collection point on land to the unloading of produce on the oil rig platform to determine the types of losses and to make recommendations to reduce losses incurred. Among the range of perishable commodities, the leafy vegetables including lettuce (*Lactuca sativa*), pak choi (*Brassica rapa chinensis*), watercress (*Nasturtium officinale*), chive (*Allium schoenoprasum*) and celery (*Apium graveolens*) were selected for evaluation since these had the highest postharvest losses. The cold chain was initiated at the caterer's facility where climacteric and non-climacteric fruits and vegetables were stored in chill rooms at 4°C–5°C and 80%–90% RH for 1 or 2 days. Overnight, the produce was loaded in cardboard containers overwrapped in shrink wrap in non-refrigerated wooden containers. These wooden containers also had frozen products such as meat, fish, butter, cheese as well as durable products such as dried grains. To maintain the freezing temperature and continue the cold chain, dry ice was used. All wooden containers with food products were transported in non-refrigerated trucks to the South-East end of the island. This journey took 2 hours. Upon arrival, the wooden boxes were placed in a temporary shed for another 2–3 hours and then loaded on special pallets onto supply boats to be taken to the off-shore oil platforms where the containers were transferred by a crane to the loading dock next to the platform kitchen. Containers were quickly emptied by the kitchen staff crew and returned to the supply vessel so that they can be returned to shore for the next delivery. The quality of the leafy vegetables was evaluated and over 65% exhibited water-soaked dark brown lesions with an obvious off-odour and were classified as unusable. The cold chain at the caterer's facility was implemented and closely monitored, but the use of dry ice to maintain the cold chain resulted in the production of carbon dioxide at levels that induced carbon dioxide injury as well as freezing injury (Mohammed, 2002).

12.3.6 COLD CHAIN MANAGEMENT OF FRESH-CUT FRUITS IN SURINAME

In Suriname, a fresh-cut fruit and vegetable caterer has successfully utilized the cold chain in the past 5 years using the lower floor of his house with an annexe to focus on the production of fresh-cut pineapple, watermelon, mango and papaya for students in the School Feeding Programme. With the acquisition of stainless steel tables, cutting devices, stainless utensils, weighing scales, cold storage rooms and refrigerated trucks, GG Enterprise Limited employed ten women from the village to prepare fresh-cut fruits as seen in Figure 12.4. The company provided their workers with hairnets, aprons and training on the significance of temperature management, quality maintenance and

FIGURE 12.4   Preparation of fresh-cut fruits in Suriname.

sanitation. The facility was subjected to food safety protocols and HACCP certified. The manager successfully incorporated, monitored and ensured effective cold chain adherence and, therefore, created an innovative framework to implement cold chain logistics for optimum quality control. Therefore, cold chain operations for fresh-cut fruits were achievable because of the introduction of cold chain logistics throughout the supply chain including distribution to school outlets using a refrigerated truck. Shifting his operations from the partial implementation of low-temperature conditions where fresh-cut fruits were prepared at room temperature then placed in a chill room at 4°C–5°C and transported in styrofoam containers in a non-refrigerated truck to schools at various locations in Suriname to transporting in refrigerated trucks held at the same temperature as the chill room resulted in a significant decrease in postharvest losses from 35%–45% to less than 2%–3% (Mohammed and Kitinoja, 2016).

12.4 CONCLUSIONS AND RECOMMENDATIONS

Based on the evidence highlighted through the case studies, it could be concluded that reducing gaps or breaks within the cold chain practices in the Caribbean is urgently needed to reduce the high rates of postharvest losses and food wastes. Caribbean producers and traders of perishable crops have to refocus on the specific goals of cold chain conditions, which could be ranked in order of priority according to safety, quality, practicality and standardization. Safety is the most important and overriding issue as perishables must be delivered to all stakeholders and be safe to consume. Perishables at the end of the cold chain should retain high-quality attributes. Cold chain conditions should be achievable and practical using current logistics and monitoring technologies. Wherever possible, the standard cold chain conditions should be practised. Stakeholders should also conduct appropriate shelf life trials under well-controlled cold chains versus cold chains that are poorly controlled. To effectively implement these guidelines, all operations must be verified by conducting random checks, frequent monitoring of temperatures, staff training, inventory controls and storage procedures.

Cold chain managers should select packaging materials (both primary and secondary) designed to minimize the risk of microbial, chemical and physical contamination; provide effective insulation; minimize product dehydration; be food-grade compliant and minimized air space within the carton. Moreover, it should have sufficient structural strength to meet the demands of cold chain storage, handling and transport, while taking into account the temperature and humidity conditions anticipated through the cold chain handling steps. Marking the product with recognizable and prominent pack size, lot identification, date code and bar code enhance the efficiency of cold chain operators with traceability, record keeping and effective stock rotation. It may be useful to include the cold chain conditions explicitly on transportation packaging. All this information should appear on two adjacent sides or two opposite sides of the packaging.

REFERENCES

Bridgemohan, P., Mohamed, M. E. S, Mohammed, M., and Felder, D. (2016). Hot peppers: IV. HPLC determination of the relative pungency and fruit quality attributes of eight (8) Caribbean hot pepper landraces. *International Journal of Research and Scientific Innovation* 3(8), 17–29.

Bridgemohan, P., Mohammed, M., Mohamed, M. S., and Bridgemohan, R. S. H. (2017). Hot pepper: VI. Effect of bio-stimulant selected agronomic practices and fruit characteristics on the relative pungency in Caribbean hot peppers. *Academic Journal of Agricultural Research* 5(10), 255–260.

International Institute of Refrigeration IIR. (2009). The role of refrigeration in worldwide nutrition. http://www.iifiir.org/userfiles/file/publications/notes/NoteFood_05_EN.pdf. Accessed 2016 September 26.

Mohammed, M., Isaac, W., Mark, N., St. Martin, C., and Solomon, L. (2014). Effects of curing treatments on physicochemical and sensory quality attributes of three pumpkin cultivars. *Acta Horticulturae 1047*, 57–62.

Mohammed, M., Isaac, W., Mark, N., and Solomon, L. (2013). Postharvest quality attributes of three pumpkin cultivars during storage. The workshop, Trinidad, March CIRSRF CARICOM UWI- McGill University Food Security Project. 15 pp.

Mohammed, M, Wilson, L. A., and Gomes, P. I. (2016). Sodium hypochlorite combined with calcium chloride and modified atmosphere packaging reduces postharvest losses of hot pepper. *International Journal of Research and Scientific Innovation 3* (10), 1–9.

Mohammed, M. (2002). Status of transportation and distribution systems for fresh fruits and vegetables produced in the CARICOM Region. UN FAO regional workshop (March 13th-15th, 2002), Jamaica. Practical elements of postharvest handling of fresh fruits and vegetables produced in the CARICOM region in order to maintain quality and safety, 23 pp.

Mohammed, M. (2013). Postharvest handling of pumpkin and hot pepper for export markets. *NAMDEVCO/ CARIRI Workshop (Trinidad) for Exporters of perishable commodities*, 11 pp.

Mohammed, M., and Craig, K. (2018). UN/FAO Food loss analysis: Causes and solutions. Case study on cassava farine, tomato and mango value chains in St. Lucia. ISBN: 978-92-5130585-0, 52 pp.

Mohammed, M., and Kitinoja, L. (2016). Gaps in the cold chain in the Caribbean. *Proceedings of the 3rd World Cold Chain Summit to Reduce Food Losses and Waste.* 1st-3rd December 2016, Singapore, 7 pp.

13 Solar-Powered Cold Storage
ColdHubs in Nigeria

Olubukola M. Odeyemi
Federal University of Agriculture, Abeokuta

Nnaemeka C. Ikegwuonu
ColdHubs Nigeria Limited

CONTENTS

13.1 INTRODUCTION

Nigeria is the most populated country in Africa with an estimate of over 200 million people. It also boasts the largest economy on the continent with the highest youth population in the world (World Bank, 2019). With the large population, agriculture remains one of the mainstays of Nigeria's economy providing a source of livelihood for most Nigerians. However, this sector faces a series of challenges which include outdated land tenure system, less developed irrigation system, limited adoption of research findings and technologies, high cost of farm inputs, poor access to credit, inefficient fertilizer procurement and distribution system, inadequate storage facility and poor access to markets. All these factors combined result in low agricultural productivity and high postharvest losses and waste (FAO, 2020).

The postharvest food loss along the value chain in Nigeria has been a major challenge over decades. Agriculture production has been increasing in the nation but so have the postharvest losses. Sawicka (2019) reported that food losses in the developing countries occur mainly at the early stages of the value chain and this can be linked to the financial management and technical constraints in the harvesting techniques as well as storage and cooling infrastructure. Specifically, in sub-Saharan Africa, around 36% of food harvested is lost, equating to an average of 167 kg per capita annually, where only 7 kg is wasted at the consumer level (FAO, 2011). In Nigeria, the annual postharvest losses along the value chain are estimated to be more than 51.3 metric tonnes valued at about US$ 10 billion (NGN 2.7 trillion), which translates into substantial economic losses (Okoruwa, 2019). The first mile of most food value chains is in the rural areas where functional cold systems are absent and up to two-thirds of overall postharvest losses occur. The food losses are much higher

DOI: 10.1201/9781003056607-16

in these rural communities because they lack even basic facilities and infrastructure for postharvest handling and are in dire need of development. More so, the adoption of modern technologies, wherever available in the early stages of supply chains (harvest, storage, transport) in developing countries, is still relatively low (Chintada et al., 2017). Generally, a farmer or other stakeholders who lack cooling or refrigeration equipment sustain high losses of perishable food products.

The cold chain infrastructure in Nigeria, where available, is concentrated mostly in or around urban areas. The Global Alliance for Improved Nutrition (GAIN) Cold Chain Capacity Mapping of Lagos, Nigeria, revealed that >200,000 m³ refrigerated spaces existed in 2018 and were powered by grid power stations. The power sector within the country generates around 4,000 megawatts of electricity most days which is insufficient to deliver reliable electric energy for the growing populace of over 200 million people. However, food production, storage and preservation are highly dependent on electric energy. Joardder et al. (2020 stated that most of these food losses occur due to the unavailability of electricity. One of the limitations for developing suitable preservative equipment for perishable produce has been this scarcity of reliable electric energy. Therefore, the need to focus on alternate sources of energy such as solar energy for powering such equipment cannot be overemphasized. The off-grid solar sector has grown tremendously over the past 10 years into a vibrant technological advancement, still developing with an annual market currently worth US$ 1.75 billion (Lighting Global, 2020).

Temperature management is important in the postharvest handling of fresh fruits and vegetables as a means of reducing food loss along the value chain. Storing these food commodities at cooler temperatures reduces the respiration rate, extends shelf life, slows down ripening and maintains the product quality. It also reduces weight loss by decreasing the rate of water loss and decay in stored produce. The ColdHubs is an automated off-grid solar-powered cold storage structure designed to extend the storage life of fruits, vegetables and other perishables to reduce food spoilage and waste. This, in turn, increases the income and improves the wellness of the major stakeholders along the horticultural value chain. In this chapter, the role of ColdHubs in filling the gaps in the cold chain in Nigeria has been explained as a case study.

13.2   POSTHARVEST LOSSES IN NIGERIA

Agriculture is the stronghold of Nigeria's economy, employing approximately two-thirds of the country's total labour force and contributing 40% to the Nigerian GDP. Nigeria is the most densely populated country in the African continent and is still struggling to achieve self-sufficiency to feed over 200 million people (Yahaya and Mardiyya, 2019). Food availability is not adequate in developing countries with an average food supply containing less than 2,000 Kcal per person, which is significantly below the average daily requirement of 2,500 Kcal per capita per day (Babatunde et al., 2010).

In Nigeria, it is estimated that about 30% – 50% of fresh fruits and vegetables are lost during postharvest handling between the farm gate and markets (Busari et al., 2015; Yahaya and Mardiyya, 2019; Kitinoja et al., 2019). While among the postharvest losses, globally it is estimated that about 25% – 30% are for animal products and 40% – 50% for roots, tubers, fruits and vegetables (FAO, 2011). Kitinoja et al. (2019) estimated 30% as the postharvest losses along the tomato value chain in Nigeria which amounts to an economic loss of US$ 446 million (N 160.7 billion) and this value could be on the increase yearly. Thus, the food loss also results in loss of earnings for the farmers and vendors. These losses were also equated to the loss in calories and nutrition. Annual losses of 116 million kilocalories, 5.65 billion g of protein, 5,356 billion IU vitamins A and 88 billion mg vitamin C were documented when 30% of tomato is lost along the value chain in Nigeria. Vitamin C loss alone could satisfy the nutritional requirements of 3 million people for 1 year in the country (Kitinoja et al., 2019).

Food loss refers to food that spills, spoils and suffers a reduction in quality such as bruising or wilting or otherwise gets lost before it reaches the consumer (Lipinski et al., 2013). Food loss occurs as an unintended result of an agricultural process or technical limitation in handling, storage, infrastructure, packaging or marketing. Furthermore, postharvest losses show up as decreased nutritional quality (loss of vitamin, development of health dangers such as mycotoxins) or decreased market value (Kitinoja and Alhassan, 2012). These fruits and vegetables are rendered unfit for consumption due to spoilage after harvesting, making them less acceptable with food safety concerns. Nutritious foods become less available and affordable for consumers (Busari et al., 2015). The lack of reliable and adequate cold chain facilities in sub-Saharan Africa is one of the main causes of postharvest losses of perishable products.

13.3   COLD CHAIN MANAGEMENT TO REDUCE POSTHARVEST LOSSES

Cold chain management along the horticultural value chain plays an important role in reducing losses. One of the most crucial points when considering reduction in the incidence and magnitude of postharvest losses is precooling which involves rapid removal of the field heat of the produce after harvest. This helps to maintain optimum produce temperature throughout the entire supply chain. Reducing food losses offers an important pathway of availing food, alleviating poverty, improving nutrition and ensuring food security along the value chain (Affognon et al., 2015). Reliable and efficient cold chains contribute not only to reduce these losses but also to improve the technical and operational efficiency of the food supply chain. However, most developing countries, including Nigeria, lack access to affordable cooling or refrigeration systems for precooling, transport, cold storage or freezing during the postharvest handling and distribution of produce (Kitinoja and Tokala, 2019). Nigeria faces acute electricity problems; the demand far outstrips the supply which is also a hindrance for an unbroken cold chain. Improving the cold storage capacity of perishable produce in Nigeria has the potential to provide significant economic benefits and overcome the socio-economic constraint of inadequate infrastructure.

13.4   COLD CHAIN DEVELOPMENT IN NIGERIA

The cold chain management along the value chain plays an important role in reducing losses in perishable crops. The lack of reliable and adequate cold chain facilities in sub-Saharan Africa is one of the main causes of losses of perishable products (FAO, 2011). One of the major challenges in Nigeria is that extent of the cold chain capacity and utilization presently is undefined. Globally, cold storage capacity reached 616 million m^3 in 2018, an increase of 2.7% since 2016. The three largest country markets – India, the United States and China – accounted for 60% of the global total of refrigerated space. On average, there are approximately $0.2\,m^3$ of refrigerated warehousing space per urban resident globally (Salin, 2018). In 2014, the Global Cold Storage Capacity Report showed that Nigeria had $10,000\,m^3$ refrigerated spaces which were one of the lowest when compared to values obtained from other countries. In 2018, an estimate of over $200,000\,m^3$ was reported indicating that the cold storage capacity has grown significantly with the potential to develop further (GAIN, 2018).

Ideally, a country like Nigeria, considering its size and population, should have at least 250 million m^3 of cold chain space which would translate to about 200,000 cold stores and hundreds of thousands of cold trucks. However, this capacity is yet to be fully harnessed. In Lagos, Nigeria, the estimated total cold chain storage capacity of $63,339\,m^3$ was recorded in 13 local government areas from 244 cold rooms facilities and 27 refrigerated logistics vehicles. A sizeable number of the cold chain facilities captured were locally built and maintained by local cold chain experts and technicians (GAIN, 2018). These cold storage spaces are reportedly used for storing frozen foods such as chicken, turkey and fish, and they rely heavily on diesel generators for their operations as alternative power due to the erratic power supply within the country. The reliance on diesel significantly increases the operational costs of the cold storage facility. None of this cold storage was reportedly used for storing fruits and vegetables.

Developing a functional cold chain for perishable foodstuffs, comparable to those in industrialized countries, is a big challenge for the Nigerian economy. This would enable developing markets to increase food supply by about 15% (Okoruwa, 2019). Efforts are being made to develop the

Nigeria and West Africa cold chain sector to improve the availability of safe and nutritious food. The Organisation for Technology Advancement of Cold Chain in West Africa (OTACCWA) was established in 2018 and became a member of the Global Cold Chain Alliance (GCCA) as an affiliate partner on 24 February 2020, for the technology advancement of the industry. This organization is working on establishing cold chain standards, logistics, warehousing and solutions to postharvest losses while preserving nutritional components of farm produce. Most importantly, policymakers in the agriculture, energy, education and food sectors should work together to promote the use of cold chain technology, improve logistics, maintenance, services, infrastructure, education and management skills and create sustainable markets for the design, use and funding of cold chains for reducing perishable food losses (Kitinoja, 2013).

13.5   CHALLENGES TO SUSTAINING COLD CHAIN IN NIGERIA

Despite all the benefits associated with the development of the cold chain, Nigeria is still faced with several challenges in sustaining the development of this sector. Okoruwa (2019) listed the constraints on investing in cold chain infrastructures in Nigeria to include the cost of investment, which may be perceived as too high with questionable benefits, and on the other hand, a complete cold chain solution requires pre-cooling, cold storage facility and refrigerating vehicles both for transportation and distribution of food which demands huge electricity consumption, which is almost unavailable. However, the adoption of innovative off-grid solutions such as the use of biomass, LPG and solar as an alternative to electricity, to effectively develop the cold storage and supply chain sector may assist in overcoming these challenges. These innovative solutions will be able to address issues of postharvest losses while improving farmer's income

13.6   COLDHUBS IN NIGERIA

There is a need to create affordable sustainable power solutions to develop the cold chain sector in Nigeria especially during the storage of produce. The power grid in Nigeria is not stable and sometimes not available, particularly in rural areas where food is primarily produced. The ColdHubs is an initiative developed by Dresden Institute for Air and Refrigeration (ILK), installed by Smallholders Foundation in Nigeria to solve the challenge of stakeholders along the value chain who do not have access and technical know-how of storing produce after harvest and during handling from the farm till it gets to the consumer. This off-grid solar-powered cold storage, with a renewable power source for refrigeration, is useful in remote communities where the power supply is limited or non-existent in Nigeria. The pilot project was implemented under the initiative "Powering Agriculture – An Energy Grand Challenge for Development" (PAEGC). This technology focused on feasible cold storage infrastructure that links smallholders with the market.

13.6.1   Justification for ColdHubs in Nigeria

The ColdHubs presents cooling to its users, which include farmers, wholesalers and retailers, by operating on a pay-as-you-store model. These business models allow users to pay for their products through technology-enabled, embedded consumer financing (Lighting Global, 2020). Farmers can store freshly harvested produce, mostly fruits and vegetables and other perishables, in cold storage without purchasing or owning to the cold storage unit. The purchase of a cold storage facility or structure has been an age-long challenge to most, especially for smallholder farmers over the years because they do not have the financial capacity to procure such facilities. More so, the cost of maintaining such a facility on-grid, wherever available in the urban centres, is high because of the unreliable nature of the power supplied and the dependence on fuel which increases the cost of operations. Each refrigerating unit is a hub that could be installed on the farms and market, within the reach of smallholder farmers and food handlers.

In 2019, 19 ColdHubs units were operational in three states (Imo, Akwa Ibom and Nasarawa) in Nigeria (Table 13.1) serving over 2,000 smallholder farmers, wholesalers and retailers (Table 13.1). Plans are underway to deploy additional 21 ColdHubs in 2020 to serve over 2,100 users.

13.6.2 ColdHubs Design and Operation

A typical ColdHubs is designed with an exterior dimension of 10 m × 10 m × 7 m with the capacity to store 2 metric tonnes of perishable produce. This capacity can withhold 150 units of plastic crates (0.54 m × 0.36 m × 0.27 m) stacked on the floor. The cold room walls are made of 120 mm insulating panels to retain low temperatures. ColdHubs combines the technologies of photovoltaic energy supply and compressive cold generation. The energy provided by solar panels mounted on the roof of the cold room is stored in high capacity batteries which allow an inverter to feed the monoblock refrigerating units. These are charged by solar panels generating about 5.7 kWh of energy. The cooling temperature within the storage chambers is adjustable between 5°C and 15°C, and the cold room door shuts efficiently to retain the cold air within the enclosed chambers. This design is rugged and can withstand the harshest weather condition prevalent in humid zones. The ColdHubs functions effectively by extending the shelf life of perishable food and making more nutritious food available to both rural and urban dwellers. It also increases the annual income of smallholder farmers by an estimated 25% and creates jobs for the local farmers and women, especially those who manage the operations and collect revenue at ColdHubs stations (Figures 13.1 and 13.2).

TABLE 13.1

ColdHubs Outlets in the Three States in Nigeria (2019)

	Imo	Akwa Ibom	Karu
Markets	Relief Market Owerri	Ibaka and Ibeno Fresh Fish Hubs	Garum Mallam Tomato Farmers Cluster, Kano
	Orlu Central Market		Kokami-Danja Farmers Cluster, Danja Katsina
	Obowo Central Goods Daily Market		Dutsin Wei Farmers Cluster and Market
	Ubomiri New Market		Kaduna and New Orange Fruits and Market
	Owerri World Bank Housing Estate Market		

FIGURE 13.1 Solar-powered ColdHubs storage. (www.coldhubs.com.)

FIGURE 13.2   Reusable Plastic crates used in ColdHubs. (www.coldhubs.com.)

TABLE 13.2

Key Performance Indicators Based on the Ownership Business Model

Internal Rate of Return (IRR) after 20 years	48%
Debt Service Ratio	1.36
Utilization rate	1.1

Source: Energypedia (2019).

13.6.3 MODE AND COST OF COLDHUBS OPERATIONS

ColdHubs Limited owns and operates all facilities within Nigeria on an ownership-based business model. The initial cost for installing a ColdHubs unit ranges between US$ 28,500 and US$ 33,500, but with this business model based on key performance indicators, an initial investment of US$ 28,500 is still profitable. The ColdHubs unit is expected to pay for itself after 3 years of operation. This business model is profitable as long as the utilization rate of the cold storage unit is above 94%. The current high productivity adds to the viability of the innovation. However, extensive use also leads to the faster discharge of the batteries which has a durability of 6 years. Replacing the batteries after every 3 years, the Internal Rate of Returns (Table 13.2) still exceeds 40% which makes the business model and innovation more viable and profitable.

Each ColdHubs unit is operated by an operator employed to monitor the day-to-day activity of the centre by ensuring proper loading and offloading of crates and stacking arrangement of produce within the storage chambers. Reusable plastic crates are provided by the centre as part of the services rendered to the users. The storage of fresh farm produce attracts a daily fee of US$ 0.25 (N 100) per reusable plastic crate per day. The cold room is used intermittently by 3–4 users who fill to maximum capacity. Records showed that 10% of the users stored their produce more than five times each month. The ColdHubs unit established in the Relief market in Imo State has 45 customers of which 30 were regular. Operations in 2018 also revealed that five units of ColdHubs reduced postharvest loss by 50% in their area of deployment. This translates to an estimate of 11,400 tonnes of food saved from spoilage while increasing the household income of 612 users from US$ 60 to US$ 120.

13.7 CONCLUSIONS

The cold chain is a requirement for a successful postharvest industry. However, refrigeration capacity is yet to be fully harnessed, resulting in huge losses of perishable food in Nigeria, due to unreliable power supply on the grid. ColdHubs, the off-grid solar-powered cold room, provides a sustainable solution to the problem of postharvest losses in horticultural produce and other perishable food in Nigeria, using solar power as an alternative power supply for cold storage. This provides cold storage capacity, helps to extend the shelf life of produce, improves the livelihood of farmers and vendors and provides jobs for those engaged in its operations.

REFERENCES

Affognon, H., Mutungi, C., Sanginga, P. and Borgemeister, C. (2015). Unpacking postharvest losses in sub-Saharan Africa: A meta-analysis. *World Development. 66*, 49–68.

Babatunde, R.O., Adejobi, A.O. and Fakayode, S.B. (2010). Income and calorie intake among farming households in rural Nigeria. Results of parametric and nonparametric analysis. *Journal of Agricultural Science. 2* (2), 135–146.

Busari, A.O., Idris-Adeniyi, K.M. and Lawal, A.O. (2015). Food security and postharvest losses in fruit marketing in Lagos Metropolis Nigeria. *Discourse Journal of Agriculture and Food Sciences. 3*(3), 52–58.

Chintada, V.G., Satyanarayana, K.V., Manyam, S.C., Srilasya, N.K. and Swati, H. (2017). Cold chain technologies – transforming food supply chains. New Delhi: The Associated Chambers of Commerce and Industry of India (ASSOCHAM). (https://www.sathguru.com/news/wp-content/uploads/2017/05/Cold-Chain-Report.pdf). Accessed 4 April 2020.

ColdHubs. (2020). ColdHubs: Solar-powered cold storage for developing countries. www.coldhubs.com. Accessed 3 March 2020.

Energypedia. (2019). ColdHubs- Solar cold rooms in Nigeria. www.energypedia.info/wiki/coldHubs_Solar_Cold_Rooms_in_Nigeria. Assessed on 17 May 2020.

FAO (2020). FAO in Nigeria: Nigeria at a glance. Food and Agriculture Organization of the United Nations.

FAO. (2011). *Global Food Losses and Food Waste – Extent, Causes and Prevention.* Eds. J. Gustavsson, C. Cederberg, U. Sonesson, R. van Otterdijk and A. Meybeck. Food and Agricultural Organization of the United Nation, Rome.

GAIN. (2018). The Global Alliance for Improved Nutrition - Nigeria Cold chain capacity mapping. Postharvest Loss Alliance for Nutrition.

Joardder, M.U., Mandal, S. and Masud, M.H. (2020). Proposal of a solar storage system for plant-based food materials in Bangladesh. *International Journal of Ambient Energy. 41*(14), 1664–1680.

Kitinoja, L. (2013). Use of cold chains for reducing food losses in developing countries PEF White Paper No. 13-03. The Postharvest Education Foundation (PEF).

Kitinoja, L. and Alhassan, H.Y. (2012). Identification of appropriate postharvest technologies for small-scale horticultural farmers and marketers in sub-Saharan Africa and South Asia-Part 1. Postharvest Losses and Quality Assessment. In *Proceedings XXXVIII IHC-IS on Postharvest Technology in the Global Markets.* Eds. M.J. Cantwell and D.P.F. Almeida. *Acta Horticulturae, 943*, 31–40. https://doi.org/10.17660/ActaHortic.2012.934.1.

Kitinoja, L. and Tokala, V.Y. (2019). Postharvest education training and capacity building for reducing losses in plant-based food crops- A critical review (2010–2017). In: *Postharvest Extension and Capacity Building for the Developing World.* Eds. M. Mohammed and V.Y. Tokala, World Food Preservation Center Book Series. CRC Press, Boca Raton, FL, pp. 29–41.

Kitinoja, L., Odeyemi, O.M., Dubey, N., Musange, S. and Gill, G. (2019). Commodity system assessment methodology studies on tomato in Sub Saharan Africa and Southern Asia. *Journal of Horticulture and Postharvest Research. 2*, 15–90.

Lighting Global. (2020). Off-grid solar market trends report 2020. www.lightingglobal.org

Lipinski, B., Hanson, C., Lomax, J., Kitinoja, L., Waite, R. and Searchinger, T. (2013). *Reducing Food Loss and Waste.* World Resource Institute, Washington, D.C.

Okoruwa, A. (2019). Advancing sustainable cold chain developments in Nigeria. PLAN-Postharvest Loss Alliance for Nutrition. GAIN-Global Alliance for Improved Nutrition. Presented at the Agrofood Conference, Agrofood Nigeria. Landmark Centre, Victoria Island, Lagos, Nigeria.

Salin, V. (2018). GCCA Global cold storage capacity report. International Association of Refrigerated Warehouses (IARW) A Global Cold Chain Alliance.

Sawicka, B. (2019). Post-harvest losses of agricultural produce. *Sustainable Development. 1*(1), 1–16. doi:10.1 007/978-3-319-69626-3_40-1.

World Bank. (2019). The World Bank in Nigeria. www.worldbank.org/en/country/nigeria/overview. Assessed 15 July 2020.

Yahaya, S.M. and Mardiyya, A.Y. (2019). Review of post-harvest losses of fruits and vegetables. *Biomedical Journal of Scientific and Technical Research. 13*(4), 10192–10200.

14 Low-Cost Cooling Technology to Reduce Postharvest Losses in Horticulture Sectors of Rwanda and Burkina Faso

Eric Verploegen
Massachusetts Institute of Technology

Mandeep Sharma, Rashmi Ekka, and Gurbinder Gill
Agribusiness Associates Inc.

CONTENTS

DOI: 10.1201/9781003056607-17

14.1 INTRODUCTION

The information presented in this chapter was derived from the work performed during Agribusiness Associates Inc. (ABA) implemented **Reducing Postharvest Losses in Rwanda** and **Improving Postharvest Practices for Tomatoes in Burkina Faso** projects under the USAID's Feed the Future initiative. During these projects, ABA partnered with the Massachusetts Institute of Technology's (MIT) D-Lab to conduct a research study on exploring the potential of non-electric evaporative cooling devices for reducing Postharvest losses (PHL) in the horticulture sector. The findings of the study are presented in this chapter.

14.1.1 Postharvest Food Losses in Rwanda

In Rwanda, agriculture serves as the backbone of the economy and is crucial to its growth and reduction in poverty. The agriculture sector accounts for 23.5% of gross domestic production and 62.4% of total employment. Various estimates say that up to 40% of food is lost in the postharvest stage (Rwanda Agriculture, 2019). While simple approaches exist to reduce postharvest losses such as improved handling of horticultural crops, there is no one isolated intervention that will prove effective at mitigating this issue.

14.1.2 Reducing Postharvest Losses in Rwanda Project

In this investigation, a wide range of horticultural crops including the four key crops – chillies, tomatoes, bananas, and sweet potatoes – were selected for the project. The project was implemented by Agribusiness Associates (ABA) in partnership with the Rwanda Agriculture Board (RAB), National Agricultural Development and Export Board (NAEB), University of Rwanda (UR), and The Postharvest Education Foundation (PEF). The 3-year project started in 2016 and ended in 2019.

During the project, research was conducted to adapt postharvest tools and technology to assist usage behaviour in Rwanda. This included research on evaporative cooling, cold rooms, solar dryers, and other technologies. The project's postharvest-related capacity building efforts focused on providing in-depth training to people as well as broad-based knowledge of postharvest. It was done through demonstrations to over a thousand people during the Agri-shows, Conferences, and Postharvest Campaigns. These events were conducted with the help of cooperatives and the training was provided by the Food Security Graduates of NAEB. Technology adoption initiatives included working with farmers, cooperatives, and private sector entrepreneurs to increase their adoption of zero energy cooling chambers (ZECC), field sheds, cold rooms, and crates. The project created partnerships with local cooperatives and assisted them in technology adoption using a cost-sharing and partnership approach along with technical assistance and on-field capacity building. In total, the project assisted in the construction of multiple collection centres and cold rooms. RAB worked with small-scale processors and equipped them with small tools. The University of Rwanda started a collection centre, sensory lab, and processing kitchen at its Postharvest Training and Services Center established up by the project.

The project worked with Cooperatives to increase the adoption of harvesting bags and crates and to build local capacity in Rwanda. Several training and events conducted by the project included (i) a cold room training for technicians; (ii) postharvest training for extension agents and lead farmers; (iii) farmers were trained through on-field extension; (iv) experts attended the Postharvest Conference in Kigali, 2018–2019 and regional horticultural experts attended the 2 day East Africa Horticulture Experts Meeting; and (v) students at the University of Rwanda received training on Postharvest Management.

Moreover, the project supported the University of Rwanda to develop the curriculum for a new Master's Program in Postharvest Management. Research and assessment efforts included assessment of cold chain and adaptive research for the postharvest extension.

14.1.3 Tomato Postharvest Losses in Burkina Faso

In Burkina Faso, tomatoes are an important crop for increasing household resilience and nutrition. However, the farming of tomatoes is largely at a subsistence level and farmers face many challenges, including poor shelf life and low bargaining power due to a supply glut in the main season. The majority of farmers lack access to appropriate seed varieties as well as proper training on growing, harvesting, postharvest storage, and marketing tomatoes. Farmers are affected by the high postharvest losses in tomato production because of its high perishability and lack of appropriate storage infrastructure. Exposure to sun and heat during delays/waits and transportation increases rotting. Farmers use only the traditional processing technique of drying tomatoes, which is done on the ground. Because of the absence of other processing techniques, farmers have not uncovered new marketing opportunities in sauces, pastes, etc.

14.1.4 Improved Postharvest Practices for Tomatoes in Burkina Faso Project

The project "Improving Postharvest Practices for Tomatoes in Burkina Faso" was aimed at improving postharvest handling, storage, processing, and marketing of tomatoes in Burkina Faso. The project was implemented in partnership with Institut de l'Environnement et de Recherches Agricoles de Burkina Faso (INERA), the main agricultural research body in Burkina Faso, and the Postharvest Education Foundation.

As part of the project, a cold room and a postharvest training centre were constructed. The results indicated that evaporative cooling can be used to increase the shelf life of fruits and vegetables by providing a more stable temperature and humidity. Training sessions were conducted for farmers and agronomists in postharvest including demonstrations for evaporative cooling. Training sessions highlighted that (i) controlling temperature from harvest to market can reduce losses; (ii) on the field, farmers should harvest in the morning and evening when the temperature is cool; (iii) collected produce should be kept in shade with good ventilation as much as possible; (iv) produce should be cooled immediately after harvest; and (v) to store for a few days, simple measures such as evaporative cooling can extend the shelf life of the produce. Farmers and extension agents from more than 35 communities were involved in this training and also included the participation of Catholic Relief Services and the Regional Directorate of the Agriculture Service.

14.2 CASE STUDY: EVAPORATIVE COOLING FOR IMPROVED FRUIT AND VEGETABLE STORAGE IN RWANDA AND BURKINA FASO

In 2019, MIT D-Lab (https://d-lab.mit.edu/), in partnership with Agribusiness Associates Inc. (http://www.agribusinessassociates.com/) and Horticulture Innovation Lab (https://horticulture.ucdavis.edu/), conducted a research study on low-cost evaporative cooling devices designed to improve the postharvest storage life of fresh produce. Most techniques for cooling and storing fruits and vegetables rely on electricity – which is unavailable or unaffordable in most rural areas in Rwanda and Burkina Faso – limiting access to effective and affordable postharvest storage options. The evaporative cooling devices that are the subject of this study function without the use of electricity and so are well suited for regions without electricity access, or where electricity-dependent cooling and storage technologies are not affordable. Effective, affordable cooling and storage technologies have the potential to prevent food loss, increase access to fresh produce, and create opportunities for additional income generation in off-grid areas and where electricity is intermittent or prohibitively expensive (Arah et al., 2016; Basediya et al., 2013; Odesola and Onyebuchi, 2009).

This project had three Postharvest Training and Services Centers in Mulindi (Center), Busoga (North), and Rubona (South). 'Improving Postharvest Practices for Tomatoes in Burkina Faso' (https://horticulture.ucdavis.edu/project/improving-postharvest-practices-tomatoes-burkina-faso) was implemented in partnership with INERA. The Project had one Postharvest Training and Services

Center in Kamboinsi at an INERA research station. Both projects had worked widely to increase the adoption of evaporative cooling devices.

14.2.1 THE CHALLENGE

Storage conditions throughout the supply chain play an important role in preventing postharvest losses for fruits and vegetables. While the optimal storage conditions vary for different fresh produce, many fruits and vegetables are best stored in a cool and humid environment to prevent rot and dehydration (McGregor, 1989).

Tomatoes are an important source of income, a key ingredient in the local cuisine, and an important crop for increasing household nutrition in both Rwanda and Burkina Faso. However, tomato farmers face many challenges, chief among them is decreased bargaining power due to a supply glut in the main season. They are also susceptible to degradation through dehydration, rot, bruising, and fungal growth. In both cases, access to improved storage could provide benefits at several stages along the value chain including on the farms directly after harvest, at farming cooperative aggregation centres, for fresh produce vendors at markets, and in consumers' homes.

14.2.2 OVERVIEW OF THE EVAPORATIVE COOLING TECHNOLOGIES STUDIED

Two classes of non-electric cooling and storage technologies were evaluated in this study:

- Zero energy cooling chambers – ZECCs
- Clay pot coolers – also known as Zeer pots

These devices function on the principle of direct evaporative cooling which causes a decrease in temperature and an increase in the relative humidity inside the storage device, conditions that increase the shelf life of many fruits and vegetables (Kader, 2005). Due to their relatively large size, ZECCs are typically used by larger producers or community groups, whereas clay pot coolers are typically used at the household level due to their simple construction and relatively small size (Figures 14.1 and 14.2).

Several variations on the common pot-in-pot design were included in this study. The following configurations were tested:

- Clay pot in a clay pot (Rwanda and Burkina Faso)
- Plastic container in a clay pot (Rwanda)
- Metal container in a clay pot (Rwanda)
- Clay pot in a plastic dish (Rwanda and Burkina Faso)
- Clay pot in a metal dish (Rwanda and Burkina Faso)
- Plastic container in a clay dish (Burkina Faso) (Figure 14.3)

14.2.3 PREVIOUS RESEARCH RESULTS

Several studies present findings indicating that the improved storage conditions provided by evaporative cooling devices lead to retaining good fruit and vegetable quality – such as weight, colour, firmness, and deterioration – resulting in extended shelf life (Basediya et al., 2013; Ambuko et al., 2017). Because evaporative cooling devices do not require electricity to function, they have the potential to be particularly beneficial for users in areas with limited or prohibitively expensive electricity access. Regardless of the context, the low energy consumption and use of simple materials make evaporative cooling devices an environmentally friendly alternative to refrigeration systems that use electricity, for short duration storage.

Reports from multiple studies in India indicate that brick ZECCs can provide temperature reductions of $10°C – 15°C$ when the ambient temperature is greater than $35°C$ and the ambient relative

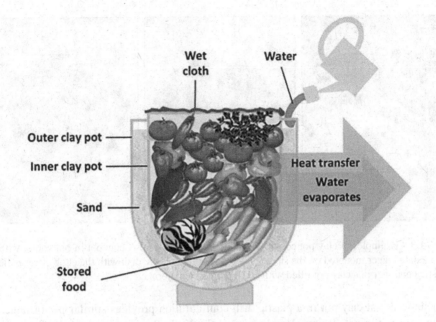

FIGURE 14.1 Diagram of a clay pot in clay pot cooler, covered by a wet cloth. (Adapted from Peter Rinker, CC BY-SA 3.0, https://commons.wikimedia.org/w/index.php?curid=33444154, Accessed 3 January 2018.)

FIGURE 14.2 Brick evaporative cooling device in Rubona, Rwanda.

humidity is less than 40% (Basediya et al., 2013; Kumar et al., 2014). In a separate study, clay pot coolers demonstrated temperature reductions of 5°C –10°C when the ambient temperature is greater than 40°C and the ambient relative humidity is less than 30% (Morgan, 2009). Research conducted by MIT D-Lab and the World Vegetable Center in Mali showed that when the ambient humidity is <40%, brick ZECCs and clay pot coolers can be expected to decrease in the average temperature and peak daily temperature between 5°C to 7°C and 8°C to 12°C, respectively. This

FIGURE 14.3 Examples of clay pot coolers included in this study. (a) A clay pot-in-pot cooler with a cloth cover and a data logger mounted on the side; (b) A clay pot in a plastic dish with the cloth cover pulled back; (c) A plastic container in a clay pot filled with chilli peppers for storage.

research showed that clay pot in a plastic dish configuration provides similar performance to the more common pot-in-pot design (Verploegen et al., 2018). Across all of the studies referenced, regardless of the ambient conditions, the brick ZECCs and clay pot coolers were shown to maintain relative humidity above 80% in the interior of the device where the fresh produce is stored. The same principle of evaporative cooling has been used with other designs and materials such as charcoal coolers (Rathi and Sharma, 1991; Nobel, 2003) and devices that use synthetic materials to hold and allow for the evaporation of water (Kitinoja, 2016). Another design – commonly referred to as a Janata cooler – consists of a metal or plastic container placed inside of a clay pot or dish, with a wet cloth covering any exposed surface of the inner container (Odesola and Onyebuchi, 2009; Roy and Khurdiya, 1985).

When assessing the potential benefits of a cooling and storage device for a given context, it is essential to consider how the storage conditions that can be achieved within the device compare to the conditions without the device. For example, the ideal storage conditions for tomatoes are between 18°C and 22°C with humidity between 90% and 95%; if the ambient conditions present an average temperature of 35°C and relative humidity of 20%, a storage device that provides conditions with an average temperature of 30°C and greater than 80% humidity can provide a significant increase in shelf life for tomatoes (McGregor, 1989).

14.2.4 STUDY DESIGN

The objective of this study was to evaluate a set of non-electric cooling and storage technologies for their suitability to improve the postharvest storage of fruits and vegetables in Rwanda and Burkina Faso and used a combination of:

- Electronic sensors to monitor the performance of the evaporative cooling devices
- Measurements of fruit and vegetable shelf life stored in evaporative cooling devices compared to storage in ambient conditions
- Structured user interviews with existing and potential users of evaporative cooling devices, fresh produce vendors, and producers of clay pots and other containers.

The research was conducted over 3 months, from February to April of 2019, at one location in the town of Kamboinse in Kadiogo Province, Burkina Faso, and three locations in Rwanda (Mulindi, Rubona, and Busogo), shown in the map below (Figure 14.4).

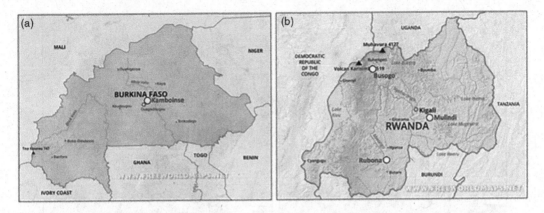

FIGURE 14.4   (a) A map of Burkina Faso; (b) A map of Rwanda. The locations where the study was conducted are labelled with white circles.

The evaporative cooling devices included in this study were selected to include a range of designs constructed from locally available materials. The clay pot coolers were assembled for this study.

14.2.4.1   Electronic Sensors Methodology

Electronic sensors developed by Sensen (https://www.sensen.co/) were installed on the ZECCs (4) and clay pot coolers (16) monitored the following parameters:

- Exterior (ambient) temperature
- Exterior (ambient) relative humidity
- Interior temperature
- Interior relative humidity
- Sand moisture

Data for each of the five parameters were recorded every 5 minutes for the 3 months of the study period. Project staff was trained on the installation and data retrieval for the electronic sensors designed for this study (Figure 14.5).

14.2.4.2   Fruit and Vegetable Shelf Life Methodology

Measurement of fruit and vegetable shelf life was conducted for a variety of fresh produce in each of the evaporative cooling devices included in this study. Fruits and vegetables that were recently harvested before being fully ripe were purchased, and 2–5 kg of each fruit or vegetable was weighed at the start of the experiment and then placed in each of the evaporative cooling devices and a container in the shade exposed to ambient conditions. The fresh produce (tomatoes, mangoes, carrots, cabbage, chilli peppers, and French beans) was weighed and visually inspected for evidence of fungal growth, rot, dehydration, bruising, discolouration, or other signs of deterioration every 2–3 days. The fresh produce was left in the chamber until signs of deterioration were observed, or the fruits and vegetables were determined to be fully ripe through visual inspection. The crops chosen are those susceptible to high losses and are grown in different experiment sites.

14.2.4.3   User Interview Methodology

Structured individual interviews were conducted with ZECC users, farmers, and households that were potential users of evaporative cooling devices (non-users), fresh produce vendors, and vendors of clay pots and other containers. Table 14.1 shows the number of each interview type conducted in Rwanda and Burkina Faso. These interviews explored:

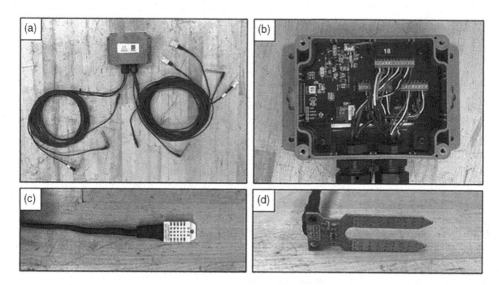

FIGURE 14.5 The sensors used for electronic data collection. (a) A full sensor data logger with sensors attached; (b) Interior of sensor control box; (c) Temperature and humidity sensor; (d) Moisture sensor.

TABLE 14.1
The Number of Each Type of Interview Conducted in Rwanda and Burkina Faso

	Interview Type				
Country	ZECC Users (Farmers)	Non-users (Farmers)	Non-users (Households)	Fresh Produce Vendors	Clay Pot and Container Vendors
Rwanda	8	55	0	10	4
Burkina Faso	0	47	32	15	15

- Types of fruits and vegetables purchased and produced
- Existing methods for fresh produce cooling and storage
- Need for improved cooling and storage technology
- Availability and cost of materials that can be used for evaporative cooling devices

Respondents were randomly selected among members of farming cooperatives whose members had received postharvest training from the Rwanda and Burkina Faso projects and households in communities near the Postharvest Training and Service Centers (PTSCs). Fresh produce vendors and vendors of clay pots, plastic containers, and metal dishes were randomly selected from markets where the farmers and households purchase these products.

14.2.5 SENSOR RESULTS

The data collected from the sensors were used to determine the temperature and relative humidity changes in the interior of the ZECCs and clay pot coolers as a function of ambient temperature and humidity and the frequency of watering. One sensor measuring the exterior (ambient) temperature and humidity was placed near the ZECC or clay pot coolers. Sensors measuring the interior temperature and humidity were located inside the ZECC or clay pot coolers, and a moisture sensor was placed in the sand layer between the two brick walls. During the period where data were being

FIGURE 14.6 Sensor placement on a brick ZECC in Mulindi, Rwanda. (a) Sensor data logger mounted on a brick ZECC; (b) Temperature and humidity sensor mounted on the interior brick wall; (c) Temperature and humidity sensor in a plastic crate containing tomatoes; (d) Moisture sensor in the top of the sand layer between the two brick walls; (e) Moisture sensor in at the bottom of the exterior brick wall; (f) Ambient temperature and humidity sensor under the ZECC shade cover.

FIGURE 14.7 Sensor placement on a clay pot cooler in Mulindi, Rwanda. (a) Sensor data logger and moisture sensor mounted on a clay pot in a clay pot containing tomatoes; (b) Ambient temperature and humidity sensor under the shade cover; (c) Temperature and humidity sensor in a plastic container in a clay pot containing chilli peppers; (d) Moisture sensor in the sand layer between a metal container and a clay pot.

collected, water was regularly added to ensure that the sand in the evaporative cooling devices remained moist (Figures 14.6 and 14.7).

14.2.5.1 Brick ZECC Sensor Data

The climates in Rwanda and Burkina Faso are significantly different, with the humidity being significantly higher in Rwanda throughout the year, including the time when this research was conducted

FIGURE 14.8 (a) ZECC in Mulindi, Rwanda, with bricks that are stacked, a cloth cover supported by a wood and straw frame, and a structure with a tightly woven straw roof providing shade for the majority of the day. (b) ZECC in Kamboinse, Burkina Faso, with bricks that are held together by mortar, a plywood cover, and a wooden and straw structure that provides partial shade throughout the day.

(average humidity of 82% and 16%, for Rwanda and Burkina Faso, respectively). The following section compares brick ZECCs with two different designs in two different climates (Figure 14.8).

14.2.5.1.1   Impact of Humidity on Cooling Efficiency

When discussing the resulting interior storage conditions (temperature and humidity), it is important to consider the interior temperature achieved in relation to the ambient conditions. Although the ambient humidity was significantly lower in Burkina Faso, when water was added at least once per week, both ZECCs provided an average interior humidity above 90%. Furthermore, both ZECCs maintained interior humidity greater than 80% throughout the day, which was consistent with results from several other studies (Verploegen et al., 2018; Basediya et al., 2013, Odesola and Onyebuchi, 2009).

The temperature decrease that can be achieved is highly dependent on the ambient humidity, where higher humidity significantly reduces the cooling that can be achieved through the evaporation of water. The minimum temperature that can be achieved through evaporative cooling at any specific time is called the "wet-bulb temperature" (LeRoy and Kuehn, 2001). The wet-bulb temperature is dependent on the ambient temperature and humidity, where higher humidity results in a wet-bulb temperature that is closer to the ambient temperature. When evaluating the performance of an evaporative cooling device, we recommend considering the "cooling efficiency," which is the observed temperature decrease divided by the maximum potential temperature decrease. The cooling efficiency is calculated as follows:

$$\text{Cooling efficiency} = \frac{(\text{Ambient temperature - Interior temperature})}{(\text{Ambient temperature - Wet bulb temperature})}$$

14.2.5.1.2   Interior Temperature and Humidity of Brick ZECCs

Figures 14.9 and 14.10 show the ambient, interior, and wet-bulb temperature over 5-day periods for the two brick ZECCs, one in Mulindi, Rwanda, and one in Kamboinse, Burkina Faso. It is important to note that even when there is not a significant decrease in the average temperature, it is beneficial to store fresh produce in an environment with a stable temperature. While the ambient temperature in Kamboinse, Burkina Faso, fluctuated by more than 20°C each day, the temperature inside the ZECC varied by less than 4°C throughout a given day. Similarly, in Mulindi, Rwanda, the ambient temperature fluctuated by more than 10°C and 15°C each day, while the temperature inside the ZECC varied by less than 3°C during a given day. Additionally, the peak daily temperature was

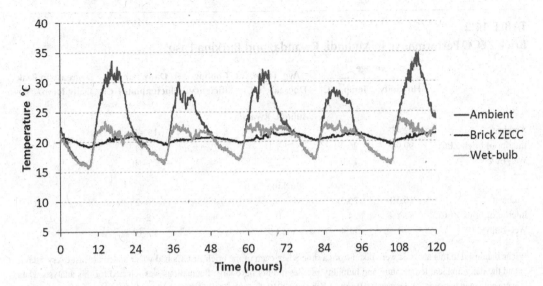

FIGURE 14.9 Typical daily ambient temperature, the interior temperature of the brick ZECC, and the wet-bulb temperature for a ZECC in Mulindi, Rwanda. The points where the wet-bulb temperature is the same as the ambient temperature corresponds to when the ambient humidity was 100%, and no evaporation can occur.

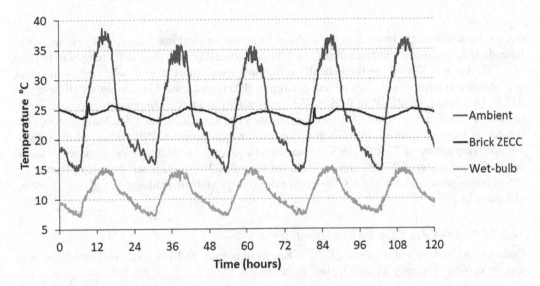

FIGURE 14.10 Typical daily ambient temperature, the interior temperature of the brick ZECC, and the wet-bulb temperature for a brick ZECC in Kamboinse, Burkina Faso.

reduced by 10°C and 15°C in Kamboinse, Burkina Faso, and 7°C – 12°C in Mulindi, Rwanda, in both cases, avoiding peak temperatures above 30°C. Such improved temperature stability and avoidance of high peak temperatures improve the shelf life of many fruits and vegetables.

As seen in Table 14.2, while the average ambient temperature was higher in Burkina Faso than in Rwanda (25.3°C and 22.6°C, respectively), the average wet-bulb temperature in Burkina Faso was lower than in Rwanda (11.4°C and 19.9°C, respectively), yielding a theoretical temperature decrease of 13.9°C and 2.7°C, for Burkina Faso and Rwanda, respectively. This difference was

TABLE 14.2

Brick ZECC Performance in Mulindi, Rwanda, and Burkina Faso[a]

	Average		Avg. Temp.	Cooling	Daily Temp.	Decrease in Peak
	Humidity	Temp. (°C)	Decrease (°C)	Efficiency	Fluctuations (°C)	Daily Temp. (°C)
Mulindi, Rwanda						
Ambient	82.4%	22.6	–	–	15 – 20	–
Interior of brick ZECC	99.9%	20.5	2.1	77%[c]	3–4	7–12
Wet–Bulb [b]	–	19.9	2.7	–	5–10	–
Burkina Faso						
Ambient	16.3%	25.3	–	–	20–25	–
Interior of brick ZECC	93.1%	24.7	0.6	4%	2–3	10–15
Wet–Bulb[b]	–	11.4	13.9	–	5–10	–

[a] The data used in this analysis were taken from a time when each of the brick ZECCs had water added at least every 5 days, and the daily ambient temperature and humidity profiles did not have large fluctuations was selected for this analysis. Data were collected between 29 January 2019 and 11 February 2019, from the brick ZECC in Mulindi, Rwanda, and from the brick ZECC in Kamboinse, Burkina Faso, between 4 February 2019 and 14 February 2019 (brick ZECCs are pictured above).

[b] The wet–bulb temperature is dependent on ambient temperature and humidity LeRoy and Kuehn (2001).

[c] Due to the small difference between the ambient temperature and the wet–bulb temperature in Rwanda, the cooling efficiency is sensitive to small variations in the interior temperature of the ZECC.

due to the significantly lower average ambient humidity in Burkina Faso (16.3%) compared to Rwanda (82.4%). While the brick ZECC in Mulindi, Rwanda, only achieved a temperature drop of 2.1°C, this was 77% of the theoretically achievable temperature drop (cooling efficiency) that was possible to achieve with the ambient conditions during that time – indicating a well-designed ZECC. In contrast, the ZECC in Burkina Faso only achieved a 0.6°C temperature decrease, yielding a cooling efficiency of only 4%. In comparison, a brick ZECC in Bamako, Mali, that was the subject of previous research (Verploegen, Sanogo, and Chagomoka, 2018) achieved an average interior temperature of 23.9°C, 3.9°C lower than the average ambient temperature when ambient conditions averaged 27.8°C and 21% relative humidity. The average wet-bulb temperature for these conditions was 14.3°C (13.4°C lower than the average ambient temperature), giving a cooling efficiency of 29%.

14.2.5.1.3   Brick ZECC Design Considerations

Considering the design of the brick ZECC in Kamboinse, Burkina Faso, there were several reasons that the cooling efficiency was not higher, including:

- The mortar was used to secure the bricks, which reduces the surface available for water to evaporate from (mortar is less permeable to water than most bricks). The mortar was used to make the ZECC permanent as the area is prone to flash floods.
- A solid wooden cover was used to provide security (the wooden cover can be locked to the ZECC wall when closed). However, this kind of cover does not absorb water and is not, therefore, an effective surface for evaporative cooling.
- The wooden structure with a loose straw roof did not provide good shade coverage, as light passes between the pieces of straw and the roof of the structure was not large enough or positioned to effectively prevent exposure to direct sunlight throughout much of the day.

FIGURE 14.11 Construction of a brick Zero Energy Cooling Chamber by cooperative members in Gakenke, Rwanda.

The brick ZECCs in Rwanda were constructed using stacked bricks, a cover directly on top of the ZECC made from a wooden frame and woven straw covered in a cloth blanket and located under thatched shade covers that extend far enough past the ZECC to protect from direct sunlight for a majority of the day. While these design features improved the cooling efficiency, they were not as critical in high humidity environments, as the wet-bulb temperature limited the cooling efficiency that could have been achieved (Figure 14.11).

14.2.5.1.4   Recommendations for Optimized Brick ZECC Design

1. Stacked bricks are commonly used and recommended for a brick ZECC of this height. If a brick ZECC taller than 1.5 m is being constructed, then it is advisable to ensure that the brick walls cannot fall over and injure anyone. Some types of reinforcement, such as a wooden and wireframe or mortar, may be necessary, along with an understanding of how these modifications will impact the performance of the ZECC.

2. A water-absorbent cover directly on the top of the ZECC should be used to allow for subsequent evaporation from its surface and provide additional cooling. Effective materials include cloth or woven straw mats supported by a wooden frame. If additional security is needed, the cover could include a sturdy open wire mesh attached to a wooden frame that can be locked to secure the produce inside the chamber. The cover should also be sufficiently thick so that it can absorb a significant amount of water and provide thermal insulation if it does become dry.

3. The ZECC should be located where direct sunlight is blocked for as much of the day as possible. If constructing a shade cover, the roof material should not allow light to pass through and be large enough and well-positioned to block the sun as it moves across the sky. If available, a tree with thick leaf coverage is also a good option. In addition to the shade provided, the transpiration of water from the leaves provides an additional cooling effect in the area under the tree. Mango trees are well suited for this, as they have thick leaf coverage that is low to the ground and wide enough to block the sun for most of the day. Additional guidance on the construction can be found in: "Evaporative Cooling Best Practices Guide" (http://d-lab.mit.edu/resources/projects/Evaporative-Cooling-Best-Practices-Guide) (Verploegen et al., 2018).

FIGURE 14.12 Examples of the 4-clay pot cooler designs studied at each of the three locations in Rwanda. (a) Clay pot in a clay pot with sensor data logger attached; (b) Plastic container in a clay pot; (c) Metal container in a clay pot; (d) Clay pot in a plastic dish. (All of the clay pot coolers are covered with wet cloth when being used and have a capacity of ~35 L.)

14.2.5.2 Clay Pot Cooler Sensor Data

14.2.5.2.1 Clay Pot Coolers in Rwanda

Four different clay pot cooler designs were tested at three locations in Rwanda (Mulindi, Rubona, and Busogo). The materials for the clay pot coolers were purchased in Kigali in triplicate, so the three versions of each design were nearly identical (Figure 14.12).

By measuring the performance of these clay pot coolers in the same environmental conditions, we gained insights into how design variations impacted the performance of the devices. Figure 14.13 shows the ambient temperature, the wet-bulb temperature, and the interior temperature of two of the clay pot coolers, the "pot-in-dish" and "pot-in-pot", and Table 14.3 shows a summary of the performance of all four devices. Similar to the data for the brick ZECC in Mulindi, Rwanda, due to the high ambient humidity – a monthly average of greater than 75% – the wet-bulb temperature was only 3.8°C lower than the ambient temperature, limiting the amount of cooling achieved through the evaporation of water. All four of the clay pot coolers studied in this portion of the research provided an average interior humidity greater than 95% and provided a decrease in the average temperature between 1°C and 3°C. In addition to the average temperature decrease, the daily peak temperature and temperature fluctuations were both reduced. On days without significant amounts of rain,[1] the ambient temperature typically fluctuated between 10°C and 15°C each day, and the temperature within the clay pot coolers had temperature fluctuations of 6°C or less on most days while reducing the peak daily temperature by 5°C – 10°C.

14.2.5.2.2 Clay Pot Coolers in Burkina Faso

Four clay pot coolers were assembled using materials found at roadside markets in Kamboinse, Burkina Faso, images of each are shown below (Figure 14.14).

The clay pot coolers in Burkina Faso showed the most significant average temperature decreases (> 4°C) of all of the devices tested in this research, particularly the small clay pot in clay pot device which provided nearly an 8°C decrease in the average temperature (Table 14.4). While this device provided the best performance, the utility for storing significant amounts of fresh produce was limited by the volume available for storage (~ 5 L) and, therefore, was best suited for household use.

Figure 14.15 shows a plot of the ambient temperature, the wet-bulb temperature, and the interior temperature of two of the clay pot coolers, the medium clay pot in a plastic dish "Pot-in-dish" and small clay pot in a clay pot "pot-in-pot" over 5 days. In addition to the temperature drop provided,

---

[1] On days with significant amounts of rain, the ambient temperature typically decreases significantly reducing the daily temperature fluctuations and the peak daily temperature.

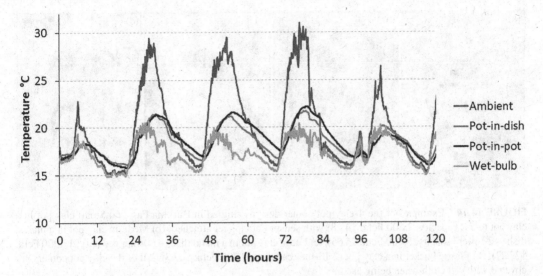

FIGURE 14.13 Typical daily ambient temperature, the interior temperature of the clay pot in a plastic dish (pot-in-dish), the interior temperature of the clay pot in pot cooler (pot-in-pot), and the wet-bulb temperature for a ZECC in Rubona, Rwanda. Only two of the four devices are shown in this figure for clarity, a summary comparing the performance of all four devices is shown below and in the Appendix. The points where the wet-bulb temperature is the same as the ambient temperature corresponds to times when the ambient humidity was 100%, and thus no evaporation can occur. The data in this figure were collected between 6 March 2019 and 10 March 2019.

TABLE 14.3
Clay Pot Cooler Performance in Rubona, Rwanda[a]

	Average		Daily			Decrease in
	Humidity (%)	Temp. (°C)	Average Temp. Decrease (°C)	Cooling Efficiency (%)	Temp. Fluctuations (°C)	Peak Daily Temp. (°C)
Ambient	76.8	20.3	–	–	12–16	–
Clay pot in clay pot	99.9	18.6	1.7	43	4–7	6–8
Plastic container in clay pot	99.6	18.5	1.8	47	4–7	6–8
Metal container in clay pot	95.7	18.1	2.2	57	4–7	6–8
Clay pot in plastic dish	99.9	18.8	1.5	39	4–7	6–8
Wet-Bulb[b]	–	16.4	3.8	–	–	–

[a] The data in this table were collected between 10 February 2019 and 13 March 2019.
[b] The wet-bulb temperature is dependent on the ambient temperature and humidity LeRoy and Kuehn (2001).

all four devices significantly reduced the ambient temperature fluctuations, from over 20°C to less than 3°C, and the peak daily temperature, from 35°C –45°C to 25°C–30°C. The three devices with a clay pot as the interior chamber provided an average interior humidity of >95%. While the plastic bucket in the clay pot device provided a significant increase in humidity, 73.8% compared to an average ambient humidity of 16.8%, it was lower than the devices with a clay pot as the interior chambers. This was likely due to the inability of water to penetrate through the plastic pail, leaving the moisture from the cloth covering the device as the only source of humidity for the fruits and vegetables stored inside. Thus, when the cloth cover becomes dry, which happened quickly during the middle of the day when ambient conditions reach 40°C and 10% humidity, the chamber started

FIGURE 14.14 Examples of the 4-clay pot cooler designs studied in Burkina Faso. (a) Small clay pot in a clay pot (~ 5 L capacity; 2,500 Fcfa; $4.28) with sensor data logger attached; (b) Medium clay pot in a plastic dish (~ 20-liter capacity; 3,300 Fcfa; $5.65); (c) Large clay pot in a metal dish (~ 60-liter capacity; 13,000 Fcfa; $22.27); (d) Plastic bucket in a clay pot (~ 10-liter capacity; 2,700 Fcfa; $4.63) (All of the clay pot coolers are covered with wet cloth when being used.)

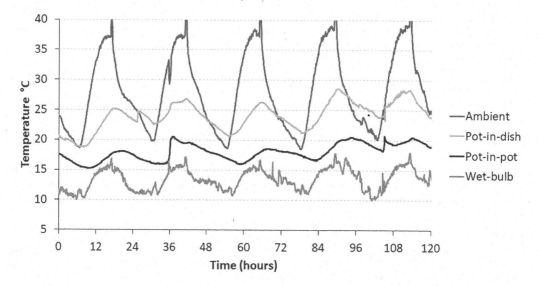

FIGURE 14.15 Typical daily ambient temperature, the interior temperature of the medium clay pot in a plastic dish (pot-in-dish), the interior temperature of the small clay pot in a clay pot (pot-in-pot), and the wet-bulb temperature. The data in this figure were collected between 4 March 2019 and 8 March 2019.

to dry out. This illustrated the importance of either designing the clay pot cooler so that the water stored in the sand layer can provide moisture to maintain a high humidity environment or to minimize the need to monitor the device to keep the cloth cover wet.

14.2.6 Shelf Life Results

14.2.6.1 Rwanda Fruit and Vegetable Shelf Life Data

Across the three regions in Rwanda (Mulindi, Rubona, and Busogo), fruits and vegetables were selected for inclusion in this study based on their relative importance among farmers in each region. For the shelf life experiments, two fruits or vegetables were tested in the same space, so care was taken to ensure that combinations selected were compatible, particularly for ethylene production and sensitivity. The fruits and vegetables studied for each region were the following:

- Busogo: cabbages and carrots
- Rubona: tomatoes and chilli peppers
- Mulindi: tomatoes and mangoes/chilli peppers and French beans

In all cases, the produce stored in the evaporative cooling devices showed improved shelf life compared to the produce stored in the shade in ambient conditions. Each experiment was continued until all of the produce showed signs of deterioration or were determined to be fully ripe through visual inspection. Although the ZECCs and clay pot coolers in each of the three locations in Rwanda were constructed using very similar materials and designs, there are significant variations in factors impacting the shelf life of the produce including (i) the source of the produce being tested and (ii) weather variations affecting the ambient conditions during a specific experiment (either due to different location or time). Thus, our analysis was focused on comparisons between produce purchased at the same time and stored at the same location in different devices.

14.2.6.1.1 Busogo: Cabbages and Carrots[2]

The evaporative cooling devices provided significant improvement in shelf life for carrots and cabbages compared to ambient conditions. When stored in ambient conditions, the carrots and cabbages began to show signs of degradation (fungal growth, rot, bruising, or dehydration) after 8 days. In comparison, when stored in any of the four clay pot coolers, both the carrots and cabbages were still edible after 18 days and retained over 85% of their original weight. The brick ZECC provided the same shelf life for carrots (shelf life > 18 days), but the cabbages began to lose weight and spoil after 12 days, which was still an improvement over the shelf life in ambient conditions (Figures 14.16 and 14.17).

14.2.6.1.2 Rubona: Tomatoes and Chilli Peppers[3]

When stored at ambient conditions, over 60% of the chilli peppers spoiled within 3 days, with only 5% of the chilli peppers remaining unspoiled by the 9th day. When stored in the brick ZECC or clay pot coolers, significant shelf life improvements were observed for the chilli peppers, with minimal degradation observed until the 12th day of the experiment when they began to rapidly degrade. In contrast, the tomatoes stored in ambient conditions did not show significant signs of degradation, primarily shrivelling, until the 9th day, with the tomatoes stored in the brick ZECC and clay pot coolers showing slightly prolonged shelf life (Figure 14.18).

14.2.6.1.3 Mulindi: Tomatoes, Mangoes, Chilli Peppers, and French Beans

In Mulindi, the primary cause of spoilage for mangoes and tomatoes in the evaporative cooling devices was bruising and fungal growth at the area of the fruit or vegetable in contact with the inner surface of the chamber. This was particularly an issue with devices where the inner pot was made of clay. This could illustrate the importance of properly cleaning the interior chamber that is in contact with the fruits or vegetables. The devices with a metal or plastic interior chamber may be able to provide benefits in this regard, due to their smoother surface, which is easier to keep clean. Further research with larger sample size is needed to validate this hypothesis.

Unlike the chilli peppers from the experiment in Rubona, when stored in ambient conditions, the chilli peppers in Mulindi did not begin to show significant signs of degradation until the 10th day of the experiment. By the third day, the French beans stored in ambient conditions showed greater

---

[2] The cabbages and carrots were stored in separate plastic containers for the ambient and brick ZECC experiments, and stored together in the four clay pot coolers

[3] The tomatoes and chili peppers were stored in separate plastic containers for the ambient and brick ZECC experiments, and stored together in the four clay pot coolers

TABLE 14.4

Clay Pot Cooler Performance in Kamboinse, Burkina Faso[a]

| | Average | | | Daily | | Decrease in |
	Humidity (%)	Temp. (°C)	Average Temp. Decrease (°C)	Cooling Efficiency (%)	Temp. Fluctuations (°C)	Peak Daily Temp. (°C)
Ambient	16.8	28.1	–	–	15–20	–
Small clay pot in a clay pot	99.9	20.3	7.8	61	3–6	12–20
Medium clay pot in a plastic dish	98.0	23.7	4.4	35	5–8	8–11
Large clay pot in a metal dish	99.9	23.7	4.5	35	5–8	8–11
Plastic bucket in a clay pot	73.8	23.5	4.6	36	7–10	6–10
Wet-Bulb[b]	–	15.4	12.7	–	–	–

[a] The data in this table were collected between 20 February 2019 and 25 March 2019.

[b] The wet-bulb temperature is dependent on the ambient temperature and humidity LeRoy and Kuehn (2001).

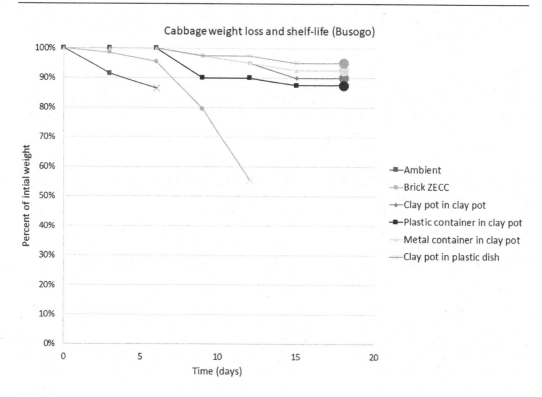

FIGURE 14.16 Weight loss and shelf life for cabbage in Busogo. The ending data points with a circle represent cabbage that was determined to be fully ripe after the experiment. The ending data points with an "X" represent cabbage that were fully spoiled at the point of the experiment indicated.

weight loss than those stored in the brick ZECC and clay pot coolers and spoiled after the 7th day. The French beans stored in the brick ZECC and clay pot coolers remained unspoiled after the 13th day and retaining 75% – 95% of their original weight. Both the chilli peppers and the French beans had less weight loss when stored in the clay pot coolers as compared to the brick ZECC (Figure 14.19 and Table 14.5).

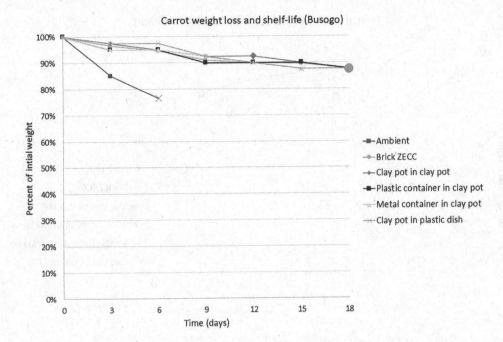

FIGURE 14.17   Weight loss and shelf life for carrots in Busogo, Rwanda. The ending data points with a circle represent carrots that were determined to be fully ripe after the experiment. The ending data points with an "X" represent carrots that were fully spoiled at the point of the experiment indicated.

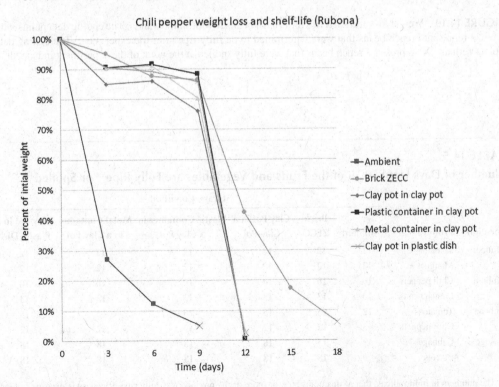

FIGURE 14.18   Weight loss and shelf life for chilli pepper in Rubona, Rwanda. The ending data points with a circle represent chilli pepper that was determined to be fully ripe after the experiment. The ending data points with an "X" represent chilli pepper that was fully spoiled at the point of the experiment indicated.

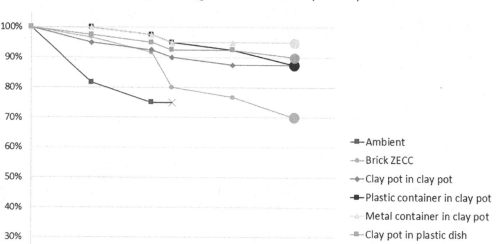

FIGURE 14.19 Weight loss and shelf life for French beans in Mulindi, Rwanda. The ending data points with a circle represent French beans that were determined to be fully ripe after the experiment. The ending data points with an "X" represent French beans that were fully spoiled at the point of the experiment indicated.

TABLE 14.5
Number of Days Until 50 % of the Fruits and Vegetables are Fully Ripe[a]* or Spoiled[b]

Location	Product		Storage Condition				
		Ambient	Brick ZECC	Clay Pot in a Clay Pot	Plastic Container in a Clay Pot	Metal Container in a Clay Pot	Clay Pot in a Plastic Dish
Mulindi	Tomatoes	12	**12**	12	**12**	**12**	12
	Mangoes	,12	12	12	12	12	12
Mulindi	Chili peppers	10	**10**	10	10	**10**	10
	French beans	10	**13**	**13**	**13**	**13**	**13**
Rubona	Tomatoes	12	15	15	15	15	15
	Chili peppers	3	12	12	12	12	12
Busogo	Cabbages	9	15	**18**	**18**	**18**	**18**
	Carrots	9	**18**	**18**	**18**	**18**	**18**

[a] The numbers in **bold** indicate that at this point 50% or more of the produce was fully ripe. Ripeness is measured objectively according to a maturity index based on the physical characteristics of the fruit or vegetable.

[b] The numbers in other than bold indicate that at this point 50% or more of the produce had spoiled.

14.2.6.2 Fruit and Vegetable Shelf Life: Key Takeaways

Across all of the regions, there is a greater incidence of degradation due to dehydration (weight loss, cracking, and shrivelling) for fruits and vegetables stored in ambient conditions, as compared to those stored in evaporative cooling devices, which provide a high humidity environment.

While these shelf life experiments indicate that evaporative cooling devices can improve fresh produce shelf life in multiple contexts, further research is needed to gain more conclusive evidence of the value that evaporative cooling devices can provide, and a deeper understanding of the factors that influence fruit and vegetable preservation in these devices.

When measuring the shelf life of fruits and vegetables, it is important to compare storage in an evaporative cooling device to that of the same fruits and vegetables in ambient conditions or other alternative methods of storage. Such comparative shelf life experiments should be conducted at the same location and time (to ensure the ambient conditions are the same), and with the same source of fruits and vegetables. Additionally, care should be taken to account for the impacts of having multiple fruits and vegetables stored together, such as ethylene production and sensitivity, fungal contamination, spreading of rot from one fruit or vegetable to another nearby (Figure 14.20).

FIGURE 14.20 Nibagwire Claudine is the President of the Dufatanye-Nyanza cooperative, in the Nyanza District of Rwanda. Claudine shared, "After being trained by the project in May 2018 on small scale post-harvest handling practices, I was motivated to build a ZECC to protect the produce against both weight loss, quality loss, and price decrease. Since then, we are saving time as we used to harvest and deliver to the market immediately fearing the losses and the full day was busy but now, we harvest and store the produce in the ZECC and continue our farm activities. We are now selling our vegetables twice a week on Mondays and Thursdays. If all the produce is not sold, we can store it back in the ZECC."

14.2.7   Interview Results

Users of evaporative cooling devices, fresh produce vendors, and producers of clay pots and other containers were interviewed to gain insights into:

- Types of fruits and vegetables purchased and produced
- Existing methods for fresh produce cooling and storage
- Need for improved cooling and storage technology
- Availability and cost of materials that can be used for evaporative cooling devices

14.2.7.1   Interviews with Clay Pot and Container Vendors

Fifteen vendors were randomly selected who sold clay, plastic, and metal containers in the Kadiogo province of Burkina Faso and were interviewed to gain an understanding of what products are available and how much they cost. All of the vendors specialize in selling either clay containers or both metal and plastic containers. Clay pot prices range from 500 to 5,000 ($0.86 to $8.60) for various size pots, with smaller pots (~ 5 to 15 L) averaging ~1,200 Fcfa ($2.06) and larger pots (~ 50 to 100L) averaging 4,200 Fcfa ($7.19). These clay pots are primarily sold to households for storing water because, in the Sahel region, it is common knowledge that storing water in a clay pot will keep it cooler than if it is stored in a plastic or metal container. Plastic dishes, buckets, baskets, and metal dishes of various sizes ranging from 30 to 70 cm in diameter can be purchased for between 600 Fcfa ($1.03) and 4,500 Fcfa ($7.71); and an average price of 2,400 Fcfa ($4.11). These containers are primarily sold to women for washing clothes, dishes, and storing food. All of the vendors selling the plastic and metal containers said they would deliver those products to their customers, but only a third of vendors reported that they would deliver clay pots to customers. This is likely because the clay pots are heavier and more fragile than plastic or metal containers. While the price is similar across the clay, plastic, and metal containers of similar size, the convenience of transporting plastic or metal containers may provide some advantages.

14.2.7.2   Interviews with Current Users of Evaporative Cooling Devices

Five farmers and three cooperative leaders who are current ZECC users in Rwanda were interviewed about what they store in the ZECC and how it compares to other storage methods. All of the ZECC users report that it provides improved shelf life compared to storage in ambient conditions. The most common fruits and vegetables stored in ZECCs were amaranthus, cabbages, carrots, and tomatoes. Other produce these respondents stored in the ZECCs includes sweet potato, sweet peppers, pineapples, chayotes, beetroots, eggplants, French beans, chilli peppers, spinach, peas, and passion fruits. Some specific shelf life improvements that were reported are shown in Table 14.6.

Users reported their ZECCs can store 35 – 200 kg, with an average storage volume of 100 kg, and cost between 30,000 and 64,000 RWF, with an average cost of 46,000 RWF.

14.2.7.3   Interviews with Potential Users of Evaporative Cooling Devices

Interviews were conducted with 57 farmers who are non-users of ZECCs in Rwanda and in Burkina Faso 47 farmers, 13 public servants, and 19 traders, artisans, or private-sector workers who are not

TABLE 14.6
Fresh Produce Shelf Life Reported by Individual Farmers

Produce	Shelf Life in Ambient Conditions	Shelf Life in brick ZECC
Carrots	2 days	4 days
Amaranthus	1 day	4 days
Cabbage	3 days	14 days
Tree tomato	6 days	10 days

TABLE 14.7

Most Commonly Grown Crops Among Farmers Interviewed in Rwanda

Produce[a]	(%)	Crop[b]	
Amaranthus	49.1	Maize	45.6%
Cabbage	47.4	Cassava	36.8%
Sweet potato	45.6	Onion	15.8%
Eggplant	42.1		
Carrot	26.3		
Tomato	22.8		
Sweet pepper	14.0		
Beetroot	7.0		
Avocado	5.3		

[a] Fruits and vegetables that are suitable for storage in evaporative cooling devices.

[b] Crops that are not suitable for storage in evaporative cooling devices due to their sensitivity to high humidity conditions.

currently using ZECCs or clay pot coolers. The farmers were randomly selected among members of cooperatives that the project works with, and the non-farmer respondents were randomly selected from the communities surrounding the farming cooperative.

14.2.7.4 Interviews with Farmers in Rwanda

All of the farmers interviewed in Rwanda are members of cooperatives, and 25% are also members of savings groups. Table 14.7 shows the most common crops grown by the farmers interviewed in Rwanda. The crops are separated by those that are suitable for storage in evaporative cooling devices and those that should not be stored in high humidity conditions.

These farmers grow the crops for a combination of personal consumption and sale. The fruits and vegetables that are sold are typically taken to local or district markets by bicycle or on foot. These markets are 1 – 6 km from the respondent's home. Many of these farmers purchase fruits – such as pineapples, mangos, bananas, and passion fruit – as well as fruits and vegetables that they do not grow for themselves.

The storage method used by the farmers interviewed in Rwanda is primarily keeping the fresh produce in a shady and well-ventilated place. Most of the respondents stated that they would prefer a ZECC or refrigerator if they could afford it, and some respondents indicated that a lack of electricity access was also a barrier to using a refrigerator. For respondents who desired a ZECC, the cost of materials is the most common reason for not having one.

14.2.7.5 Interviews with Farmers in Burkina Faso

The primary harvest season for tomatoes in Burkina Faso is December through March, which is grown for both personal consumption and sale. The farmers interviewed in Burkina Faso sell an average of 18,000 kg of tomatoes per week with an average of ~ 15% tomatoes spoiling before they can be sold. All of the farmers reported selling their tomatoes right after harvest for prices ranging from between 7,000 Fcfa ($12) and 25,000 Fcfa ($43) per 100 kg, depending on the market conditions. 50% of the farmers interviewed are members of farming cooperatives or savings groups. In addition to tomatoes, most farmers (79%) also grow onions, along with some farmers growing eggplants (28%), cabbages (19%), carrots (15%), and chilli peppers (11%). None of the farmers interviewed used either refrigerators or ZECCs for vegetable storage, as they were sold soon after harvest to avoid large amounts of spoilage. For fruit and vegetable storage in their homes, these farmers used straw or plastic baskets with an average capacity of 6 kg, costing an average of 1,000

Fcfa ($1.70). All of the respondents would prefer a refrigerator but cited the lack of electricity or the high cost as the reason for not currently having a refrigerator. Because these farmers sell their crops immediately after harvest, there may not be a significant value that ZECCs can provide; however, further investigation could provide insights into potential benefits that improved storage could provide including a reduction in the need to travel to the market every day to sell the harvest and improved negotiating position with vendors purchasing the produce from farmers.

14.2.7.6   Interviews with Non-Farmers in Burkina Faso

Among the heads of households interviewed that are not farmers, 40% were traders, 40% work in public service, and the remain 20% with various private-sector jobs. 5% are members of savings groups. Someone from these households travels an average of 2 km to the local market to purchase fruits and vegetables including tomatoes (96%), onions (75%), cabbages (50%), eggplants (28%), and chilli peppers (22%). There was a wide range in the frequency of travelling to the market to purchase fruits and vegetables, ranging from every day (25%) to once per week (16%), with an average of three trips to the market per week among this respondent group. 46% of the households use a refrigerator for storage with an average capacity of 12 kg and 60% of the households use straw or plastic baskets with an average capacity of 7 kg (some households reported using both a refrigerator and baskets for fruit and vegetable storage). The refrigerators, costing an average of 220,000 Fcfa ($375), provided an average shelf life of 13 days for tomatoes, eggplants, and cabbages; while the straw and plastic baskets, costing an average of 1,200 Fcfa ($2), provided an average shelf life of 3 days for tomatoes, eggplants, and cabbages. One respondent within this group reported using a clay pot to store tomatoes, providing a shelf life of 7–10 days. All of the respondents that are currently using straw and plastic baskets would prefer a refrigerator but cited the cost as being the major barrier to purchasing one. Among the respondents that are currently using refrigerators, a majority reported that there is no other method of storage that they would prefer, with two respondents saying that they would prefer a solar refrigerator to avoid power cuts, and one woman having recently heard about ZECCs said she would prefer this storage method due to the lack of electricity consumption. Further research would be required to determine what benefits-improved shelf life would provide to these households to determine a clay pot cooler capable of storing 5 – 10 kg of tomatoes that can provide enough value to justify its cost (Table 14.8).

14.2.7.7   Interviews with Fruit and Vegetable Vendors

14.2.7.7.1   Fresh Produce Vendors in Rubona, Rwanda

Interviews were conducted with ten randomly selected fresh produce vendors at the Nyanza market in Rubona. The most common fruits and vegetables being sold were carrots, eggplants, tomatoes, and French beans. Most vendors purchase their fresh produce directly from farmers who come to the market to sell their produce. Some vendors also travel to other larger markets to purchase produce at bulk prices. The vendors sell their produce at mark-ups of 50% – 100% from their purchase price. The price that vendors pay for the produce and their selling price typically varies by a factor of 2, where the price in the peak harvest season is typically half of the low season. Most vendors sell a variety of fruit and vegetable types and at volumes anywhere between 5 and 50 kg of a given

TABLE 14.8
Current Storage Methods Used by Non-farmers in Burkina Faso

Storage Method	% using	Average Capacity	Average Cost	Average Fruit and Vegetable Shelf Life
Refrigerator	46%	12 kg	220,000 Fcfa ($375)	13 days
Straw or plastic basket	60%	7 kg	1,200 Fcfa ($2)	3 days

fruit or vegetable per day, and report losing between 10% and 50% of the fruits and vegetables that they purchase per day.

All of the vendors interviewed are currently storing their fruits and vegetables where they are exposed to ambient conditions under the shade of the market stall, in baskets, sacks, or on tables. Over half the vendors said they would prefer to store their fruits and vegetables in a cold room or shared warehouse, and the remaining 40% responded that they were not aware of any other options for improved storage.

14.2.7.7.2   Fresh Produce Vendors in Kadiogo, Burkina Faso

Fifteen randomly selected fresh produce vendors from three markets were interviewed, all of whom primarily sell tomatoes, as well as some who also sell cabbages, eggplants, and onions. Tomatoes are purchased by the vendors from nearby wholesalers in 100 kg boxes for 7,500 – 60,000 Fcfa ($12.85 – $102.78), depending on the season. Most vendors sell 35 – 200 kg of tomatoes per day, with a margin between 2,000 and 5,000 Fcfa ($3.43–$8.56) per 100 kg box (mark-up ranging from 7% to 57%). These vendors reported the tomato spoilage ranging from 1.5 to 10 kg per day (3% – 12% of vegetables sold), with an average of 4 kg per day of tomatoes that spoil. Daily tomato spoilage of 4 kg per day results in the loss of 370 to 1,600 Fcfa ($0.63–$2.73) per day, which is a significant portion of the sales compared with the margin calculated. Additionally, given the prices for clay pots and other containers reported from vendors at markets, a clay pot cooler with a capacity of 35 – 70 kgs can be constructed for under 10,000 Fcfa ($17.13). For vendors that need to store more than 70 kg of tomatoes per day, more than one clay pot cooler may be needed. If these clay pot coolers can reduce tomato spoilage by 50%, they would pay for themselves in 2–8 weeks depending on the season.

14.3   POTENTIAL OF EVAPORATIVE COOLING DEVICES TO BENEFIT FRESH PRODUCE VENDORS

Clay pot coolers or small brick ZECCs could provide single vendors storage for one or more compatible fruit and vegetable types. One key consideration regarding the practicality for vendors to use clay pot coolers for tomato storage at markets is the portability of the devices. If the markets have permanent structures, the clay pot coolers can be stored overnight or permanent ZECCs could be constructed. However, in cases where vendors move to different locations, the clay pot coolers would have to be transported each day.

The markets where the vendors interviewed in Rwanda and Burkina Faso sell their produce are permanent, allowing for the possibility of constructing a brick ZECC to meet the storage needs of a single vendor, or for a group of vendors, with a need to store similar fruits and vegetables. Such a permanent structure could serve as a shared warehouse allowing for produce that did not sell on a given day to be stored overnight. While the shelf life experiments in this and other studies show that a well-designed and maintained ZECC can provide significant shelf life improvements, organizing the construction, use, and security of a shared storage facility requires planning and coordination. Alternatively, individual vendors could construct clay pot coolers for storing smaller amounts of fresh produce near where they are selling their produce, and if they have access to secure storage space, the clay pot coolers would not have to be transported away from the market each day.

For vendors that purchase fruits and vegetables in suitable boxes or crates (where there are holes for ventilation in the side and bottom of the container), there is the potential to cover the container with a wet cloth to produce some measure of an evaporative cooling effect. This is a potential area for future research to evaluate the performance of a device with such a design (Figure 14.21).

14.4   SUMMARY

The results of this study indicate that low-cost evaporative cooling devices, such as clay pot coolers and brick ZECCs, can improve fruit and vegetable storage shelf life by providing a stable storage

FIGURE 14.21   Tomatoes in a crate used for transportation and storage.

environment with reduced temperatures and high humidity. These conditions reduce water loss and spoilage in the produce studied (tomatoes, mangoes, carrots, cabbage, chilli peppers, and French beans) and could have important benefits for households, farmers, and fresh produce vendors. Specific key results are listed below.

Limited cooling in high humidity environments: It is well understood that evaporative cooling is highly sensitive to ambient humidity. In this research, the wet-bulb temperature (the minimum temperature that can be achieved through evaporative cooling) was compared to the interior temperature of the devices to determine the cooling efficiency. The results showed that in Rwanda, while the cooling efficiency of clay pot coolers was ~40%– 60%, the average temperature decrease was only 1°C – 2°C. In contrast, the low humidity environment in Burkina Faso allowed for temperature decreases of 4°C – 8°C to be achieved with cooling efficiencies of 35% –60%.

Impacts of design on performance: Some specific results demonstrate the impact of evaporative cooling device design on performance:

- A brick ZECC that has a reduced surface area available for water evaporation and was exposed to direct sunlight for a significant portion of the day showed a low cooling efficiency (4%) compared to brick ZECCs with more optimized designs that achieve cooling efficiencies ranging from 29% to 77%.
- Even when evaporative cooling devices do not provide significant decreases in the average temperature, brick ZECCs, and clay pot coolers provide other benefits including increased interior humidity, decreased temperature fluctuations, and decreased peak daily temperatures, all of which are expected to improve fruit and vegetable shelf life.

- The four-clay pot cooler designs in Rwanda showed similar performance in terms of the interior temperature and humidity, although greater variations in performance may be observed in lower humidity environments where greater cooling can be achieved.
- The clay pot coolers in Burkina Faso showed the most significant average temperature decreases (> 4°C), particularly the small clay pot in clay pot device, which provided nearly an 8°C decrease in the average temperature. While this device provides the best performance, the volume available for fruit and vegetable storage is limited (~ 5 L).

Improvements in fresh produce shelf life: Shelf life improvements are dependent on the ambient conditions, the specific fruit, and vegetable types, and the time they were harvested in the maturation cycle. Given this context, specific results from this research include:

- Carrots and cabbages saw greater benefits than the tomato experiment.
- Chilli peppers saw a greater shelf life improvement in Rubona than Mulindi.
- Besides ethylene compatibility (sensitivity and production), fungus, or rot from one fruit or vegetable type can accelerate deterioration in another type.
- Contact with the walls of the inner chamber can cause degradation if the surface is not clean.

Benefits reported by ZECC users: Farmers who have constructed brick ZECCs were satisfied with their investment and reported shelf life improvements ranging from ~two times the shelf life in ambient conditions (tree tomato) to a shelf-life improvement of over four times the shelf life in ambient conditions (cabbage).

Potential for evaporative cooling devices to provide benefits to fresh produce vendors: Fresh produce vendors in Rubona, Rwanda, and Kadiogo, Burkina Faso suffer high rates of spoilage. Because the vendors surveyed sell at daily markets with permanent locations (as opposed to travelling each day to sell at different weekly markets), there is the potential to construct brick ZECCs or store clay pot coolers at these locations to help the vendors avert food spoilage and financial losses.

REFERENCES

Ambuko, J., Wanjiru, F., Chemining'wa, G. N., Owino, W. O. and Eliakim, M. (2017). Preservation of postharvest quality of leafy amaranth (*Amaranthus* spp.) vegetables using evaporative cooling. *Journal of Food Quality.* https://doi.org/10.1155/2017/5303156

Arah, I. K., Ahorbo, G. K., Anku, E. K., Kumah, E. K. and Amaglo, H. (2016). Postharvest handling practices and treatment methods for tomato handlers in developing countries: A mini-review. *Advances in Agriculture, 2016.* https://doi.org/10.1155/2016/6436945

Basediya, A.L., Samuel, D. V. K. and Beera, V. (2013). Evaporative cooling system for storage of fruits and vegetables-a review. *Journal of Food Science and Technology, 50*(3): 429–442.

Kader, A. A. (2005). Increasing food availability by reducing postharvest losses of fresh produce. *Acta Horticulturae, 682*: 2169–2176.

Kitinoja, L. (2016). Innovative Approaches to Food Loss and Waste Issues. Frontier Issues Brief for the Brookings Institution's Ending Rural Hunger project.

Kumar, A., Mathur, P. N. and Chaurasia, P. B. (2014). A study on the zero-energy cool chamber for the storage of food materials. *International Research Journal of Management Science and Technology, 5*(7): 66–70. https://doi.org/10.32804/IRJMST

LeRoy, J. T. and Kuehn, T. H. (2001). Psychrometrics. *2001 ASHRAE Fundamentals Handbook (SI).* 6.1–6.7.

McGregor, M. B. (1989). *Tropical Products Transport Handbook.* USDA Office of Transportation, Agricultural Handbook. No. 668. p. 148.

Morgan, L. (2009). *Clay Evaporative Coolers Performance Research.* Practical Action, Sudan.

Nobel, N. (2003). *Evaporative Cooling: Practical Action Technology, Challenging Poverty.* Bourton, UK. Retrieved online from www.practicalaction.org. (Accessed on April 1, 2019).

Odesola, I. F., and Onyebuchi, O. (2009). A review of porous evaporative cooling for the preservation of fruits and vegetables. *The Pacific Journal of Science and Technology, 10* (2): 935–941.

Rathi, R. and Sharma, B. (1991). *Few More Steps Toward Understanding Evaporating Cooling and Promoting Its use in Rural Areas.* A Technical Report. Delhi, India.

Roy, S. K. and Khurdiya, D. S. (1985). *Zero Energy Cool Chamber.* India Agricultural Research Institute: New Delhi, India. Research Bulletin. Volume 43.

Rwanda Agriculture. (2019). *The World Bank Web site.* Retrieved January 14, 2020, from http://www.data. worldbank.org/country/rwanda.

Verploegen, E., Rinker, P. and Ognakossan, K. E. (2018). *Evaporative cooling best practices.* Copyright © Massachusetts Institute of Technology (Accessed on April 1, 2019).

Verploegen, E., Sanogo, O. and Chagomoka, T. (2018). *Evaluation of Low-Cost Vegetable Cooling and Storage Technologies in Mali.* Copyright © Massachusetts Institute of Technology (Accessed on April 1, 2019).

15 CoolBot™ Cool Rooms for Small-Scale Value Chain Systems

Neeru Dubey
Ernst and Young, India

CONTENTS

15.1 INTRODUCTION

With the world moving from food security to nutritional security, the growing awareness for nutrient-rich diet is increasing the demand for horticultural crops. Compounding this is the reality of the COVID-19 pandemic which has raised the demand for nutritional and immunity-boosting food products. Further, perishable food needs to be maintained at low temperature and high humidity, throughout the supply chain. When compared to cereals, the horticultural crops are many times more perishable because of their postharvest physiological characteristics which can be attributed to their natural biological processes. If a proper cold chain is not maintained along the supply chain, the ripening process followed by senescence and deterioration of the horticulture produce happens at a faster rate and ultimately leads to food losses.

In India, the need for small-scale value chain systems arises from the fact that about 86.2% of farmers are small and marginal farmers owning just two hectares or less than two hectares of land, which represents only 47.3% of the total cropped area (Anonymous, 2018). These farmers cannot afford to build mechanical cold storage because of the cost and technical limitations. In developing countries, the cold chain system is characterized by poor storage facilities (FAO, 2011), poorly developed and inefficient infrastructure which have caused the highest amount of deterioration in harvested produce over the years (Kitinoja and Cantwell, 2010). Most small-scale farmers are operating below the poverty line and are unable to obtain the highly sophisticated facilities to maintain a cold chain at the farm level. As per the industry estimates, about 104 million metric tonnes (t) of high-value produce is distributed and transported between various cities of India every

DOI: 10.1201/9781003056607-18

year. Out of this, about 100 million t of perishable produce is transported through non-reefer mode while only 4 million t is distributed through reefer mode (Sivaraman, 2016). The main reasons for low adoption are economic feasibility as well as temperature specificity for each horticulture produce.

In this chapter, an attempt is made to describe the development of a small-scale value chain system that makes the facilities available for farmers/farmer groups and retailers at an afford-able price so that the commodities produced are not lost and be able to sell at a price which is beneficial for the farmers. The CoolBot™ cold storage system hereby discussed is cold storage and temperature-controlled transport system, which is technically and economically feasible, is relevant to this large group of farmers. The CoolBot™ was developed by Ron Khosla of 'Store It Cold' (storeitcold.com) as an easy and affordable way for small-scale farmers to cool produce on their horticultural lands. The Horticulture Innovation Laboratory, USA, has verified and tested cool rooms furnished with CoolBot™ in three different countries viz. India, Uganda and Honduras.

15.2 COOLBOT™ EQUIPPED COOLING SYSTEM – THE CASE OF INDIA

The training for the CoolBot™ system was provided at the University of California, Davis, through the project 'Cool room and cool chain transport for small scale farmer' funded by Horticulture Collaborative Support Project (HortCRSP) and United States Agency for International Development (USAID) in June 2010. After the training, the group from Amity University, Uttar Pradesh, continued with the project and decided to install the system at Amity university farm with local insulating materials to increase its applicability is increased for the local farmers. The team leader Professor Michael Reid of UC Davis and the team comprising of people from India, Honduras and Uganda tested a cooling device, the CoolBot™. The control functions of the air conditioner unit were moulded to bypass the thermostat and reach low air temperatures without the development of ice on the evaporator coils as ice formation on the coils would restrict airflow, thereby slowing the cooling process and then stopping it altogether, so standard air conditioners have thermostats set to prevent temperatures falling to where ice might accumulate. CoolBot™ is equipped with a thermostat system that has a microcontroller and sensors to control a standard room air conditioner to produce cooling temperatures below its standard set, converting a simple well-insulated room into a cool room. Experiments included testing a range of low cost and locally available insulation materials that can be utilized in insulating cool rooms, testing of the CoolBot™ in combination with a room air conditioner. The team also explored some systems for short-distance transportation and low-cost novel insulating materials, e.g., inverted earthen containers, wheat straw, mud and rice husk. The most important component of any cooling system is insulation, therefore, Amity University focused on developing a cooling system that was made of insulating materials that are easily available in the village and easily constructible and affordable to the farmers and retailers.

Insulation consisting of mud and rice husk with 0.91-m-wide walls provided an excellent bar-rier for cold air. The capacity of the room was 8 t for which the dimensions were 3.6 m × 3.65 m. For strength and mechanical support, a layer of bricks was provided in the wall surrounded by a wide mud wall (2 m). The mud was combined with rice husk in the ratio of 2:1. As most of the heat exchange takes place from the door, roof and floor, the door of the cool room was insulated with 60 mm polyurethane foam (PUF). The roof was insulated with expanded polystyrene material of density 20 kg per cubic meter and thickness 0.15 m, with a slope on one side to facilitate drainage of rainwater. The roof was covered with a polyethylene sheet which acted as a vapour barrier. An ante-chamber of 1.5 m × 1.8 m was also constructed to control the airflow and break the pattern of hot air in and cold air out. For protection from extreme heat and excessive rain, the cold room was painted with heat-reflective and water-resistant paint.

15.2.1 System Installation

The cooling room was constructed as per the details mentioned above. An energy-efficient 2 KWA window room air conditioner unit with a digital display was fitted in the cool room. All the gaps near the air conditioner unit door and roof were sealed with PUF spray foam sealant and the CoolBot™ was installed near the air conditioner fixed to the wall. It also modifies a retrofitted insulated room into a cool room using a cheap, easily available, room air conditioner, which substantially reduces the cost of a cool storage environment for perishable horticultural produce.

The unit had three labelled wires emerging from the unit. One wire measures the temperature of the room, the second wire which is labelled as a frost sensor is stuck into the cooling fins of the air conditioner unit to control the ice formation on the cooling units. The third wire was connected to the temperature sensor of the air conditioner unit. These two wires are attached and wrapped with aluminium foil to ensure a good thermal connection. The wires should be aligned parallel with each other. The CoolBot™ on its front has three buttons. The first button is set to the required room temperature and the second button to the desired frost temperature taking care that its temperature is lower than the temperature set for the room and third is the reset button (Figure 15.1).

15.2.2 Construction Cost

The capacity of the cold room is 8t with 25% free space for air circulation. The room size is 144 m² with a height of 9 m sloping on one side for rainwater drainage. The external dimension is 18 m × 18 m and the roof dimension is 19 m × 19 m. The construction cost of the cool room was USD 1028, significantly less than the cost of conventional cold storage (Table 15.1).

The cost may vary depending upon the site location, ongoing labour charges and other seasonal and local variations. The project is specially designed for small and marginal farmers and retailers.

FIGURE 15.1 (a) Construction stage. (b) Installed system. (c) Data recording (d) CoolBot™ cool room.

TABLE 15.1

Comparative Costs of Installing a Commercial Refrigeration System and a CoolBot™ Controlled Room Air Conditioner

	Conventional 2 ton Refrigeration System[a]		[b]CoolBot™ Controlled Room Air Conditioner	
	USD	INR	USD	INR
Refrigeration equipment	4,000	2,20,000	678	37,300
Construction cost + labour	4,000	2,20,000	1520	83,600
Total	8,000	4.40,000	1028	1,20,900

[a] Based on estimates by Thompson et al. (2002).
[b] Based on actual expenses at Amity University cool room.

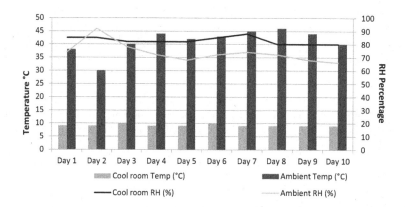

FIGURE 15.2 Comparative temperature and relative humidity in a cool room and ambient condition.

15.2.3 COMPARATIVE TEMPERATURE IN COOLBOT™ COOL ROOM

A set of experiments proved the ability of CoolBot™ cool room to sustain the set temperature along with relative humidity. The experiments were conducted on many important vegetable crops in different seasons, i.e., cauliflower (*Brassica oleracea* var. *botrytis*, tomato (*Solanum lycopersicum*), cabbage (*Brassica oleracea* var. *capitata*), okra (*Abelmoschus esculentus*), chilli (*Capsicum annum*), French beans (*Phaseolus vulgaris*), mango (*Mangifera indica*), cucumber (*Cucumis sativus*) and coriander (*Coriandrum sativum*).

The temperature differences between the cool room and outside temperatures showed that the CoolBot™ cool room was able to maintain a temperature below 10°C when the outside temperature fluctuated between 25°C and 46°C through the study period (Figure 15.2).

15.2.4 QUALITY PARAMETERS IN COOLBOT™ COOL ROOMS

The experiments conducted at Amity University showed that CoolBot™, which enabled the room to maintain a low temperature and relative humidity (> 80%), was able to maintain the quality of the horticulture produce for the extended storage period. This system provides the farmers access to a very important resource, i.e., storage at safe low temperatures at a lower price.

In general, the weight losses increased successively with the increase in the time interval in both conditions. Physiological losses in the weight of produce during storage is attributed to the loss of

TABLE 15.2
Quality Parameters of Some Horticulture Commodities Stored in the CoolBot™ Cool Rooms

Product Commodity	Storage Life (Days)		Physiological Loss in Weight (PLW) (%)		Pulp Temperature (°C)	
	Ambient	Cool Room	Ambient	Cool Room	Ambient	Cool Room
Cabbage	4–5	21	30	5	26	9.14
Brinjal	4–5	10	40	8	25	10
Tomato	6	20	42	4	34	9.26
Green chilly	6	21	25	7	32	9
Cauliflower	7	21	35	6	26	9.24
French beans	2–3	15	37	8	26	9
Coriander	1–2	12	53	8	24	9
Okra[a]	3–4	7	44	7	30	13
Mango	2–3	25	40	8	35	10
Cucumber	3–4	10	37	3.75	34	12

Source: Saran et al. (2012).
[a] Okra: There was a problem of skin blackening.

moisture and respiration (Saran et al., 2012). The physiological weight losses in the produce are generally attributed to water loss through transpiration. This weight loss results in shrivelling and wilting, which decrease market value and consumer acceptability.

Data mentioned in Table 15.2 showed that the pulp temperature in the CoolBot™ cool room was lower significantly than the pulp temperature of the commodity kept in ambient storage. The pulp temperature has a direct relationship to the postharvest storage life of the commodity. The lower the inner core temperature is, the relatively greater is the storage life of the commodity. For most of the crops, the pulp temperature remained between 9°C and 13°C in the CoolBot™ cool room, while at the ambient condition, the pulp temperatures ranged between 25°C and 35°C.

Sensory scores for all the commodities were higher when stored in a cool room whereas the samples at ambient temperatures showed the very mean ratings which can be attributed to the enhanced spoilage and loss of keeping quality. The firmness of the horticultural produce is one of the most important physical parameters which is used to analyse the enhancement of maturity and ripening and is most commonly considered as a key criterion to determine the degree and stage of fruit ripening. Besides, it is an important marketing criterion for produce and determines its market value (Figure 15.3).

15.2.5 CoolBot™ Cool Room-Field Testing

As the agribusiness sector in India continues to evolve and more sophisticated buyers start to link with rural area production bases, affordable cold storage units will certainly become more prominent in rural areas. The type of units that could be used by both the farmers and buyers will be in demand. Amity University Uttar Pradesh, Noida, has done an extensive amount of research and tested the CoolBot™ technology and in partnership with Agribusiness System International under their Sunhara India Project. The CoolBot™ system was constructed at a horticulture retail market (locally called mandi) in the district of Sultanpur in Uttar Pradesh. The construction was done in early 2012 and the data were recorded in May–July showed that when in the hot summer months, the ambient temperature ranged from 30°C to 40°C the temperature inside the CoolBot™ cool room was recorded between 6°C and 10°C, even with interruptions in electrical supply.

Visual commodity comparison

FIGURE 15.3   Comparative visual analysis produce stored in a cool room and ambient condition.

The commodities stored were a variety of fruit and vegetable crops including guava, mango, citrus, muskmelon, watermelon, okra, cauliflower, chilli, cabbage, potato, bitter gourd, bottle gourd, potato, amaranthus, coriander, tomato and spinach. The storage fee charged on a per crate basis was ~USD 0.07 and the retailers sold their produce only as per the market demand. The losses in the horticulture commodities stored inside the cool room were reduced from 25% to 30% to as low as 2%.

The retailers at the *mandi* at Sultanpur earned profit with the CoolBot™ cool room storage infrastructure and their total earning was approximately USD 335 and USD 220 monthly (total earning and net earnings, respectively), with 25 days of operational period and only half capacity utilization. If the capacity utilization is enhanced on a per-day basis and if the full capacity utilization is done while adding more members to the group, then earnings could be increased to the tune of USD 400–540 per month.

15.2.6   ADVANTAGES OF THE COOLBOT™ COOL ROOM TECHNOLOGY

The CoolBot™ cool room contributes to reducing postharvest losses by retarding the rates of physiological activities and thereby extending the storage and marketing life of the fresh produce. The temperature of the cool room is maintained at the range of 2°C –16°C. It can be used for the storage of all fresh horticulture commodities. It is an economically viable, income-generating and affordable storage option for farmers, farm women, self-help groups, cooperatives and market groups. The technology was pilot tested for storage of different vegetables and shelf life enhancement. Postharvest loss mitigation and management strategies specific to the agro-climatic conditions need to be developed. Different aspects related to the scaling up of CoolBot™ cool room concerning constructional details and storage capacity for different regions need to be considered.

The most important envisaged benefits of technology are increased profits through:

- Easily applicable, low-cost technology-cold storage accessible to small and marginal farmers, retailers, private farmers markets and cooperative markets
- An increase in the postharvest storage life of the produce – by quickly lowering the inner core temperature of the produce and removing the field heat after harvest enhances life by

decreasing the rate of biological processes, metabolic activities and diminishing microbial load.

- Gives farmers an advantage to leverage market situations in case of bulk productions and leverage the market prices to their benefit.
- Substantial cost reduction when compared to the existing industrial cooling solution which is not only four times more expensive but consumes up to 30% more energy as compared to a CoolBot™ enabled system
- Compatible with a normal household window air-conditioner and mini-split ACs, as it does not impact the compressor life and does not put extra load on the compressor.
- And above all, the increased employment opportunity and thus increased standard of living, nutrition, health and education.

15.2.7 PROBLEMS ENCOUNTERED

As with all the technologies, during the usage of technology, we encountered few problems as described below:

- The mud and rice husk insulation were effective in controlling the temperature of the commodity but needed constant repairing in case of thunderstorms, hailstones and very heavy rain. After every monsoon season, the walls needed repairs.
- For future constructions more sturdy insulating materials with higher R-value are to be used.
- The relative humidity is maintained at around 60%–80%. If a humidifier is installed, it will prove more effective.
- The sensors of the CoolBot™ are very sensitive and are damaged easily. They need to be frequently replaced.

15.3 RECOMMENDATIONS

The CoolBot™ ™ cold storage technology is operationally enhancing postharvest storage life and maintaining the nutritional and biochemical quality of many fruits and vegetables. Other published studies have shown it to be effective for the storage of milk, wine, flowers and other products requiring cold temperatures for storage.

15.4 ADOPTION AND SCALING UP

It is recommended for adoption by smallholder farmers to reduce the high postharvest losses. Farmers will also be able to store their produce for relatively longer period after the harvest especially during the market glut and will be able to sell the produce when the prices go up hence increasing their profit margin.

The technology with few modifications can easily be utilized for the transport of horticultural commodities in an insulated vehicle along with solar panels and utilizing solar energy to power the air conditioner. Training and close collaboration with industry, local and regional markets, farmer cooperatives and wholesale purchasers through the below steps is required for wide adopt the CoolBot™ technology.

- i. Education:
  Enhanced training in the technology to reach directly to the farmers and educate them on this simple, easy, affordable technology and in their language.
- ii. Investment:
  Research investment to develop innovative options suitable for a range of potential users and to identify entrepreneurs willing to invest in the CoolBot™.

iii. Capacity building:

Since temperature management is critical in reducing the high postharvest losses in most perishable commodities, capacity building should be enhanced among smallholder farmers on the best possible ways to reduce temperature abuse on harvested produce.

15.5 CONCLUSIONS

Considering all the observations and quality parameters, it can be easily inferred from the results that the CoolBot™ cool room can be constructed at a very low cost when compared to conventional cold storage. The cold room can be built easily with locally available materials to maintain the optimum storage temperatures for the horticulture commodities even in extreme weather conditions. The CoolBot™ cool room successfully enhances the postharvest storage life and preserves the postharvest parameters of the commodity to sustain their marketability and provide farmers with the benefits to leverage the market price fluctuations. The visual appearance of the commodities was excellent after the completion of the storage period and was fresh, firm, marketable and consumable. This technology can provide a sustainable low-cost solution to the farmers, farmer groups and retailers. Low cost, easily affordable and easier technology options can easily be utilized at the village and semi-urban areas for storage, transport and processing of horticultural produce, which can not only add value to the produce but also generate extra income for the farmers and farm women thereby providing impetus to entrepreneurship.

REFERENCES

Anonymous. (2018). Small and marginal farmers own just 47.3% of crop area, shows farm census. Live Mint. October.2018.

FAO (2011). Global food losses and food waste- extent, causes and prevention, Rome, Italy.

Kitinoja, L. and Cantwell, M. (2010). Identification of appropriate postharvest technologies for improving market access and incomes for small horticultural farmers in Sub-Saharan Africa and South Asia. WFLO Grant Final Report, 323p.

Saran, S., Dubey, N., Mishra, V., Dwivedi, S. K. and Raman, N. L. M. (2012). Evaluation of shelf life and different parameters under CoolBot cool room storage conditions. *Progressive Horticulture*, *45*(1): 115–121.

Sivaraman, M. (2016). Government's role in India's ailing cold storage sector. Centre for Public policy research. December.2016, 26p (No. id: 11505).

Thompson, J. F., Mitchell, F. G., Rumsey, T. R., Kasmire, R. F. and Crisosto, C. H. (2002). *Commercial cooling of fruits, vegetables, and flowers* (No. 635.046 C734c 2002). California, US: University of California, Division of Agriculture and Natural Resources.

16 Policy, Strategies, Investments and Action Plans for Cold Chain Development

Lisa Kitinoja
The Postharvest Education Foundation

Divine Njie
Food and Agriculture Organization of the United Nations

CONTENTS

16.1 INTRODUCTION

Proper utilization of the cold chain during fresh produce handling reduces food losses as well as improves incomes by promoting agricultural value chain development. The successful implementation of the cold chain requires effective collaboration between the academic, industry, public and private sectors. Cold chain development is a process that may take decades and go through different phases or stages of development. In the 1880s, the cold chain initially developed in the United States with the use of ice in insulated railroad cars, loaded with fresh foods intended for distant wholesale markets. Over many years, this has changed to refrigerated road transport, with reefer truckloads of foods intended for individual retail vendors who manage their private distribution networks (i.e., supermarket chains or Big Box stores). Investors in early refrigerated (ice) transport systems and wholesale markets were slowly pushed out by reefer truck operation companies and supermarkets chains. In India and China, the same changes are already occurring, and with access to modern communication methods and business practices, many developing countries may bypass some of the steps in this progression.

DOI: 10.1201/9781003056607-19

The use of a cold chain creates new jobs (skilled and unskilled, rural and urban) along the agricultural value chain and in the agricultural support sector. Public policies and investments should strive to support these jobs in packaging, packinghouses, cooling facilities, freezers, cold transport, distribution centres and all the engineering, construction, logistics, management, maintenance and repair operations related to designing, building and operating these cold chain facilities and support services.

Cold chains are often developed and utilized first for the **export** of higher value commodities and food products, which may appear to favour large-scale growers or processors. But once cold chains are in place, they can also be utilized for domestic handling and marketing, thus improving nutrition and food safety while increasing food availability for the local population. Ideally, reducing food losses along the agricultural value chain will allow growers to earn more and provide lower market prices for consumers.

As local and global resources become more scarce and more expensive (i.e., land, water, fertilizers, fuels, other inputs), preventing food losses will become even more cost-effective than it is at today's resource prices.

16.2 ASSESSING REQUIREMENTS FOR COLD CHAIN DEVELOPMENT

Each region, sub-region or country will have its own set of perishable foods and horticultural crops, local resources, existing infrastructure along the value chains, level of development and past attempts at agricultural development. The following factors will need to be taken into consideration in assessing requirements for cold chain development:

- The objective for assessments: Comparing the current situation to what is desired.
- Food system approach: identifying all the key products to focus for cold chain development.
- Value chain approach: Include all the actors and decision-makers along the specific value chain or commodity system (during planning, production, postharvest, processing and marketing).
- Key elements: Identify foods of priority interest (due to volume, economic value or cultural benefits); determine current levels of economic and nutritional losses, quality problems and food safety issues; identify bottlenecks and missing links; identify cold chain technologies that can solve the priority problems.
- Cost-benefit considerations: What are the costs of the cold chain technologies needed for solving the problems? What are the expected benefits for individual actors, the investors or society at large? What are the potential profit and the estimated rate of return on investment?
- Sustainability perspectives: Considering the triple-bottom-line of sustainability – economic, social and environmental so that it contributes to the SDGs

There are many models and methods used for conducting assessments of the agricultural value chain, including rapid rural appraisals, commodity systems assessments and value chain analysis. Publications that provide examples, guidelines for conducting assessments and sample questionnaires include da Silva et al. (2009), FAO (2015a, b), Fearne et al. (2007), LaGra et al. (2016), Shepherd (2007), Shewfelt and Prussia (1993), Ton et al. (2007) and Weinberger et al. (2008). The cold chain needs assessments can be conducted by potential investors, private companies, development projects and other organizations.

16.3 COST-BENEFIT CONSIDERATIONS

The use of the cold chain will be deemed a desirable investment for any given food product, based upon its market value and whether the cost of implementing and maintaining the cold chain (including pre-cooling or freezing, cold storage, cold transport and cold retail marketing) is less than the expected loss of value if the cold chain is not utilized. It is also possible to build cold chains as part of development projects, charity efforts or relief work.

It is possible that using a cold chain for lower value foods will never be cost-effective. For this reason, there are many examples of cold chain investments for the fast-food sector (i.e., potatoes for frozen *French fries*) and high-value horticultural export crops. It is less common to see cold chain investments made for handling and storage of lower value but important staple foods such as bananas in Rwanda and cabbage in Ghana, or for less highly regarded foods such as out of season fruits and vegetables in India.

Every cost-benefit calculation and determination of potential ROI must begin with the market value of the food product. Since food prices fluctuate with the seasons and can differ from year to year, there is no perfect way to predict actual financial returns. Instead, a range of costs and benefits can be used to evaluate the potential of any given horticultural crop in any given locale. If costs are determined to be higher, or market value is found to be lower than the ranges required for success, then it is unlikely that investments in the cold chain will be cost-effective or that a long-term agribusiness based upon that crop will be sustainable. However, if government policy is to promote the cold chain, it can introduce policy instruments to:

- incentivise adoption,
- subsidize energy costs, and
- ensure that externalities, such as environmental pollution, which could constitute a major problem in cold chains, are paid for.

Table 16.1 summarizes the associated costs and expected benefits for some of the cold technologies presented in other chapters of this book. These estimated capital costs and energy use comparisons can be used as guidelines when assessing which cold technologies may be the most cost-effective for which produce. Local prices of energy and the cost of importing or constructing needed equipment also will be important for deciding factors.

16.4   CREATING NATIONAL STRUCTURES FOR COLD CHAIN DEVELOPMENT

A combination of the public sector and private sector efforts are required for promoting cold chain development in any nation. A **cold chain task force** at the national level can bring together stakeholders in the government and private industry and create a venue for communication on priorities and strategies for mutually beneficial efforts in cold chain development.

16.4.1   PUBLIC SECTOR

An appropriate framework is required to ensure coordination of the different Ministries and public-sector institutions that handle food systems issues. These include agriculture, industry, transportation, commerce, nutrition, food safety, energy, environment, emergency response and so on. Mechanisms are also required to ensure the coherence of the policies addressing these issues. Furthermore, the staff of these institutions must be mobilized and receive education to be made more aware of the current high levels of food losses, the importance of the use of the cold chain for reducing losses, protecting food quality and safety, the benefits of the cold chain for the population in terms of improved nutrition, food security and job creation as well as the specific roles that their services might play in addressing these issues.

The proportion of urban population in the world is expected to reach 68% in 2050 (UNDESA, 2018) and the cold chain will be increasingly necessary to supply urban dwellers with perishable food products. Urban policies have traditionally not considered food-related issues. In this regard, it is important to ensure that (i) urban planning actions and policies take into consideration food-related matters, especially as concerns cold chain development and (ii) there is coordination between the policies and actions put in place at national and sub-national levels.

TABLE 16.1
Costs and Benefits for Selected Cooling Technologies

Cooling Technology	Estimated Capital Cost $US[a]	Energy Use per MT (kWh)	Expected Benefits
Shade cloth	Under $100	0	Can reduce temperatures by 10°C
Evaporative cooling	$200–400	0.7 kWh	Can reduce temperatures to 2°C above ambient dew point temperature
Night air ventilation	$300–400	0	Can reduce storage temperature to the range of night air temperature
Zero-energy cool chamber (ZECC)	$200 for 100 kg capacity $1,000 for 1 MT capacity	0	Can reduce temperatures to 2°C above ambient dew point temperature
Insulated package or pallet covers	$200–300	0	Can maintain cold during transport (with estimated 1°C rise per hour)
Portable forced air cooler	$600–1000	55 kWh 35 kWh for target temperature of 12°C	Can rapidly reduce temperature up to 2°C above cold room temperature
CoolBot controller with one ton A/C unit	$1500	55 kWh 35 kWh for target temperature of 12°C	Can reduce temperature to 0°C
Solar chiller (ice bank)	$2,000	0	Can hold cold food products indefinitely if recharged each day
Use of ice	$6,000–10,000	66 kWh	1 kg of ice will lower the temperature of 3 kg of the product by 28°C
Retrofit 20 reefer for cold room storage	$5,000–10,000	55 kWh 35 kWh for target temperature of 12°C	Quick cold storage with minimal construction cost
Owner built small-scale cold room	$5,000–10,000	55 kWh 35 kWh for target temperature of 12°C	2 tons of refrigeration can cool 1 MT of horticultural produce from 27°C to a target temperature of 2°C in 6–8 hours
Hydro-cooler – shower or spray type	$10,000–20,000	80–110 kWh 35–100 kWh for target temperature of 12°C	Small diameter produce will cool to 1°C in less than 10 minutes, larger produce (melons) will require up to 1 hour
Hydro-cooler – immersion type	$10,000–20,000	110–150 kWh	Cooling is more rapid than shower type hydro-cooler
Pre-fabricated small-scale cold storage room	Over $20,000	55 kWh 35 kWh for target temperature of 12°C	2 tons of refrigeration can cool 1 MT of horticultural produce from 27°C to a target temperature of 2°C in 6–8 hours
Vacuum cooler	Over $20,000	30 kWh	The cooling rate to 1°C is very rapid (15–20 minutes); about 5 times faster than forced-air cooling
Intermediate scale cold rooms	Over $20,000	55 kWh	5 tons of refrigeration can cool 3 MT of horticultural produce from 27°C to a target temperature of 2°C in 6–8 hours
Refrigerated (reefer) transport	Over $20,000	55 kWh 35 kWh for target temperature of 12°C	Reefer can maintain product temperature during transport

(Continued)

TABLE 16.1 (*Continued*)

Costs and Benefits for Selected Cooling Technologies

Cooling Technology	Estimated Capital Cost $US[a]	Energy Use per MT (kWh)	Expected Benefits
Large scale cold storage facility	$1,500 per m^2	55 kWh 35 kWh for target temperature of 12°C	High-income potential due to the large potential volume

[a] *Source*: Modified from Winrock (2009). Energy costs per MT will increase whenever cooling equipment or facilities are not utilized to their full capacity. Cost estimates for energy use are not provided due to the substantial volatility of energy prices (e.g., diesel, propane, kerosene, and electricity) and their geographic variability.

There is also an international dimension. Regional coordination, e.g., to ensure cross-border trade, can be an important factor. In Africa, the African Continental Free Trade Agreement has been implemented to address these needs and reduce transportation delays and border crossing issues.

16.4.2 PRIVATE SECTOR PARTNERS

The cold storage sector includes those companies that consolidate finished goods for distribution, export or value addition. Cold storage operators typically offer services to suppliers (packinghouses or processing companies) and provide services to buyers through consolidating orders, picking products for distribution and preparing products for export. Cold storages provide a critical link in the integrated value chain and can be independently operated or part of a private company network.

Refrigerated trucking operations are gaining ground in many countries and have a vested interest in promoting the development and use of cold technologies in other parts of the value chain. Reefer services will achieve much better results when the food products being loaded are pre-cooled and when their deliveries are made to well-managed cold storages and marketed via refrigerated retail displays.

In India, McDonald's was the first company to establish an integrated cold chain. Working with RK Foodland, a large food service company in Mumbai in the late 1990s, McDonald's established modern facilities for food processing under refrigerated conditions, cold storage (chilled and frozen) and cold transport. RK Foodland managed the entire cold chain and distributed chopped onions, sliced tomatoes, fresh cut lettuce and frozen French-fried potatoes to local McDonald's restaurants daily.

16.4.3 COLD CHAIN ASSOCIATIONS

GCCA, the Global Cold Chain Alliance, launched in 2007, serves as the focused voice of the international cold chain industry. GCCA is a platform for communication, networking and education for each link of the cold chain and is an umbrella organization consisting of four core partners and many affiliates, strategic and supporting partners (https://www.gcca.org/about/partners). The five core partners are cold chain associations managed from GCCA headquarters in the United States and are closely aligned in personnel, programs and services. The Global Cold Chain Alliance currently represents 1,233 member companies in over 65 countries.

Many countries have formed their own local cold chain associations during the past decade, and a few are listed below:

- Organization for Technology Advancement of Cold Chain in West Africa (OTACCWA)
- Bangladesh Cold Storage Association

- Brazil: Associação Brasileira da Indústria de Armazenagem Frigorificada
- Egyptian Cold Chain Association
- Asosiasi Rantai Pendingin Indonesia (ARPI)

In addition to GCCA, there is also the Global Food Cold Chain Council (www.foodcoldchain.org) and the Cold Chain Federation (www.coldchainfederation.org.uk).

16.5   DEFINING NATIONAL STRATEGIES AND SETTING PRIORITIES

The public sector, private sector and cold chain associations need to work together to define and set national priorities, based upon the findings of cold chain assessments, and as guided by the overall strategy of the government. The lead organization can be a private investor, development project or public sector agency.

Research assessments should be conducted at the national level in each country on the economic, social and environmental aspects of reducing postharvest losses and food waste via the use of the cold chain. These comprehensive studies will reveal the key commodities and food products that would most benefit from the use of the cold chain in each country, the barriers to adoption of local and regional cold chain technologies, as well as the socio-economic aspects of making changes in the food handling systems and key challenges for government policymakers.

16.6   DEVELOPING ACTION AND INVESTMENT PLANS

Each country's cold chain task force and cold chain working groups must develop their action plans and make recommendations regarding priority investments in the cold chain and in supporting logistics systems. A fully functioning integrated cold chain for a specific food product should be the overarching goal, and missing pieces of the cold chain should be filled in before beginning any new ventures. A model system will serve to provide an ongoing source of information and feedback, as well as a site for practical training for those who are interested in learning the skills they will need to work in the field. Over-reaching, by planning and attempting to implement too many competing or over-lapping cold chains at once, should be avoided.

Cold chain development is a complex goal, and governmental agencies or private sector businesses that are interested in investing in the area should avail themselves of the resources provided in this document. It is desirable as well that they reach out to the cold chain specialists who work via professional trade organizations in the many regions of the world and conduct feasibility studies based upon their local food products, capital and operating costs and market prices.

The government may need to invest in public goods, such as roads and electricity, to promote private sector investments. Furthermore, it may need to put in place the policy incentives, economic instruments and legal and regulatory measures that are required for cold chain development.

The development of the cold chain in any country must take into account the needs of the product beginning with the producer and ending with the ultimate consumer. If any piece of the cold chain is missing, then the entire chain is at risk. It is possible that the cold chain will not be cost-effective for a given food product, and the desire to reduce food losses and increase incomes for growers and marketers may come instead from food processing technology such as aseptic packaging or solar drying, which create products that are shelf stable at ambient temperatures.

16.7   CASE STUDIES OF COLD CHAIN INFRASTRUCTURE
REQUIREMENTS AND GOOD PRACTICES

The requirements for a successful cold chain extend across the entire value chain and can become very complicated depending upon the nature of the product and ultimate customer preferences.

TABLE 16.2
Three Scales of Typical Cold Chains for Fresh Produce

	Small-Scale Intended for Local Markets	Medium-Scale Intended for Regional Markets	Large-Scale Intended for Export Markets
Fresh foods	Harvest	Harvest	Harvest
	Packing and pre-cooling	Field packing	Transport to packinghouse
	Reefer transport	Transport to pre-cooler	Packaging
	Direct retail marketing by grower/shipper	Pre-cooling	Pre-cooling
	Final consumer	Temporary cold storage	Cold storage
		Reefer transport to modern retail	Reefer transport to the perishable distribution centre
		Refrigerated display	Dock unloading
		Final consumer	Cold storage
			Loading
			Dispatch to tarmac
			Loading in aircraft for export
			Flight
			Reefer transport to the perishable distribution centre
			Dock unloading/loading
			Reefer transport to modern retail
			Refrigerated display
			Final consumer

The following generalized examples of three cold chains cover the range of size and scope of cold chains found in developing countries.

The small-scale cold chain examples are shorter value chains with fewer actors and typically do not include cold storage if local market demand is well matched to the harvests. Medium-scale cold chains include cold storage as a buffer to better control the flow of fresh or frozen foods to regional or urban markets. If demand does not match production, the foods can be kept in cold storage for a few weeks or months, and the marketers can wait until market prices improve by avoiding selling produce during a period of over-supply (Table 16.2).

16.8   GENERAL CASES OF COLD CHAIN DEVELOPMENT

During the past decades, dozens of countries have attempted to jump-start their economic development by promoting food exports and/or by developing local perishables handling and storage capacity. Most early efforts were stand-alone projects, such as building cold storage facilities for high-value vegetable crops exports near an airport in Kenya. In many cases, these expensive efforts fell short when they were not supported by an integrated cold chain, and they became infamous examples of 'white elephants'. There are instances of packinghouses in Upper Egypt and Indonesia constructed in the mid-2000s that are not yet connected to electric power and large-scale cold storages in South America and Africa that have been abandoned due to the too high costs of operation. There are poorly constructed concrete cold storages in India, in multi-story buildings where stairs are the only means of hand carrying individual containers.

The following case studies are presented as examples of past and present efforts in cold chain development for fresh horticulture produce, and in each case, the lessons learned regarding good practices are summarized. Case studies include private investments, public initiatives and PPPs (public-private partnerships).

16.8.1 COLD CHAIN DEVELOPMENT IN KENYA

Efforts to provide pre-cooling and cold storage facilities for Kenya's fresh produce industry were funded by the Japan International Cooperation Agency (JICA) under the Horticultural Crops Development Authority (HCDA) during 2001–2005. Many problems were encountered, from proper siting to financial constraints, and the facilities remained underutilized (Kitinoja, 2010). In 2007, a new project was launched by JICA in Kenya, the Smallholder Horticulture Empowerment Project (SHEP), which included training of frontline extension staff, capacity building for farmers on group formation and management, market surveys for small farm produce and rural infrastructure development linked to these activities.

During 2010–2011, the Government of Kenya, with donor support, developed 38 produce markets with cold storage. It has been reported that due to a lack of modern storage facilities the horticultural industry loses over Ksh 30 billion (US$370 million) every year. "The reason we are experiencing huge losses of fresh produce is that our markets do not have modern storage facilities like cold rooms, which can store the produce before they are exported," explains Fresh Produce Exporters Association of Kenya (FPEAK) chief executive Dr Stephen Mbithi (http://www.new-ag. info/en/news/newsitem.php?a=1705).

Research studies and assessments revealed that perishability, price fluctuations, lack of storage and lack of transportation were the leading causes of food losses, and the Ministry of Agriculture Permanent Secretary predicts that the Ksh 80 billion generated in 2011 by the horticultural industry could reach a potential of Ksh 240 billion with improved distribution and cold chain development (Okulu, 2011).

16.8.2 COLD CHAIN DEVELOPMENT IN GHANA

Early cold chain assessment studies undertaken in Ghana in the early 2000s determined that investment in infrastructure would be very costly, and therefore, it would be very difficult for every industry player to build cooling facilities on their own. The absence of cooling facilities was shown to impact negatively on the market image of produce exported from the country as well as reducing the potential foreign exchange earnings from exports with high rates of losses due to perishability. It was proposed that it would be easier for Governments to build cool chain infrastructure for common use by industry, managed by industry specialists at economically viable rates, where each sector benefits – both private sector and government. An example of this in Ghana is the new Millennium Development Authority (MiDA) Perishable Cargo Centre at the Kotoka International Airport (KIA).

Thousands of small-scale farmers drawn from 30 districts have been trained by private extension agencies such as Agroteque and the Adventist Development and Relief Agency (ADRA) on good agricultural practices. Packinghouses were constructed in various regions and equipped with pre-cooling equipment (via forced air cooling or hydro-cooling systems). The funding for this program is from the US Millennium Challenge Corporation (MCC), but the execution was carried out by Development Alternatives, Inc. (DAI) and private sector partners in Ghana during 2006–2012.

When the Perishable Cargo Centre facility broke ground in January 2011, the Chairman of MiDA, Prof. Samuel Sefa-Dedeh, stated that "this $2.5m facility at KIA together with the postharvest and value chain enhancing investments by MiDA is also intended to enable Ghanaian horticultural exports to address food safety concerns and meet International Certification Standards required for effective competition at the higher end of markets in the European Union (EU)" (http://mida.gov. gh/site/?p=2626).

This MiDA project follows up on previous projects implemented by the United States Department of Agriculture (USDA) and the United States Agency for International Development (USAID), mainly in the establishment of packinghouses and cold facilities. The construction of Shed 9 at the Tema Port has proved to be sustainable over time. The formation of Farmer-Based Organizations (FBOs) has enabled farmers to gain access to credit, but the ability of farmers to repay their loans remains a significant challenge.

16.8.3 Cold Chain Development in the Philippines

Nestor B. Fongwan (as governor of the Province of Benguet) told reporters that he requested of President Gloria Macapagal-Arroyo during one of her visits in the province to provide the farmers with facilities to extend the shelf life of fresh produce. The aim was to minimize postharvest losses of crops and increase the income of farmers. "The President asked me, how she can help the farmers of Benguet? I told her we need cold chain facilities to preserve the freshness of the harvests. When she asked me how much it will cost, I said around eight million pesos. Immediately after, she gave a check amounting to four million pesos, and the remaining amount was shouldered by the Department of Agriculture," Fongwan recalled. (Eight million pesos=$US 182,000) (http://www.philmech.gov.ph/?page=News&action=detailsandCODE01=FB10070003)

The cold chain program for Benguet includes reefer trucks (6) cold storage chambers (2) and plastic crates for packing produce. The Benguet cold storage project provides services to farmers, including rentals of refrigerated trucks. The rental rate is P5,600 ($US 128) for 20 foot and P3,500 ($US 80) for 16 foot or 10-foot refrigerated trucks (on a 24-hour basis with a driver, excluding fuel). Space in cold storage is rented at P700 ($US 16) per day per chamber with a capacity of five metric tons. The plastic crates are rented at 50 centavos ($US 0.011) if the farmers also use the trucks or the cold storage, or 75 centavos ($US 0.017) as a stand-alone fee. Initial costs were higher than returns, but users give the project positive evaluations, and the Philippines Center for Postharvest Mechanization and Development (PhilMech) is preparing to expand investments in this kind of project into other provinces.

The Philippine Center for Postharvest Development and Mechanization (PhilMech) was asked by the government to propose several new cold chain-related development projects valued at over $US33 million, for improved distribution of fish, meats, fruits and vegetables (http://www.global-coldchainnews.com/?p=638). Future projects are being planned to include pre-coolers, packinghouses, cold rooms and cold trucks, and services will be marketed using the catchphrase "It pays to be cool". Ledda (2011) points out "refrigeration is a relative term", so the cold chain is more likely to be a cool chain for many of the Philippines-grown fruits and vegetables.

16.8.4 Cold Chain Development in Cameroon

Assessment studies in Cameroon, undertaken by FAO (2015a) resulted in the following general perceptions of logistics systems:

- Small firms: Lacked an understanding of wider logistics management and tended to associate it with only infrastructure and transport issues. For 86% of managers, logistics was limited to the possession and management of infrastructure.
- Medium-sized firms: Believed that proper logistics management was crucial for the success of the enterprise but had limited understanding of logistics.
- Large firms: Had limited understanding of logistics as relating only to aspects such as transport, maintenance of equipment and buildings (Table 16.3).

TABLE 16.3

Lessons Learned in Recent Case Studies of Cold Chain Development

Country and Types of Fresh Produce	Investors in the Cold Chain/Actors in the Value Chain	Lessons Learned
Kenya Fresh produce for domestic markets	Gov't of Kenya HCDA JICA (2001–2007) Regional market authorities	The benefits of training efforts had longer-term positive results than the large, difficult to manage, stand-alone infrastructure projects (i.e. cold storage facilities). Better siting of the Nairobi Horticultural Centre would have made it more convenient for growers and traders to use. Due to problems securing land, the Nairobi Horticultural Centre was constructed on a plot some distance from the airport, instead of within the grounds of Nairobi International Airport as originally planned. The centre's location was far from the area where exporters gathered, making it inconvenient to use. Contracts must be clearly spelt out and strictly honoured for growers to be able to begin to trust and have faith in the agreements made with the HCDA managers.
Ghana Fresh fruits and vegetables for export	MiDA MCC DAI Local private agri-business companies (mainly exporters)	In developing countries where the private sector has not fully emerged, the government must take the lead and provide the framework of cold chain development for the private sector to execute. MCC is working to find ways to work more directly with countries to ensure that they fully understand the responsibilities implicit in the written guidance, especially when it comes to consultation with civil society and coordination with donors. The MCC should take responsibility for communicating its approach regarding local leadership to other stakeholders. The evidence of past project failures in Ghana has lead current project implementers to learn from the mistakes of past horticultural development projects and work to avoid creating unrealistic expectations. MCC should appreciate how other donors' models of fostering country ownership through long-term capacity building in Ghana (i.e. DFID, USDA, USAID) can be strong complements to the MCC's more hands-off approach.
Philippines Temperate vegetable crops	PhilMech/Bureau of Postharvest Research and Extension (BPRE) Province of Benguet	Initial costs of establishing a cold chain will likely outweigh returns, so the government must be willing to take a long term view. By the 4th year of operation, the cold storage project was close to full utilization of capacity and was beginning to see a positive economic return and become self-sufficient.
Cameroon Cereals, oilseeds, fruits and vegetables, meat, fish and seafood	26 Private sector companies	No matter the size of the firm (small, medium or large) or the level of investment, there was limited understanding of cold chain logistics.

These case studies provide a range of examples of the issues found in developing countries and provide some of the activities that were part of national cold chain development action plans. Successful cold chain development policy and investments require regular assessment, re-evaluation and formative active.

16.9   CONCLUSIONS

For more details and additional case studies, FAO has published several studies on cold chain development and logistics in different regions:

- Firm-level logistics systems for the agri-food sector in sub-Saharan Africa (FAO, 2015a). http://www.fao.org/3/a-i5017e.pdf
- Logistics in the horticulture supply chain in Latin America and the Caribbean (FAO, 2015b). http://www.fao.org/3/a-i4792e.pdf
- Logistics Systems Need to Scale Up Reduction of Produce Losses in the Latin America and Caribbean Region (Fonseca and Vergara 2014). http://www.fao.org/fileadmin/user_upload/save-food/PDF/Articles/logistics_systems.pdf
- Developing the cold chain in the agri-food sector in sub-Saharan Africa (FAO and IIR, 2016). http://www.fao.org/3/a-i3950e.pdf. This is an IIR/FAO policy brief that was generated from studies conducted in a selection of countries.

On the whole, to improve the logistics and overall performance of agri-food firms involved in cold chain development to reduce food losses and improve market access, attention needs to be given to critical areas such as (i) policy implementation; (ii) infrastructure and facility improvement; (iii) networking and linking of agri-food firms in vertical and horizontal chains; (iv) training services; (v) improving information connectivity and (vi) improving access to financial sources (FAO, 2015a).

Promoting better linkages and improved communication among the many actors in the agro-industrial sector would serve to reduce duplication of efforts and increase the adoption of cost-effective cold chain development practices. Effective communication, coordination and collaboration among research, extension and industry personnel involved in the postharvest system and the development of the cold chain are the foundations of any successful plan. In most cases, the solutions to existing food loss, quality and food safety problems will require the use of already available information and the application of available proper cold chain technologies and logistics rather than conducting new research or developing new technologies.

REFERENCES

da Silva C. A., D. Baker and A. W. Shepherd. (2009). Agro-Industries for Development. CABI, UN FAO and UNIDO.

FAO and IIR. (2016). Developing the cold chain in the agrifood sector in sub-Saharan Africa. http://www.fao.org/3/a-i3950e.pdf.

FAO. (2015a). Firm-level logistics systems for the agrifood sector in sub-Saharan Africa – Report based on appraisals in Cameroon, Ghana, Uganda and the United Republic of Tanzania, by Gebresenbet, G. and Mpagalile, J. Rome, Italy.

FAO. (2015b). Logistics in the horticulture supply chain in Latin America and the Caribbean – Regional report based on five country assessments and findings from regional workshops by Fonseca, J. M. and Vergara, N., Rome, Italy.

Fearne, A., D. Ray and B. Vorley (Eds.) (2007). *Regoverning Markets: A Place for Small Scale Producers in Modern Agrifood Chains?* Abingdon, UK: Gower Publ., 248 p.

Fonseca, J.M. and N. Vergara (2014). Logistics Systems Need to Scale Up Reduction of Produce Losses in the Latin America and Caribbean Region. In: Mohammed, M. and J.A. Francis (Eds.) *Proceedings of 3rd International Conference on Postharvest and Quality Management of Horticultural Products of Interest for Tropical Regions. Acta Horticulturae.* 1047.

Kitinoja, L. (2010). Appropriate Postharvest Technologies for Improving Market Access and Incomes for Small Horticultural Farmers in Sub-Saharan Africa and South Asia. WFLO project final report for the Bill and Melinda Gates Foundation. 318 pp.

LaGra, J., L. Kitinoja, and K. Alpizar (2016). Commodity Systems Assessment Methodology for Value Chain Problem and Project Identification: A first step in food loss reduction. San Jose, Costa Rica: IICA. 246 pp. http://repiica.iica.int/docs/B4232i/B4232i.pdf.

Ledda, A.G. 2011. Cold chain technology: A technique for better competitive market advantage. August 31 2011 https://zamboanga.com/z/index.php?title=Nueva_Ecija_News_August_2011#Cold_chain_technology:_A_technique_for_better_competitive_market_advantage http://archives.pia.gov.ph/?m=110&item=201108 http://www.pia.gov.ph/?m=1&t=1&id=51285

Okulu, L. 2011. Kenya: State to put up fresh produce markets countrywide. Nairobi Star (11 Aug, 2011) http://allafrica.com/stories/201108120537.html

Shepherd, A. (2007). Approaches to linking producers to markets. FAO agricultural management and finance occasional paper. 67 p.

Shewfelt, R.L., and S.E. Prussia (Eds). (1993). *Postharvest Handling: A Systems Approach*. San Diego, CA: Academic Press.

Ton, G., J. Bijman, and J. Oorthuizen (2007). *Producer Organisations and Market Chains. Facilitating Trajectories of Change in Developing Countries*. Wageningen: Wageningen Academic Publishers.

UNDESA. (2018). United Nations Department of Economic and Social Affairs. https://www.un.org/development/desa/en/news/population/2018-revision-of-world-urbanization-prospects.html. Accessed on 28 September 2020.

Weinberger, K., Genova Ii, C., and Acedo, A. (2008). Quantifying postharvest loss in vegetables along the supply chain in Vietnam, Cambodia and Laos. *International Journal of Postharvest Technology and Innovation*, 1(3), 288–297.

Winrock International. (2009). Empowering Agriculture: Energy Options for Horticulture. US Agency for International Development 79 pp.

Disclaimer: The views expressed in this publication are those of the author(s) and do not necessarily reflect the views or policies of the Food and Agriculture Organization of the United Nations.

17 Gaps in the Research on Cooling Interventions for Perishable Crops in Sub-Saharan Africa and South Asia

Lisa Kitinoja
The Postharvest Education Foundation

Deirdre Holcroft
The Postharvest Education Foundation
and
Holcroft Postharvest Consulting

CONTENTS

17.1 INTRODUCTION

The Ceres2030 Project was developed and managed by a partnership between the International Food Policy Research Institute (IFPRI), International Institute for Sustainable Development (IISD) and Cornell University Global Development, with the objective of identifying sustainable solutions to end hunger. Eight topics were selected, and teams of researchers conducted scoping reviews and/or evidence syntheses to provide decision-makers with the information needed to help meet the targets set by Sustainable Development Goal 2 of ending hunger by 2030 (https://www.un.org/sustainabledevelopment/hunger/).

The Ceres2030 (https://ceres2030.org/) postharvest team reviewed the evidence-base to determine the interventions that small-scale producers and associated value chain actors in Sub-Saharan Africa (SSA) and South Asia (S. Asia) can adopt to reduce postharvest losses along food crop value chains. This involved systematically screening 12,907 research studies conducted from the 1970s through 2019 on 22 selected crops to identify the relevant research. Of these, 334 studies satisfied the predetermined inclusion criteria and provided evidence on the interventions for reducing postharvest losses. This research was published by *Nature Sustainability* as an open access article (Stathers et al., 2020).

The list of the 334 studies can be found in Table 17.1 of the supplementary data associated with this publication (https://doi.org/10.1038/s41893-020-00622-1). The supplementary and extended

DOI: 10.1201/9781003056607-20

(https://doi.org/10.1038/s41893-020-00622-1) information also includes the classification system used and specific interventions for each crop. The Ceres2030 postharvest database will be available online (https://phceres2030.net/) and can be searched by crop, country and postharvest intervention stage.

The 22 focal crops from five crop groups chosen for this study were:

- Cereals: maize, rice, sorghum, wheat
- Legumes: bean, cowpea, pigeon pea, chickpea, groundnut
- Roots and tubers: cassava, potato, sweetpotato, yam
- Vegetables: cabbage, onion, tomato, leafy vegetables
- Fruits: plantain, banana, mango, papaya, citrus.

The study was limited to the 57 countries of SSA and S. Asia. Twenty-five of the countries had no studies that fit the criteria for inclusion.

This chapter presents the evidence on the specific postharvest interventions associated with the cold chain i.e., precooling, cold storage and cold transportation for the 13 perishable crops (roots, tubers, fruits and vegetables) in the target countries, hereafter referred to as 'cooling' interventions.

17.2 OVERVIEW OF THE DATA ON PERISHABLE CROPS

As of October 2020, the database had 1,565 postharvest interventions of which 780 (49.8%) were on perishables. Of these, 311 interventions were for roots/tubers, 301 for fruits and 168 for vegetables (Table 17.1). The individual studies included data for multiple interventions and sometimes for more than one crop.

Data on postharvest interventions for perishable crops were identified for only 21 of the 57 countries included in the evidence review (35%). India alone accounted for 53% (413) of the total interventions on perishable crops, and 63% of the studies of interventions for fruits. In Africa, Nigeria was the country with the highest number of studied interventions for perishable crops (68), followed by Ghana (51) and South Africa (39) (Table 17.1).

17.3 COLD CHAIN INTERVENTIONS

In October 2020, we searched the Ceres2030 postharvest database for studies related to any aspect of the cold chain, i.e., evidence on interventions related to precooling, cold storage and cold transport in all the countries of SSA and S. Asia. Table 17.2 provides a summary for each perishable crop, with the total number of studies and the number of studies on cooling interventions identified in the evidence review. Studies included data for multiple crops and/or multiple interventions plus control treatments. Only 96 cooling interventions were identified in the database (12.3% of the total). The vast majority of these cooling intervention studies focused on only four crops: potato, banana, mango and tomato. For four crops (cassava, sweetpotato, yam and plantain), there were no cooling interventions found in the database.

Only 22 precooling interventions, including control treatments, were studied for the 13 perishable crops, which accounts for less than 3% of the total interventions. There were 74 interventions on cold storage which included evaporative cooled structures, rooms cooled with a CoolBot™ (an air-conditioner with a controller to override the frost sensor) and commercial cold rooms at various set points, which was 9.3% of the interventions for perishables crops. There were no studies on cooling during transportation.

17.4 CROP-SPECIFIC RESULTS

The cooling interventions with their relative effects on postharvest losses are presented for each crop in Tables 17.3–17.6. Of the four root/tuber crops, cooling interventions were only evaluated

TABLE 17.1

The Number of Postharvest Loss Reduction Interventions Studied for the Different Three Categories of Perishable Crops in 32 Countries Located in SSA and S. Asia

Country	Total for All Crops	Roots/Tubers	Fruits	Vegetables	Total for Perishables	Percent Perishables (%)
			Number of Interventions			
Afghanistan	2	0	0	0	0	0.0
Bangladesh	34	4	12	6	22	64.7
Benin	46	2	0	0	2	4.3
Burkina Faso	13	0	0	0	0	0.0
Cameroon	12	12	0	0	12	100.0
Cote d'Ivoire	22	19	0	0	19	86.4
DRC	4	4	0	0	4	100.0
Eritrea	11	0	0	0	0	0.0
Ethiopia	67	4	15	11	30	44.8
Gambia	3	0	0	0	0	0.0
Ghana	129	39	0	12	51	39.5
Guinea	2	2	0	0	2	100.0
India	541	130	190	93	413	76.3
Kenya	98	16	3	0	19	19.4
Malawi	29	0	0	0	0	0.0
Mali	3	0	0	0	0	0.0
Mauritius	6	3	0	3	6	100.0
Mozambique	13	8	0	0	8	61.5
Nepal	27	4	5	10	19	70.4
Niger	22	0	0	3	3	13.6
Nigeria	107	49	8	11	68	63.6
Pakistan	71	4	22	3	29	40.8
Sierra Leone	3	0	0	0	0	0.0
Somalia	4	0	0	0	0	0.0
South Africa	39	2	34	3	39	100.0
Sri Lanka	29	0	7	3	10	34.5
Sudan	15	0	2	7	9	60.0
Tanzania	66	4	3	3	10	15.2
Togo	28	0	0	0	0	0.0
Uganda	42	9	0	0	9	21.4
Zambia	8	0	0	0	0	0.0
Zimbabwe	69	0	0	0	0	0.0
Totals	**1,565**	**311**	**301**	**168**	**780**	
Percentages	**100.0%**	**19.9%**	**19.2%**	**10.7%**	**49.8%**	

in 13 studies on potato. Storage in a cold room tended to minimize losses more than evaporative cooled structures, which, in turn, were more effective than ambient storage (Table 17.3). However, cold storage alone did not reduce sprouting effectively. No studies were looking at the entire cold chain.

In cabbage and leafy vegetables, two studies on the CoolBot™ or zero energy cool chamber (ZECC) demonstrated benefits in reducing quantity loss (predominantly water loss) and quality loss (predominantly decay) (Table 17.4). Despite the short shelf life and perishability of leafy vegetables, which are an excellent source of nutrition, there were no studies on precooling, storage in

TABLE 17.2

Key Crops and Postharvest Intervention Studies in SSA and S. Asia

Crop	No. of Interventions/Crop	No. Cooling Interventions	No. and Type of Cooling Interventions	Cooling Interventions as Percent of Total for Each Crop (%)
Cassava	20	0	–	0.0
Potato	153	22	9 cold room 13 evaporative cool room	14.3
Sweetpotato	53	0	–	0.0
Yam	85	0	–	0.0
Plantain	0	0	–	0.0
Banana	46	19	9 precooling 3 cold room 7 evaporative cool room	41.3
Mango	106	19	13 precooling 1 cold room 2 cold + pesticide 2 CoolBot cold room 1 evaporative cool structure	17.9
Papaya	16	5	5 evaporative cool structure	31.3
Citrus fruits	133	9	4 cold room 5 evaporative cool structure	6.8
Cabbage	7	1	1 CoolBot cold room	14.3
Onion	97	4	4 cold room	4.1
Tomato	59	15	2 cold room 1 CoolBot cold room 12 evaporative cool structure	25.4
Leafy vegetables	5	2	2 evaporative cool room	40.0
Total	**780**	**96**		**12.3**

TABLE 17.3

Cooling Interventions Evaluated on Potato to Reduce Postharvest Loss

Study	Treatment	Quantity Loss (%)	Quality Loss (%)	Sprouting (%)	Weeks of Storage	References
11753	Ambient storage with CIPC			20.0	8.6	Mehta and Singh
11753	Cold room			100.0	6.4	(2015)
11753	Cold room with CIPC			50.0	6.4	
11753	Ambient storage with CIPC			100.0	8.6	
12004	Cold room (10°C) with CIPC	1.2		14.8	8.6	Malik et al. (2008)
12004	Cold room (2°C–3°C; no CIPC)	0.3		0.0	8.6	
12336	Cold storage	2.6	0.0		17.1	Venugopal et al.
12336	Evaporative cooled structure	3.6	0.0		17.1	(2017)
12336	Ambient storage in jute bags in the field	53.0	42.0		17.1	
12336	Cooling with night air ventilation	11.8	3.6		17.1	
12336	Traditional storage in a pit	27.6	5.4		17.1	
3649	Cold storage in bags	4.8	1.0		20.0	Verma et al.
3649	Ambient storage in bags	19.1	29.0		20.0	(1974)
3956	Cool storage (15°C)	23.0	7.0		18.6	Thomas et al.
3956	Cold storage (2°C–4°C)	4.6	8.2		18.6	(1979)
3956	Cool storage (15°C)	15.3	1.0		17.1	

(Continued)

TABLE 17.3 (*Continued*)
Cooling Interventions Evaluated on Potato to Reduce Postharvest Loss

Study	Treatment	Quantity Loss (%)	Quality Loss (%)	Sprouting (%)	Weeks of Storage	References
3956	Ambient storage (27°C–32°C)	52.7	22.4		13.6	
3956	Cool storage (15°C) after irradiation	9.4	1.2		17.1	
3956	Ambient storage (27°C–32°C) after irradiation	50.4	30.8		13.6	
3956	Cool storage (15°C) after irradiation	14.1	10.8		18.6	
3956	Ambient storage after irradiation, unwashed	69.2	47.8		18.6	
3956	Ambient storage (27°C–32°C) washed with sanitizer	73.9	44.3		18.6	
3956	Ambient storage (27°C–32°C) unwashed	75.8	42.8		18.6	
5105	Ambient storage	17.2	6.4		14.0	Mehta and Kaul
5105	Evaporative cooled structure	9.1	1.9		14.0	(1991)
5550	Control (no cooling)	15.4	1.9		12.9	Kumar et al.
5550	Cool room, evaporative cooled	20.2	3.4		12.9	(1995)
6274	Cold storage	7.7			12.0	Patel et al. (2001)
6274	Traditional storage – heap	21.5			12.0	
6274	Traditional storage – kachcha ventilated rooms	20.4			12.0	
6274	Traditional storage – rustic diffused light stores	13.8			12.0	
6551	Cold storage (2°C–4°C)	3.5	1.6		21.0	Wong Yen Cheong
6551	Cold storage (8°C–10°C)	3.2	2.5		12.0	and Govinden
6551	Ambient storage	34.3	16.0		12.0	(2001)
6750	Cold storage in bags	8.0			21.0	Kumar and Gupta
6750	Ambient storage indoors on trays	18.0	80.0		13.0	(1999)
6750	Evaporative cooled (electric desert cooler)	17.5	3.0		19.0	
6750	Ambient with passive draft	22.5	3.0		19.0	
6750	Radiatively cooled	22.0	8.0		16.0	
6750	ZECC (brick/sand)	14.0	3.0		17.0	
6830	Evaporative cooled store+CIPC	9.1		20.3	16.0	Mehta and Ezekiel
6830	Evaporative cooled store+CIPC applied twice	6.9		5.9	16.0	(2002)
6830	Evaporative cooled store (no CIPC)	11.1		100.0	16.0	
6830	Traditional storage in a heap; CIPC	15.3		32.7	16.0	
6830	Traditional storage in a heap; CIPC applied twice	9.0		11.4	16.0	
6830	Traditional storage in a heap (no CIPC)	14.3		100.0	16.0	
6830	Traditional storage in a pit; CIPC	8.2		13.4	16.0	
6830	Traditional storage in a pit (no CIPC)	10.7		100.0	16.0	
6830	Traditional storage in a pit; CIPC applied twice	5.1		4.5	16.0	
8267	Ambient storage	18.3			10.7	Mishra et al.
8267	Evaporative cooled store	12.5			10.7	(2010)
8825	Bamboo iceless refrigerator	13.6	4.8		17.0	Anjali and Kumar
8825	Traditional storage in a heap	15.5	5.3		17.0	(2011)
8825	ZECC	10.0	3.5		17.0	

CIPC, sprout inhibitor; ZECC, zero energy cool chamber.

TABLE 17.4

Cooling Interventions Evaluated on Cabbage, Leafy Vegetables, Onion and Tomato to Reduce Postharvest Loss

Study	Treatment Details	Quantity loss (%)	Quality Loss (%)	Sprouting (%)	Vit. C (mg/100g)	Weeks of Storage	Shelf Life (days)	References
Cabbage								
9434	Ambient storage (25°C–35°C)	30.0				3.0		Saran et al. (2013)
9434	Coolbot cold room (10°C)	5.0				3.0		Dari et al. (2015)
Leafy vegetables								
1074	Ambient storage in a room	44.0				0.4	2	
1074	ZECC (kiln fired brick)	5.3				0.4	5	
1074	ZECC (mud brick)	6.1				0.4	5	Nabi et al. (2013)
Onion								
10089	Ambient in cement room (34°C–38°C; 50%–60% RH)	98.1		100.0		16.0		
10089	Cold room (0°C–1°C; >75% RH)	6.1		9.6		16.0		
10089	Ambient in mud room (27°C–31°C; 60%–70% RH)	92.7		91.4		16.0		Jolayemi et al. (2018)
10646	Cold room (4°C–5°C)				4.0	8.0		
10646	Warm storage (45°C–50°C)				4.0	8.0		
10646	Ambient storage (28°C–30°C)				8.0	8.0		
4157	Ambient (20°C)		9.0			8.0		Musa et al. (1973)
4157	Cold room (5°C)		2.3			8.0		
4157	Traditional storage structure ('cottage')		9.5			8.0		
4157	Drying before storage (ventilated oven 35°C–40°C)		7.3			8.0		
Tomato								
11332	Ambient storage	3.0				0.7	5	Woldemariam and Abera (2014)
11332	Evaporative cooled bamboo jute cooler	0.5				0.7	21	
11332	Evaporative cooled pot in pot cooler	0.7				0.7	19	

(Continued)

TABLE 17.4 (Continued)
Cooling Interventions Evaluated on Cabbage, Leafy Vegetables, Onion and Tomato to Reduce Postharvest Loss

Study	Treatment Details	Quantity loss (%)	Quality Loss (%)	Sprouting (%)	Vit. C (mg/100g)	Weeks of Storage	Shelf Life (days)	References
12753	Cool room (12°C)	4.7				2.9		Nkolisa et al. (2018)
12753	Evaporative cooled structure (fan and pad)	8.5				2.9		
12753	Ambient storage in room	10.5				2.9		
1838	ZECC Pusa	3.8	29.5			1.7		Goswami et al. (2008)
1838	ZECC modified Pusa with a stronger structure	5.5	30.8			1.7		
1838	ZECC modified with wind tunnel	14.2	28.5			1.7		
1838	Ambient storage in room	16.5	23.7			1.7		
2439	Ambient storage	17.8	90.1			2.3		Woldetsadik et al. (2011)
2439	Evaporative cooled structure (fan and charcoal)	8.4	19.1			2.3		
5649	Evaporative cooled structure with fans (active)	7.0				7.1		Garg et al. (1997)
5649	Ambient storage	20.0				3.6		
5649	Evaporative cooled structure without fans (passive)	12.0				6.0		
5649	ZECC (brick/sand)	14.0				7.1		
8267	Ambient storage	21.5				4.3	7	Mishra et al. (2010)
8267	Evaporative cooled structure	10.2				4.3	14	
8724	Ambient storage in CFB carton	7.0				1.3		Jain et al. (2005)
8724	Ambient storage in wooden box	7.2				1.3		
8724	ZECC in CFB carton	8.7				1.3		
8724	ZECC in bamboo basket	9.1				1.3		
8724	ZECC in wooden box	9.6				1.3		
9434	Ambient storage (25°C–35°C)	42.0				3.0		Saran et al. (2013)
9434	Coolbot cold room (10°C)	5.0				3.0		

ZECC, zero energy cool chamber; CFB, corrugated fibre board.

conventional cold rooms or cold transportation. Cold storage reduced quantity loss and sprouting in onions, although 2 of the 3 studies showed that some traditional storage structures were effective and affordable means of extending onion shelf life (Stathers et al., 2020; Table 17.4). Losses of tomato were reduced and the shelf life extended by evaporative cooled structures or cold rooms across all eight of the studies. However, despite the contribution of vegetables to human nutrition, there were no studies on the effect of these interventions on vitamin C or carotenoid content, except for one study on onions, and neither were there any studies on cold transportation (Table 17.4).

All the fruits selected for the review were chilling sensitive, i.e. are injured by low but non-freezing temperatures, as is common in crops of tropical or subtropical origin, making them well adapted to evaporative cooled storage structures. Five studies showed that the reduction of quantity and quality losses of banana (optimum storage temperatures of 13°C–15°C) was achieved by the use of evaporative cooled structures, cool storage and even cellar storage (Table 17.5). Precooling of banana, by hydrocooling or forced air cooling, was not very effective unless followed by cold storage, demonstrating the need for an integrated cold chain.

Losses in quantity and quality of citrus were reduced by a range of interventions from commercial cold rooms to evaporatively cooled structures in five studies when compared to ambient storage (Table 17.5). Quality loss of mango was reduced in five studies by storing the fruit at 13°C–15°C compared to ambient conditions (Table 17.6). Hydrocooling mangoes before storage had beneficial effects on both quantity and quality loss, especially when combined with fungicides to control decay. Storing papaya in an evaporative cooled structure in one study reduced quantity and quality losses compared to ambient storage regardless of packaging (Table 17.6).

17.5 CONCLUSIONS

There is an obvious need to study these cold chains and cooling interventions in more countries over a wider range of perishable crops and include all the value chain actors involved in precooling, storage and transport and distribution.

While precooling and cold storage are important components of the cold chain, not one of these studies evaluated the effect of cold transportation or an integrated cold chain from harvest to consumption. At most the studies evaluated precooling alone or precooling followed by cold storage.

One possible reason cold chain-related postharvest interventions have not received a lot of attention is the cost of the experimental setup. The costs cited in the studies (converted to US$ at the average exchange rate of the year of study) are about US$2,000 for cold rooms cooled evaporative (20 MT capacity) or with a modified air-conditioner or CoolBot™ (8 to 10 MT capacity) (Table 17.6). Experiments often require more than one cold room (if testing different temperature regimes or packaging treatments in conjunction with cold storage, for example). Smaller ZECC units are more affordable but limited in their storage capacity.

However, when calculations of costs and benefits are performed, investments in the cold chain provide a positive return on investment. Saran et al. (2013) reported that the cost of a CoolBot™ equipped cold room in India was about 25% of the cost of a 2-tonne commercial cold room of the same capacity. Ambuko et al. (2018) found their cold-stored mango fruits remained saleable for up to 35 days which was 23 days more than those stored at ambient room conditions. They recommend that the CoolBot™ could be promoted in Kenya for adoption as a low-cost cold storage alternative to conventional cold rooms.

The synthesis of this evidence resulted in only 334 studies for the selected food crops in 57 countries that met the inclusion criteria for the study. Although more than 475 million of the world's 570 million farms are smaller than 2 hectares, and populations are predominantly rural in low-income countries, most of the published research (>95%) was not relevant to these small-scale farmers. Several laboratory studies implied that postharvest interventions would be beneficial in developing countries, but there was a lack of scale-up testing. To clarify, those recommendations were based on very small sample size and were never tested on a large enough scale or under real-world conditions.

TABLE 17.5

Cooling Interventions Evaluated on Banana and Citrus to Reduce Postharvest Loss

Study	Category	Quantity Loss (%)	Quality Loss (%)	Weeks of Storage	Shelf Life (Days)	References
Banana						
1838	ZECC Pusa	7.1	47.5	1.7	13.0	Goswami et al. (2008)
1838	ZECC modified Pusa with a stronger structure	7.8	53.0	1.7	11.0	
1838	ZECC modified with wind tunnel	13.7	37.2	1.7	13.0	
1838	Ambient storage in room	18.6	31.7	1.7	14.0	
3378	Cool storage (13°C) in MAP with ethylene absorber	6.2		5.0		Kudachikar et al. (2007)
3378	Cool storage (13°C) in MAP	6.5		3.0		
3378	Cool storage (13°C)	7.6		2.1		
6109	Ambient storage; 3/4 ripe	18.0		2.3		Nagaraju and Reddy (1995)
6109	Ambient storage; ripe	18.0		1.9		
6109	Evaporative cooled structure; 3/4 ripe	5.0		2.3		
6109	Evaporative cooled structure; ripe	8.0		2.3		
7741	Ambient storage (26–30°C)	6.9	21.2	1.1	8.0	Deka et al. (2006)
7741	Cool storage (12°C)	1.0	0.0	1.1	24.0	
7741	Cellar (semi-underground; 22°C–25°C)	4.5	15.5	1.1	9.0	
8919	Cold storage	21.3	19.0		19.0	Doshi and Sutar (2010)
8919	Ambient storage	23.1	28.0		10.0	
8919	Hydrocooling (dip) then cold storage	30.3	19.0		24.0	
8919	Hydrocooling (dip) then ZECC	9.0	7.0		16.0	
8919	Hydrocooling (dip) then ambient storage	20.9	29.0		10.0	
8919	Hydrocooling (drench) then ambient storage	21.1	21.0		10.0	
8919	Hydrocooling (drench) then cold storage	20.7	10.0		24.0	
8919	Hydrocooling (drench) then ZECC	7.9	0.0		19.0	
8919	Forced air cooling then ambient storage	22.3	13.0		24.0	
8919	Forced air cooling then ZECC	22.1	17.0		10.0	
8919	Forced air cooling then cold storage	7.6	13.0		19.0	
8919	Wax then cold storage	24.2	0.0		24.0	
8919	Wax then ambient storage	7.6	0.0		10.0	
8919	Wax then ZECC	5.4	0.0		19.0	
8919	ZECC	10.2	0.0		16.0	

(Continued)

TABLE 17.5 (*Continued*)

Cooling Interventions Evaluated on Banana and Citrus to Reduce Postharvest Loss

Study	Category	Quantity Loss (%)	Quality Loss (%)	Weeks of Storage	Shelf Life (Days)	References
Citrus						
11571	Ambient storage+kinetin	36.2	34.4	4.0		Kaur et al. (2014)
11571	Ambient storage+kinetin	34.6	32.2	4.0		
11571	Ambient storage	50.1	46.7	4.0		
11571	Cold storage (3°C–4°C)	16.2	25.3	8.6		
11571	Cold storage (3°C–4°C)+kinetin	12.2	17.7	8.6		
11571	Cold storage (3°C–4°C)+kinetin	11.1	11.7	8.6		
1838	ZECC Pusa	3.6	6.9	1.7	31.0	Goswami et al. (2008)
1838	ZECC modified Pusa with a stronger structure	3.8	8.6	1.7	31.0	
1838	ZECC modified with wind tunnel	6.5	7.4	1.7	23.0	
1838	Ambient storage in room	13.2	8.2	1.7	16.0	
6835	Ambient storage (15°C–31°C; 19%–55% RH)	25.2	13.5	3.4		Bhardwaj and Sen (2003)
6835	ZECC (11°C–22°C; 90%–95% RH)	16.2	5.3	3.4		
7244	Cold storage (10°C)	9.0	0.7	12.9		Ladaniya (2004)
7244	Cold storage (6°C)	9.1	1.2	12.9		
7244	Cold storage (8°C)	8.5	0.5	12.9		
7562	Evaporative cooled structure (13°C–19°C)	2.0	1.0	2.0		Chopra et al. (2004)
7562	Ambient storage in room	9.0	10.0	2.0		

ZECC, zero energy cool chamber; MAP, modified atmosphere packaging.

TABLE 17.6

Cooling Interventions Evaluated on Mango and Papaya to Reduce Postharvest Loss

Study	Category	Quantity Loss (%)	Quality Loss (%)	Weeks of Storage	Shelf Life (Days)	References
Mango						
11159	Cool storage (13°C) with fungicide (Biosafe)		9.0	3.6	26.0	Pujari et al. (2016)
11159	Ambient storage		76.0	2.1	13.0	
11159	Ambient storage with fungicide (Biosafe)		29.0	2.1	16.0	
11159	Ambient storage with fungicide (carbendazim)		31.0	2.1	15.0	
11159	Cool storage (13°C) with fungicide (carbendazim)		22.0	3.6	26.0	
11159	Cool storage (13°C)		44.0	3.6	22.0	
11180	Cold storage with Coolbot (12 days)			1.7		Ambuko et al. (2018)
11180	Cold storage with Coolbot (35 days)			5.0		
11180	Ambient storage (12 days)			1.7		
5581	No precooling		31.1		9.2	Puttaraju and Reddy (1997)
5581	Hydrocooling		4.5		11.5	
5855	Hydrocooling (12°C)	17.5	18.0	2.1	14.3	Kapse and Katrodia (1997)
5855	Hydrocooling (12°C)+fungicide (Bavistin)	16.2	10.0	2.1	15.3	
5855	Hydrocooling (12°C)+fungicide (Dithane)	16.7	14.0	2.1	15.0	
5855	Hydrocooling (16°C)	19.9	68.0	2.1	12.3	
5855	Hydrocooling (16°C)+fungicide	18.0	38.0	2.1	14.0	
5855	Hydrocooling (16°C)+fungicide	18.5	42.0	2.1	13.0	
5855	Hydrocooling (8°C)	16.1	0.0	2.1		
5855	Hydrocooling (8°C)+fungicide	14.2	0.0	2.1		
5855	Hydrocooling (8°C)+fungicide	15.0	0.0	2.1		
5855	No precooling	26.0	100.0	2.1	10.7	Singh et al. (2003)
6783	Hydrocooling	10.9		1.3		
6783	No precooling	12.3		1.3		
8446	Ambient storage	80.0		2.1		Odey et al. (2007)
8447	Evaporative cooled structure	50.0		2.1		
8448	Storage pit lined with plantain leaves	90.0		2.1		

(Continued)

TABLE 17.6 (*Continued*)
Cooling Interventions Evaluated on Mango and Papaya to Reduce Postharvest Loss

Study	Category	Quantity Loss (%)	Quality Loss (%)	Weeks of Storage	Shelf Life (Days)	References
Papaya						
11521	Ambient storage, wooden crate	18.3	73.6	1.3	8.0	Azene et al. (2014)
11521	Ambient storage, wooden crate lined with banana leaves	14.7	63.9	1.3	7.0	
11521	Ambient storage, wooden crate lined with HDPE	5.3	48.6	1.3	9.0	
11521	Ambient storage, wooden crate lined with LDPE sheet	6.2	45.8	1.3	9.0	
11521	Ambient storage, wooden crate lined with newspaper	11.2	62.5	1.3	9.0	
11521	Evaporative cooled, wooden crate	4.4	23.6	1.3	13.0	
11521	Evaporative cooled, wooden crate lined with banana leaves	3.6	16.7	1.3	15.0	
11521	Evaporative cooled, wooden crate lined with LDPE	1.9	5.6	1.3	18.0	
11521	Evaporative cooled, wooden crate lined with HDPE	2.8	5.6	1.3	18.0	
11521	Evaporative cooled, wooden crate lined with newspaper	3.9	15.3	1.3	15.0	

HDPE, high density polyethylene; LDPE, low density polyethylene.

Our review of the evidence base also indicates a serious gap due to the lack of economic, gender and environmentally focused studies.

Despite the contribution of fruits and vegetables to human nutrition, there was only one study that considered the effect of temperature on vitamin C (onion), and no studies that measured carotenoid content. In all of these perishable crops, there were no studies on the effect of an integrated cold chain, i.e. precooling followed by storage, transportation and marketing at low temperatures, despite the widespread adoption of the cold chain for exports from these countries or adoption by commercial farms.

In conclusion, we propose that the cold chain be a stronger focus for future research as it will impact produce quality and nutritional value of root/tubers, vegetables and fruits as well as reducing postharvest losses.

ACKNOWLEDGEMENTS

The authors thank the postharvest team leader, Dr. Tanya Stathers of NRI (UK) and the *Ceres2030 postharvest team* for their inputs and reviews of the original article and development of the database on which this chapter is based.

REFERENCES

Ambuko, J., Karithi, E., Hutchinson, M., Wasilwa, L., Hansen, B. and Owino, W. (2018). Postharvest shelf life of mango fruits stored in a Coolbot™ cold room. *Acta Horticulturae* 1225:193–198.

Anjali, C. and Kumar, S.A. (2011). Effect of evaporative cooling chamber on storage of potato. *Journal of Dairying, Foods and Home Sciences* 30(2):131–133.

Azene, M., Workneh, T.S. and Woldetsadik, K. (2014). Effect of packaging materials and storage environment on postharvest quality of papaya fruit. *Journal of Food Science and Technology* 51:1041–1055.

Bhardwaj, R.L. and Sen, N.L. (2003). Zero energy cool-chamber storage of mandarin (Citrus reticulata Blanco) cv. 'Nagpur Santra'. *Journal of Food Science and Technology (India)* 40(6):669–672.

Chopra, S., Kudos, S.K.A., Oberoi, H.S., Baboo, B., Ahmad, K.U.M. and Kaur, J. (2004). Performance evaluation of evaporative cooled room for storage of Kinnow mandarin. *Journal of Food Science and Technology (India)* 41(5):573–577.

Dari, L., Owusu-Ansah, P., Nenguwo, N. and Afari-Sefa, V. (2015). Comparative study of the performance of kiln-fired brick and mud block zero energy cool chambers in the Northern Region of Ghana. *International Journal of Tropical Agriculture* 33(3):2037–2041.

Deka, B.C., Choudhury, S., Bhattacharyya, A., Begum, K.H. and Neog, M. (2006). Postharvest treatments for shelf life extension of banana under different storage environments. *Acta Horticulturae* 712:841–850.

Doshi, J.S. and Sutar, R.F. (2010). Studies on pre-cooling and storage of banana for extension of shelf life. *Journal of Agricultural Engineering* 47(2):14–19.

Garg, S., Gupta, A.K. and Kumar, A. (1997). Storage of tomatoes in evaporatively cooled chambers. *Journal of Research, Punjab Agricultural University* 34(3):320–327.

Goswami, S., Borah, A., Baishaya, S., Saika, A. and Deka Bidyut, C. (2008). Low cost storage structures for shelf-life extension of horticultural produces. *Journal of Food Science and Technology (India)* 45(1):7074.

Jain, P.K., Naidu, A.K. and Raut, R.L. (2005). Effect of oil emulsion, packaging and storage conditions on shelf life and quality of tomato fruit. *Udyanika* 11(2):84–86.

Jolayemi, O.S., Nassarawa, S.S., Lawal, O.M., Sodipo, M.A. and Oluwalana, I.B. (2018). Monitoring the changes in chemical properties of red and white onions (*Allium cepa*) during storage. *Journal of Stored Products and Postharvest Research* 9(7):78–86.

Kapse, B.M. and Katrodia, J.S. (1997). Studies on hydrocooling in Kesar mango (*Mangifera indica* L.). *Acta Horticulturae* 455:707–717.

Kaur, N., Kumar, A. and Monga, P.K. (2014). Beneficial effects of kinetin on shelf life of kinnow mandarin. *International Journal of Agricultural Sciences* 10(1):66–69.

Kudachikar, V.B., Kulkarni, S.G., Vasantha, M.S. and Aravinda Prasad, B. (2007). Effect of modified atmosphere packaging on shelf-life and fruit quality of banana stored at low temperature. *Journal of Food Science and Technology (India)* 44(1):74–78.

Kumar, D., Kaul, H.N. and Singh, S.V. (1995). Keeping quality in advanced potato selections during non-refrigerated storage. *Journal of the Indian Potato Association* 22(3/4):105–108.

Kumar, S. and Gupta, A.K. (1999). Studies on non-refrigerated storage of potatoes. *Journal of Research, Punjab Agricultural University* 36(3–4):242–251.

Ladaniya, M.S. (2004). Response of 'Kagzi' acid lime to low temperature regimes during storage. *Journal of Food Science and Technology (India)* 41(3):284–288.

Malik, T.P., Kumar, J., Bhatia, A.K. and Singh, V.P. (2008). Effect of 3-chlorophenyl isopropyl carbamate (CIPC) on shelf life of potato cultivars. *Haryana Journal of Horticultural Sciences* 37(3/4):365–366.

Mehta, A. and Ezekiel, R. (2002). Evaluation of non-refrigerated storage methods for short-term on-farm storage of potatoes. *Journal of the Indian Potato Association* 29(3–4):292–295.

Mehta, A. and Kaul, H.N. (1991). Effect of sprout inhibitors on potato tubers (*Solanum tuberosum* L.) stored at ambient or reduced temperatures. *Potato Research* 34:443–450.

Mehta, A. and Singh, B. (2015). Effect of CIPC treatment on post-harvest losses and processing attributes of potato cultivars. *Potato Journal* 42(1):18–28.

Mishra, B.K., Jainz, N.K., Kumar, S., Doharey, D.S. and Sharma, K.C. (2010). Shelf life studies on potato and tomato under evaporative cooled storage structure in southern Rajasthan. *Journal of Agricultural Engineering* 46(3):26–30.

Musa, S.K., Habish, H.A., Abdalla, A.A. and Adlan, A.B. (1973). Problems of onion storage in the Sudan. *Tropical Science* 15(4):319–327.

Nabi, G., Rab, A., Sajid, M., Farhatullah, Abbas, S.J. and Imran, A. (2013). Influence of curing methods and storage conditions on the post-harvest quality of onion bulbs. *Pakistan Journal of Botany* 45(2):455–460.

Nagaraju, C.G. and Reddy, T.V. (1995). Deferral of banana fruit ripening by cool chamber storage. *Advances in Horticultural Science* 9(4):162–166.

Nkolisa, N., Magwaza, L.S., Workneh, T.S. and Chimpango, A. (2018). Evaluating evaporative cooling system as an energy- free and cost- effective method for postharvest storage of tomatoes (*Solanum lycopersicum* L.) for smallholder farmers. *Scientia Horticulturae* 241:131–143.

Odey, S.O., Agba, O.A. and Ogar, E.A. (2007). Spoilage of freshly harvested mango fruits (*Mangifera indica*) stored using different storage methods. *Global Journal of Agricultural Sciences* 6(2):145–148.

Patel, C.K., Sasani, G.V., Patel, S.H. and Patel, N.H. (2001). Comparison of some traditional storage methods of potato. *Journal of the Indian Potato Association* 28(1):152–153.

Pujari, K.H., Joshi, M.S. and Shedge, M.S. (2016). Management of postharvest fruit rot of mango caused by *Colletotrichum gloeosporiodes*. *Acta Horticulturae* 1120:215–218.

Puttaraju, T.B. and Reddy, T.V. (1997). Effect of precooling on the quality of mango (cv. 'Mallika'). *Journal of Food Science and Technology (India)* 34(1):24–27.

Singh, R., Sharma, R.K., Kumar, J. and Goyal, R.K. (2003). Effect of oil emulsions and precooling on the shelf life of mango cv. Amarpali. *Haryana Journal of Horticultural Science* 32(1&2):54–55.

Stathers, T.E., Holcroft, D., Kitinoja, L., Mvumi, B., English, A., Omotilewa, O., Kocher, M., Ault, J. and Torero, M. (2020). Evidence synthesis: Interventions for reducing crop postharvest losses in food systems in sub-Saharan Africa and South Asia. *Nature Sustainability* 3, 821–835. https://doi.org/10.1038/s41893-020-00622-1.

Saran, S., Neeru, D., Vigya, M., Dwivedi, S.K. and Raman, N.L.M. (2013). Evaluation of CoolBot cool room as a low cost storage system for marginal farmers. *Progressive Horticulture* 45(1):115–121.

Thomas, P., Srirangarajan, A.N., Joshi, M.R. and Janave, M.T. (1979). Storage deterioration in gamma-irradiated and unirradiated Indian potato cultivars under refrigeration and tropical temperatures. *Potato Research* 22:261–278.

Venugopal, A.P., Viswanath, A. and Ganapathy, S. (2017). Development of night time on-farm ventilated potato storage system in Nilgiri Hills of Southern India. *International Journal of Processing and Post Harvest Technology* 8(1):37–43.

Verma, S.C., Sharma, T.R. and Verma, S.M. (1974). Effects of extended high-temperature storage on weight losses and sugar content of potato tubers. *Indian Journal of Agricultural Sciences* 44(11):702–706.

Woldemariam, H.W. and Abera, B.D. (2014). Development and evaluation of low cost evaporative cooling systems to minimise postharvest losses of tomatoes (Roma vf) around Woreta, Ethiopia. *International Journal of Postharvest Technology and Innovation* 4(1):69–80.

Woldetsadik, K., Workneh, T.S. and Getinet, H. (2011). Effect of maturity stages, variety and storage environment on sugar content of tomato stored in multiple pads evaporative cooler. *African Journal of Biotechnology* 10(80):18481–18492.

Wong Yen Cheong, J.K.C. and Govinden, N. (2001). Quality of potato during storage at three temperatures. *Proceedings of the Third Annual Meeting of Agricultural Scientists*, Reduit, Mauritius, 17–18 November 1998.

18 Summary and Conclusions

Vijay Yadav Tokala and Majeed Mohammed
The Postharvest Education Foundation

CONTENTS

SUMMARY

This book comprises three sections and 17 chapters from the experts and practitioners of cold chain development in developing countries. The book provides a scenario of cold chain management in different parts of the world. It starts with the discussion on the importance of the cold chain as well as the different options and innovations of cooling systems. The following chapters include the case studies, success stories, capacity building activities and opportunities in cold chain development.

Section I titled *Cooling and Cold Chain* has two chapters, chapter by Kitinoja, Tokala, Mohammed and Mahajan elucidated the importance of the cold chain and its current global status. They have examined the cold chain market size, status and global outlook for the period 2019–2025 and discussed the implications of an unbroken cold chain and the essentiality of proper refrigeration to maintain produce quality during storage and to reduce postharvest food losses and wastes. The authors emphasized that maintaining safe low temperatures along the food supply chain would slow down physiological activities of the produce, minimize the nutritional degradation as well as ensure food safety by impeding the growth of harmful microorganisms. Innovations in the cold chain were identified and related to sensors and data analysis. The chapter also addressed how connecting the sensor system to mobile networks provided ways to reach out and control the temperature from virtually anywhere on the globe by allowing centralized, or even outsourced computing resources, to reduce infrastructure costs.

In Chapter 2, Bridgemohan and co-authors elaborated on suitable temperature regimes of selected perishable commodities along the cold chain. The potential improvements in the cold chain provided by real-time temperature monitoring and the effects of temperature and relative humidity on the selected perishables along the cold chain were also reviewed. The most critical weaknesses that needed to be corrected for better produce quality and safety were identified. Potential management systems to improve the cold chain, based on the measurement of perishable product temperature, were discussed, and challenges related to the implementation of such systems were highlighted as well as the proposal of relevant research projects for global and inclusive improvements to the cold chain.

Section 2 has six chapters on various cooling systems as related to types, methods, efficiency, requirements, benefits and applications. Sharma and others in Chapter 3 explained the development of traditional storage methods for the rural farmers to maintain the quality and the longevity of horticultural produce cost-effectively. They gave a comprehensive overview of on-farm storage structures such as cellar storage, trench/pit, clamp, windbreak, barn, night ventilation, evaporative cooling, ZECC (Zero energy cool chamber) and clay pot/pot-in-pot refrigerator. The popularity of these systems among the farmers with respect to storage effectiveness and economic benefits for short-term storage of perishable commodities was discussed. The authors argued that conventional/ traditional storage systems, which provided an alternative to other expensive storage systems,

DOI: 10.1201/9781003056607-21

required no special skills to operate and justified their application as being suitable for rural stakeholders in the fresh produce industry in India and other developing countries.

Kitinoja and Wilson in Chapter 4 articulated how improved cooling and cool storage methods allowed producers, traders and marketers to handle relatively higher volumes of produce and also attained extended shelf life of produce. Both authors explained how the selection of mechanical refrigeration versus non-mechanical cooling and their relationship to the level of sophistication of the value chain. Further, the importance of simple on-farm practices for providing cooling, such as shade, ice or evaporative cooling systems in reducing the postharvest losses and increase shelf life in a cost-effective was discussed. Accordingly, they emphasized that relatively high-cost cooling methods, such as forced air cooling, hydrocooling or refrigerated transport, are economically feasible only when the product's market value is relatively high and when the facilities are used efficiently.

In Chapter 5, Peters explored in detail that optimum, the sustainable and climate-friendly cold chain can be introduced only if the entire system is designed cohesively and with an integrated system approach from the outset, rather than piecemeal technology deployments, which solve singular aspects of the problem but ignore the end outcomes and bigger picture goals. He gave an overview on what is required alongside technology and who will need to be engaged to drive change, including policy, creative and collaborative models for finance and business, training and capacity building as well as live end-to-end demonstrations for all stakeholders. Furthermore, the strategic and step-change approaches being developed were identified that could allow packhouses and cold chain businesses to expand their services and create sustainable community cooling hubs, make use of renewable power and develop new revenue streams by providing cooling to a portfolio of district services. Equally an intelligent approach to developing smart cold chains for food distribution capability that can provide the architect to other supply chains and rural cooling demands presents the greatest economic opportunity to developing a lasting cold chain legacy that is both resilient and sustainable.

Ramadhani and others, in Chapter 6, discussed the issues related to availability, affordability and access to cooling and packing facilities and associated challenges to most small- to medium-scale farming enterprises in Tanzania. They explored how the limited use of cooling facilities and poor handling practices reduced produce compliance for high-value markets. Furthermore, a detailed account of the design and establishment of the mini-pack house using marine shipping containers to undertake produce sorting, grading, packaging, precooling and storage were highlighted.

Chapter 7 by Holcroft and others covered different types of clean cold chain technologies and some successful case studies. They explained how current refrigeration and cold transport technologies utilized either diesel fuel or electricity and the expansion of such technologies for cold chain development would result in increased production of greenhouse gases and pollution. Accordingly, they discussed the need to develop a cold chain in an environmentally sustainable manner because of the huge potential for using clean technologies for cooling and cold chain management.

In Chapter 8, Kerbel focused on the rising demand for fresh perishable goods, and the need for efficient transportation of high-demand produce in temperature-controlled insulated containers with a refrigerated unit attached. The chapter included information on the importance and scope of refrigerated transportation, challenges encountered during refrigerated transport and methods to improve the efficiency of refrigerated transport. The author provided several successful case studies on improvised refrigeration transportation.

Section 3 comprises nine chapters with the overall theme based on cold chain development and capacity building. Chapter 9 written by Brondy, Randell and Jaco discussed different approaches for capacity building for cold chain development and articulated how the development of cold chain and successful management is being driven by consumer demand, adequate access to funds as well as policy regulations. They argued that the limited availability of well-trained technical professionals and lack of knowledge on quality standards as major hindrances in the development of the cold chain industry in most of the developing countries. The Global Cold Chain Alliance (GCCA) was identified as the leading trade association for temperature-controlled logistics which provided guidelines for cold chain development in 75 different countries that are consistent with international best practices and standards.

In Chapters 10 and 11, the historical perspectives, progress and status of the cold chain in India were reviewed in detail by Kohli. The cold chain development and supporting policies were tracked since the advent of the Green Revolution in the 1980s, where India witnessed a multi-fold increase in production across the agricultural sectors and was transformed from a starvation zone to an exporter of food. Despite the surplus production, the author explained that undernutrition and hunger remain at a significant proportion, due to inadequate and inefficient supply logistics. This lack of proper logistics along the supply chain resulted in postharvest losses and wasted resources. The future trajectory of India's cold-chain development was identified as being dependent on keeping abreast of cold chain developments and the deployment of strategies and resources to foster a holistic balance and a supply chain orientation.

Bridgemohan and Mohammed elaborated on the opportunities and challenges of cold operations in the Caribbean, in Chapter 12. They examined gaps within the cold chain of various outlets in the Caribbean and highlighted the causes of postharvest losses, explored methods to measure losses at critical loss points and recommended the required corrective actions to achieve a fully integrated cold chain management system to reduce postharvest losses and wastage.

In Chapter 13, Odeyemi and Ikegwuonu described the negative implications of the lack of refrigeration facilities in addition to an unreliable power supply on the electricity grid in Nigeria, which often resulted in high postharvest losses of perishable commodities. Upon installation of ColdHubs, a solar-powered off-grid cold room, the problem of postharvest losses in perishable produce was reduced and not only improved the livelihood of farmers and vendors but also provided employment for those engaged in its operations.

Verploegen and others, in Chapter 14, investigated the use of a combination of electronic sensors, fruit and vegetable shelf life measurements, and structured user interviews to gather information about users' needs for improved postharvest storage, current methods of postharvest storage and the performance of the evaporative cooling devices to address challenges of postharvest fruit and vegetable storage in Rwanda and Burkina Faso. The ZECCs and clay pot coolers, which were generally used by horticulture farmers, households as well as farmer groups and cooperatives, were evaluated. The authors explained the results of their experiments where they used different sensors to monitor temperatures and humidity. They provided insights on how the selected devices relied on the evaporation of water to create a cooling effect, and the outcome of their performance is significantly affected by the ambient temperature and humidity of the environment in which they operated.

In Chapter 15, Dubey discussed the effectiveness of CoolBot™ room, a low-cost insulated construction, in maintaining the storage temperature, extend shelf life and enhance marketability, compared to conventional cold storage. The core temperatures of the commodities kept in the CoolBot™ room were significantly lower than that of the produce kept in ambient conditions. Such technology options were recommended to be utilized in villages and semi-urban areas for storage, transport and marketing of horticultural produce to generate added value and provide extra income impetus to entrepreneurship among stakeholders in the fresh produce industry.

Kitinoja and Njie highlighted key policy issues and strategies while providing case studies in Kenya, Ghana, The Philippines and Cameroon. The investments in the best practices for better linkages and improved communication among the actors in the agro-industrial sector to reduce duplication of efforts and increase the adoption of cost-effective cold chain development practices were discussed in Chapter 16. The authors emphasized the need for effective communication, coordination and collaboration among research, extension and industry personnel involved in postharvest systems and the development of the cold chain as the foundation of any successful plan. Critical areas proclaimed to improve the logistics and overall performance of agri-food firms involved in cold chain development would require policy implementation, infrastructure and facility improvement, networking and linking of agri-food firms in vertical and horizontal chains, training services, improving information connectivity as well as access to financial sources.

Chapter 17 was articulated by Kitinoja and Holcroft. Their chapter presented the evidence on the specific postharvest interventions associated with the different stages of the cold chain, for the 13 perishable crops (roots, tubers, fruits and vegetables) in Sub-Saharan Africa and South Asia. Their evidence-based review indicated that a serious gap existed in the research on cooling interventions for perishable crops and this is due to the lack of economic, gender and environmentally focused studies. Both authors indicated that for perishable crops there were no studies on the effect of an integrated cold chain, i.e. precooling followed by storage, transportation and marketing at low temperatures, despite the widespread adoption of the cold chain for exports from Sub-Saharan Africa and South Asia countries or adoption by commercial farms.

CONCLUSIONS

The global food losses are result from a lack of necessary infrastructure, improper food safety handling procedures and insufficient training for the personnel working in the cold chain. However, cold chain systems are crucial to the growth of global trade in perishable products and the worldwide availability of food products. Fresh produce requires an uninterrupted cold chain due to its perishable nature. By controlling parameters of temperature, humidity and atmospheric composition, along with utilizing proper handling procedures, cold chain service providers can increase the product life of fresh foods for days, weeks or even months. These services allow fresh products to hold their value longer, increasing their transportability and providing opportunities that expand their market reach.

The exact structure of each cold chain varies significantly depending upon the product and customer requirements. However, the goal of a properly designed cold chain system is to safely move temperature-sensitive products in a way that reduces waste, maintains the quality and integrity of the product and limits chances for microbial contamination. A complete cold chain system includes postharvest precooling, processing, temperature-controlled warehouse or storage, refrigerated transport between locations and retail or distribution.

Cold chain management is a requirement to achieve a global reduction of postharvest losses and food waste. Despite recent advances in cold chain expansion in several developing countries, initiatives for improvement must be accelerated. Efforts to foster improved infrastructure, generation of information to propagate cold chain maintenance and applications for the fresh produce industry in developing countries have to be strengthened. The increasing capacity for cold chain growth in developing countries is driven by a greater reliance on the cold chain to meet increasing trade and consumption of better quality perishable foods, adherence to international standards of quality and food safety and the potential retail boom and emerging middle class in these countries. The cold chain industry in developing countries needs to be recognized and characterized by a higher return on investment compared with other businesses. This book seeks to increase capacity building, enhance relevant technical information on the benefits and the applications of cold chain management in developing countries.

Index

Note: **Bold** page numbers refer to tables; *italic* page numbers refer to figures and page numbers followed by "n" denote endnotes.

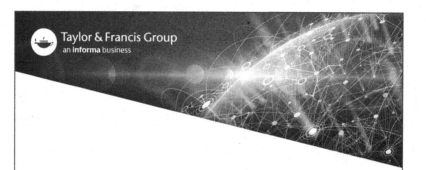